APPLIED HYDROCARBON THERMODYNAMICS

VOLUME 2

SECOND EDITION

Volume 1

Volume 2

Gulf Publishing Company
Book Division
Houston, London, Paris, Tokyo

APPLIED HYDROCARBON THERMODYNAMICS

VOLUME 2

SECOND EDITION

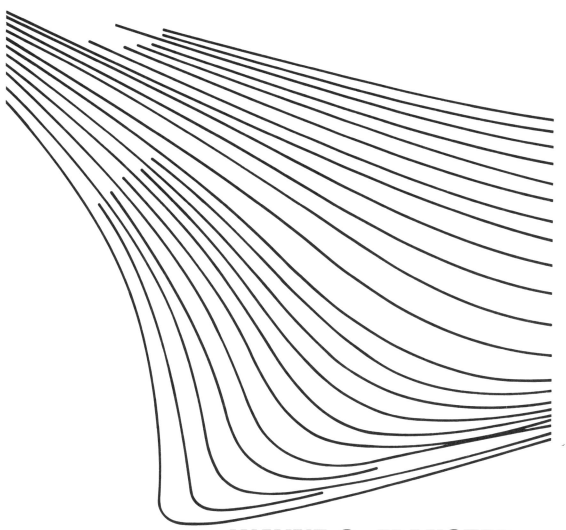

WAYNE C. EDMISTER

Applied Hydrocarbon Thermodynamics

VOLUME 2, SECOND EDITION

First Edition, October 1974

Second Edition, January 1988

Library of Congress Cataloging-in-Publication Data
(Revised for vol. 2)

Edmister, Wayne C.
  Applied hydrocarbon thermodynamics.

  Includes bibliographies and indexes.
  1. Hydrocarbons. 2. Thermodynamics. I. Lee, Byung Ik. I. Title. II. Title: Hydrocarbon thermodynamics.
TP248.H9E3 1984   661'.81   83-22654
ISBN 0-87201-855-5 (v. 1)

ISBN 0-87201-858-X (v. 2)

Series ISBN 0-87201-859-8

D
661.81
EDM

Contents

Scope of This Book. Hardware and Software. Basis of Discussion. Program Requirements. Component Properties. Solution Methods. Multi-component phase equilibria. Vapor-Liquid Equilibria. Bubble Points and Dew Points. Equilibrium Flash Vaporizations. Retrograde Phenomenon. Useful Features of the Mixture Phase Diagram. Binary Interaction Coefficients. Isenthalpic and Isentropic Processes. Other Thermodynamic Processes. References.

Data Resources. Analytical Distillations. Equilibrium Flash Vaporization. Empirical Correlations. EFV Correlations. ASTM—TBP—EFV Relationships. Pressure Effect on EFV Curves. EFV Vapor and Liquid Properties. References.

Conversion of ASTM Distillation Assays. Phase Equilibria for Sub-Atmospheric Pressures. Pressure Effect on 30% and 50% Points. Vacuum Phase Diagram Construction. Consistency Between Atmospheric and Vacuum EFV Correlations. References.

Dedication

To my wife Margaret, whose verses are more entertaining than my prose, for years of playing second fiddle while the computer and I burned.

Acknowledgments

I would like to express my appreciation to—

Robert L. Montgomery, Ph.D., who wrote Chapter 24, co-wrote Chapter 25, and contributed to Chapters 18 and 22. His activities related to hydrocarbon thermodynamics have included a Ph.D. thesis on the heat capacity of sulfur (Oklahoma State University, 1975), measurements of heats of combustion during a post-doctoral fellowship at Rice University (1976–1978), and work at the M. W. Kellogg Company, Houston, Texas (1977–1982) on evaluating and estimating thermochemical data, including the estimation methods described in Chapter 24 of this book. Dr. Montgomery is presently with Martin Marietta Astronautics Group in Denver, Colorado.

Carlos Alberto Dantes Moura, who made valuable contributions to Chapters 11 and 18 and to the pure component data bank of the FORTRAN computer program. He is a process engineer with Petrobras in Rio de Janeiro, Brazil.

Kaiser Engineers for letting me use their computer in Oakland before I got my own.

Faculty colleagues and students at visiting-professor courses in Argentina, Brazil, China, England, Puerto Rico, South Korea, Trinidad, USA, Venezuela, and West Germany since 1972, for helping keep my technical interests alive during my "retirement."

Foreword

This volume completes the second edition of Professor Edmister's *Applied Hydrocarbon Thermodynamics*. The objective of the second edition has been the same as for the first edition; namely to provide a source of practical information and tools for process engineers to use in thermodynamic property prediction, simulation, and design calculations.

Although the overall objective of the second edition is the same as that of the first edition, the information and tools are now different. Process engineers use computer hardware and software more than they use charts and slide rules in their work today. With this in mind the author has developed computer programs and data banks for vapor liquid equilibria, reaction equilibria, and other process calculations, all based upon the same basic theory, as presented in Volume 1 of this second edition.

Of special interest to petroleum refining process engineers will be Professor Edmister's pseudo component technique and computer method for characterizing petroleum fractions in phase equilibrium calculations, in which discrete and pseudo components get the fugacity-type K-value treatments.

Another unique part of this work includes material on the thermochemical properties of petroleum fractions in Chapters 24 and 25, where a modified method of corresponding states was applied to calculate the heats of combustion, formation, and reaction, and free energy of formation for petroleum fractions.

This volume will prove valuable to engineering students and practitioners in chemical and petroleum process development and design.

John J. McKetta
Chairman
Editorial Committee
Hydrocarbon Processing
Houston, Texas

Preface

Volume 1 of this second edition of *Applied Hydrocarbon Thermodynamics* contains ten chapters of fundamentals, properties data, and charts. The theoretical concepts and mathematical derivations included in Volume 1 prepare the way for the development of the ultimate thermodynamic tool—computer programs for applying thermodynamics to hydrocarbon process simulation and design. Such programs are the objective of this second edition of Volume 2.

Programs were first prepared and tested on a mainframe computer and then downloaded to a personal computer for further development. The capacity, speed, and convenience of the micro, or PC, computer have increased greatly during the past few years, making it ideal hardware for applications of thermodynamics. FORTRAN programs for this were written in a user-interactive format. Complete source codes are not included, but an executable version of the programs is available separately. The software development work is described in this volume, with examples of the subprograms, input data, and results.

Special features of interest are a data bank for 213 hydrocarbons and associated gases; computer breakdown of petroleum fractions into 11 or 21 pseudo components from ASTM or TBP assays of the oil; acceptance of narrow petroleum cuts as hypothetical components in the feed; bubble and dew point temperature or pressure calculations; alternate flash calculations to find unknown temperature, pressure, or vapor fraction; enthalpy or entropy balance flashes; adiabatic/polytropic compression and expansion design calculations for gases; compressi-ble fluid flow calculations; enthalpies at alternate datum states; heats of chemical reactions and chemical reaction equilibria; and thermochemical properties and absolute enthalpies of petroleum fractions.

To capsulize this volume: Chapter 11 introduces the reader to thermodynamic properties and process simulation; Chapters 12 and 13 contain the graphical phase equilibria correlations for petroleum fractions from the previous edition; Chapters 14 and 15 repeat the integral technique and pseudo-components of petroleum fractions from the previous edition, with additional examples; Chapters 16 and 17 give background information on computer predictions and then describe the subroutines of the computer algorithm; Chapter 18 defines the three alternate zero enthalpy datum states, gives a computer program for calculating constants for converting to each datum, and gives the conversion constants for all 213 hydrocarbons in Block Data; Chapters 19 and 20 present gas compression and calculation methods and compressible fluid flow through orifices and pipes. Included are some first edition material and new computer programs. Chapters 21 and 22 cover heats of reaction and reaction equilibrium, with some first edition material repeated and a new computer program presented; Chapter 23 covers the construction of *H-X* diagrams with information repeated from the first edition, plus new items; Chapters 24 and 25 present all new material on enthalpies of combustion and formation and free energies of formation, plus alternate zero enthalpy states for petroleum fractions.

Wayne C. Edmister
San Rafael, California

1

Introduction—Thermodynamic Properties and Process Simulation

Scope of this Book

Volume 1 of this two-volume set presented the basic thermodynamic tools (charts, tables, and equations) needed to make phase and energy calculations in hydrocarbon process design. It included theoretical concepts and mathematical derivations, and thus prepared the way for this volume's development of the ultimate thermodynamic tool—a computer program for actually applying thermodynamics to hydrocarbon processing.

Hardware and Software

The thermodynamic properties and vapor-liquid equilibrium programs described in this second volume of *Applied Hydrocarbon Thermodynamics* are in FORTRAN 77 for mainframe and microcomputers (IBM compatible).

The FORTRAN source code files were first prepared, compiled, linked, and tested on the mainframe. Then the source code files were down-loaded to the micro, where they were again compiled, linked, and run. This microcomputer work was done with the MicroSoft FORTRAN compiler.

Complete FORTRAN source code versions of the programs are not included in this book, but compiled and linked versions are available in executable form. I have

tried to describe the programs clearly enough to allow the reader to prepare his or her program if desired.

Basis of Discussion

Some of the important basic topics related to vapor-liquid equilibrium calculations (designated VLE throughout this text for convenience) were covered in Volume 1's Chapters 3, 9, and 10. For example, the criteria for phase equilibria were discussed in Chapter 3, while the definition and usage of the VLE K-values were presented in Chapter 9, including several methods for making vapor-liquid equilibrium calculations. In Chapter 10, the K-value relationships with other properties, such as the vapor pressure and fugacity, were reviewed and some graphical K-value correlations were presented.

The manual calculation methods discussed in Chapters 9 and 10 were primarily intended for background information on VLE calculations and K-values. Even with the over-simplified assumptions normally used, the manual methods are very tedious, time consuming, and of limited accuracy, particularly for multicomponent mixtures. Accordingly, such manual methods are seldom used in engineering practice now, being replaced by the more accurate rigorous methods.

This volume discusses computer methods for making K-value predictions and VLE calculations for hydrocarbon mixtures and petroleum fractions, with or without

1

associated gases. Special emphasis is given to a newly developed VLE program, named EQUIL.

Program Requirements

A workable program for making VLE and thermodynamic properties calculations for complex mixtures needs

1. A data bank for the discrete components of interest.
2. Methods for converting petroleum fractions into pseudo-components and predicting their characterizing properties.
3. Methods for making equilibrium calculations for single and multiple stage processes.
4. Methods of converging the iterative calculation inherent in the calculations.
5. A program for creating input data files and an output file handling and printing system.
6. A user-interactive system to permit operating the program conveniently from a keyboard.

Component Properties

Properties of hydrocarbons were tabulated in Chapter 7 of Volume 1. These data have been put into a data file for convenient reading by the program. This file, identified as BLOCK DATA, is an integral part of EQUIL, the computer program. Similar properties for the pseudo components of petroleum fractions must be found by empirical equations. The technique for breaking petroleum fractions into pseudo components and finding the necessary property values are given in Chapters 14 and 15 of this volume, following graphical equilibrium flash vaporization correlations for petroleum in Chapters 12 and 13.

Solution Methods

Methods of converging to unknown temperatures, pressures, vapor/liquid fractions, enthalpies, and/or entropies are covered in Chapter 16, as are the eleven possible phase equilibria processes with specified and unknown variables. Chapter 16 also presents the six source code files containing more than 50 subprograms that make up the master program that has been the objective of this work. Descriptions of these subprograms are presented in Chapter 17. Chapter 18 covers alternate zero enthalpy datum states for pure components, Chapter 25 for petroleum components.

Other topics covered in this volume include flow of compressible fluids in Chapter 19; gas compression and expansion calculations in Chapter 20; heats of chemical reactions in Chapter 21; chemical reaction equilibria in

Chapter 22; enthalpy-composition diagrams in Chapter 23; thermochemical properties of petroleum fractions in Chapter 24; and absolute enthalpies of petroleum fractions in Chapter 25.

The rest of this chapter describes several phenomena of vapor liquid equilibria and reviews basic fugacity relationships, thus establishing the basis for the preparation of the actual programs.

Multicomponent Phase Equilibria

This section reviews the phase behavior of multicomponent mixtures and describes retrograde condensation/vaporization phenomena, different VLE processes (bubble and dew points, isothermal or constant temperature and pressure flash, constant pressure and vapor fraction flash, constant temperature and vapor fraction flash, isenthalpic flash, and isentropic flash), and methods of solving the complex mathematical expressions.

An understanding of the nature and types of phase equilibria, commonly encountered in hydrocarbon processing calculations, is important to having a working knowledge of the calculation procedures. Hydrocarbon mixture separations into the many constituents, such as pure components, gasoline, naphtha, kerosene, diesel oil, gas oil, fuel oil etc., by vaporization, condensation, and distillation are the hydrocarbon processes of interest here.

As distillation consists of a series of equilibrium stages, accurate K-values and VLE calculations are critical in designing hydrocarbon separation processes, as was illustrated in Volume 1's Chapter 9. In such process design work, it is sometimes necessary to predict three-phase equilibria for vapor-liquid-liquid (VLLE), in addition to VLE, for hydrocarbon mixtures containing non-hydrocarbons, such as water, carbon dioxide, and/or hydrogen sulfide.

Another phase equilibrium state frequently encountered in hydrocarbon processing is that of hydrates forming at low temperatures from water and some hydrocarbons, and/or CO_2 and H_2. Though less frequent, there are other types of phase equilibrium encountered in hydrocarbon processing—vapor-solid, liquid-liquid, vapor-liquid-solid, vapor-solid-solid, liquid-solid-solid, vapor-liquid-liquid-solid, and vapor-liquid-solid-solid. This chapter deals only with vapor-liquid phase equilibrium.

Vapor-Liquid Equilibria

Figure 11.1 is a typical pressure-temperature diagram for a mixture of fixed composition, in which the curves *B* and *D* represent the bubble-point and dew-point loci,

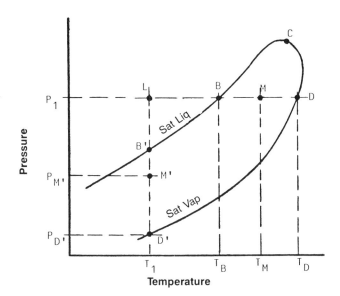

Figure 11.1. A typical *P-T* diagram for a mixture of given composition.

respectively, and point *C* is the critical point of the mixture. The envelope formed by the bubble-point and dew-point curves is commonly called the "phase boundary" or "saturated" phase envelope. Liquid that exists along the bubble-point curve *B* is "saturated liquid" because the liquid cannot retain any more molecules in the liquid state when there is an infinitesimal change in the pressure or temperature toward the inner area of the phase envelope. Similarly, the vapor along the dew-point curve *D* is "saturated vapor."

On the other hand, liquid outside the bubble-point curve is "subcooled liquid" and vapor outside the dew-point curve is "superheated vapor." For example, the point *L* (at P_1 and T_1) represents a subcooled liquid. This liquid becomes saturated at point *B*, if the temperature is increased to T_B at constant P_1. At this point, the first tiny bubble of vapor is about to appear in the liquid, yet the composition of the liquid remains unchanged.

However, if the temperature is increased further, the liquid starts to vaporize, forming a vapor with a different composition from that of the liquid. Consequently, the liquid composition also changes as the vaporization proceeds with the increase in temperature. For example, at point *M* (P_1 and T_M), inside the phase envelope, both vapor and liquid phases coexist and their compositions are different, not only from each other but also from the original liquid *L*. As the temperature increases, the amount of vapor also increases until the last tiny drop of liquid is about to disappear at T_D. At this point, the composition of the vapor (saturated) becomes identical to that of the original liquid *L*, and the composition of the last drop of liquid is completely different from the original liquid. If the temperature is increased further, the va-

por becomes "superheated," and the composition remains unchanged. Similar phenomena occur when the pressure varies along *L B' M' D' V'* at constant temperature T_1.

The purpose of VLE calculations is to determine the conditions and the phase compositions at such points as *B, M, D, B', M',* or *D'*. The phase equilibrium calculations along the bubble-point curve (*B* or *B'*) and along the dew-point curve (*D* or *D'*) are commonly called *bubble-point calculations* and *dew-point calculations*, respectively. Those VLE calculations made at conditions inside the two-phase envelope are called *flash calculations*.

Bubble Points and Dew Points

It should be obvious that there are two types of bubble-point calculations for a mixture—one is to find the bubble-point temperature and the composition of the coexisting equilibrium vapor at a given pressure, e.g., point *B* in Figure 11.1; the other calculation is to find the bubble-point pressure and the composition of the coexisting equilibrium vapor at a given temperature, e.g., point *B'*. Similarly, the point *D* represents the dew-point temperature calculation at P_1, and point *D'* represents the dew-point pressure calculation at T_1.

From a mathematical point of view, the bubble- or dew-point calculation is made to determine ($n + 1$) unknowns from the following ($n + 1$) equations for a mixture of *n* components:

$$\overline{f}^{L}_i(T,P,\overline{X}) = \overline{f}^{V}_i(T,P,\overline{Y}), \ i = 1,2,\ldots N \qquad (11.1)$$

$$\sum_1^N y_i = 1.0 \text{ for bubble point}$$

$$\left(\text{or } \sum_1^N x_i = 1.0 \text{ for dew point}\right) \qquad (11.2)$$

where T = temperature
   P = pressure
   $\overline{X}$ = liquid mole fraction, vector
   $\overline{Y}$ = vapor mole fraction, vector

The fugacity identity, Equation 11.1, was discussed in Chapters 3 and 10, whereas Equation 11.2 was given in Chapter 9 as Equation 9.42 or 9.43. For bubble-point temperature calculations, the $n + 1$ unknowns are the temperature *T* and the vapor composition $y_i (i = 1,2,\ldots N)$. Similarly, for dew-point pressure calculations, the ($n + 1$) unknowns are the pressure and the liquid composition, i.e., the *P* and x_i values.

Equilibrium Flash Vaporizations

Perhaps the most familiar flash is the so-called isothermal flash, which is also isobaric, as represented by M and M' in Figure 11.1. The isothermal flash calculation is made to find the vapor fraction (V/F) and the vapor and liquid compositions. The relationships of the vapor and liquid compositions are illustrated by the pressure-temperature diagram in Figure 11.2 for a two-component mixture. In this diagram, Comp 1 and Comp 2 represent the pure component vapor pressures for components 1 and 2, respectively, and the three-phase envelopes I, II, and III represent three different compositions of the two component mixtures. Each envelope is similar to that in Figure 11.1.

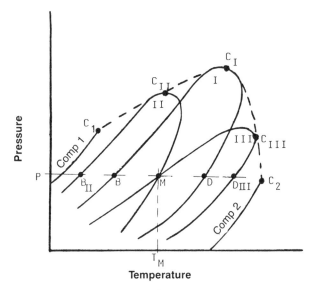

Figure 11.2. A typical diagram for several compositions of a binary system.

When the mixture I is brought to the condition given by M (at P and T) its vapor phase composition is the same as that of mixture II, and its liquid phase composition is the same as that of mixture III. Although Figure 11.2 is for a two-component mixture, the same relationship also holds for multicomponent mixtures.

Mathematically, the isothermal flash calculation is made to find ($2n + 1$) unknowns from a set of ($2n + 1$) equations for an n-component mixture. These ($2n + 1$) unknowns are the vapor fraction (V/F), and the $y_i (i = 1,2,..N)$ and $x_i (i = 1,2,..N)$ values. The ($2n + 1$) equations include Equations 11.1 and 11.2, and the following n material balance equations that were discussed in Chapter 9.

$$z_i = (V/F)y_i + (1 - V/F)x_i, \ i = 1,2,...N \qquad (11.3)$$

where F = total moles of given feed mixture
         V = total moles of equilibrium vapor
         z_i = feed mole fraction

It should be obvious from Equations 11.1 through 11.3 that the equations may also be solved for T, y_i, and x_i from the known values of P, V/F, and z_i. This type of flash calculation is often called "constant vapor-fraction flash" or "constant liquid-fraction flash."

As discussed earlier, the vapor fraction (V/F) is uniquely determined at each temperature between B and D at constant pressure between B and D at constant pressure P_1 in Figure 11.1, with its value varying from 0 at point B to 1 at point D. Indeed the constant vapor fraction flash calculation at constant P is identical to the bubble-point temperature calculation when $V/F = 0$, and is identical to the dew-point temperature calculation when $V/F = 1$. It should be noted that Equation 11.3 becomes $z_i = x_i$ when $V/F = 0$, which is the bubble-point requirement, and also reduces to $z_i = y_i$ when $V/F = 0$, which is a required condition for the dew point.

Retrograde Phenomenon

The one-to-one relationship of V/F to the temperature at constant pressure does not always hold true. Figure 11.3 illustrates an example of a phase condition with two temperatures for a given V/F at constant pressures. Note first that there are two bubble points, B_1 and B_2, at a given pressure P_B between the critical pressure P_C and the maximum pressure P_{MAX}. If the temperature is increased from point B_1 at constant P_B, the vapor fraction (V/F) gradually increases from $V/F = 0$ at B_1 to the max-

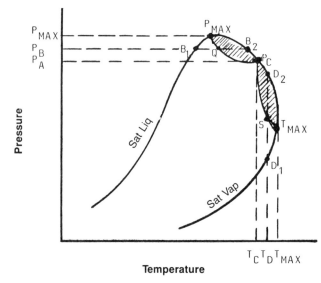

Figure 11.3. Mixture P-T diagram showing boundary curve points in critical region.

imum value of V/F (smaller than unity) at point Q, then decreases with a further increase in temperature, finally becoming $V/F = 0$ again at B_2, i.e., to completely condensed.

This phenomenon, appropriately called *retrograde condensation*, occurs at all pressures in the shaded area between P_C and P_{MAX}. Because of the occurrence of retrograde condensation, there are two identical values of V/F corresponding to two different temperatures at constant pressure; one between B_1 and Q, and the other between Q and B_2. In other words, there exist two sets of valid solutions for T, y_i, and x_i for a given value of V/F and P between P_C and P_{MAX}. Therefore, once a solution is obtained in this region, it is important to identify the location of the solution by making another flash calculation (either isothermal flash at P_B and $T = T + \Delta T$, or constant vapor fraction flash at P_B and $V/F = V/F + \Delta(V/F)$) in the vicinity of the original solution to check whether or not V/F increases with the increase in temperature. If it does, the solution is clearly between B_1 and Q; if it does not, the solution may be assumed to be between Q and B_2. It is obvious that the specified value of V/F must not exceed the maximum value in making flash calculations in the retrograde region.

A phenomenon called *retrograde vaporization,* similar to the retrograde condensation, also occurs in the shaded area between T_C and T_{MAX} on the dew-point curve side of the phase diagram, Figure 11.3. For instance, if the pressure is increased from D_1 at constant T_D, the amount of liquid also increases from D_1 at $L = 0$ to the maximum value at S, then decreases with a further increase in pressure, finally becoming $L = 0$ again at D_2, i.e., to completely vaporize at D_2. This retrograde vaporization occurs at all temperatures in the shaded area between T_C and T_{MAX}. Like in the case of retrograde condensation, there are two sets of valid solutions for T, y_i, and x_i with a given V/F value and with T between T_C and T_{MAX}; one between D_1 and S, and the other between S and D_2. In this case, a procedure similar to the one just described for retrograde condensation may be followed to find the location of the desired solution.

This retrograde condensation (or vaporization) region, which is also loosely called the critical region, is a difficult area for VLE calculations, i.e., the solution often fails to converge, or converges to a trivial solution for which $y_i = x_i$ for all components, primarily resulting from the numerical sensitivity of the calculations to the variables (T, y_i, x_i, or P, y_i, x_i) and also from the poor accuracy of the thermodynamic models used in the computation methods.

As shown in the shaded area of Figure 11.4, the retrograde condensation (or vaporization) region is located just below the critical locus, which is the locus of the critical points, connected by a dashed line on Figure 11.2. This locus can be easily obtained for a two-compo-

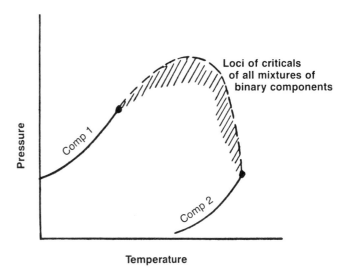

Figure 11.4. Mixture P-T boundary curve showing retrograde condensation and vaporization regions.

nent mixture by drawing a series of phase envelopes between the two vapor pressure curves of the two pure components and then connecting their critical points. Such critical loci are also shown as "convergence pressure curves" in Figure 10.13 and Figures 10.18 through 10.25. The region bounded by the two pure component vapor pressure curves and the critical locus, as shown in Figures 11.2 and 11.4, is called the "two-phase region" for this system.

The relative locations of critical points on the liquid-vapor phase boundary curves are shown in Figure 11.5, which is from Rowlinson (15). Critical points shown are C, the true critical point; CT, the cricondentherm point; and CB, the cricondenbar point. The names "cricondentherm" and "cricondenbar" were first proposed in 1934 by Sage et al. (16) to define the points of maximum temperature and maximum pressure on the phase boundary diagram.

Three mixture types in this illustration are I, normal proportions of light and heavy components; II, lighter component dominating; and III, heavier component dominating. The dashed lines through the critical point are the loci of other mixtures of the same components. From these diagrams, it can be seen that the true critical point, C, is located in between CB and CT for mixture Type I, whereas C is located on the BP side for Type II and on the DP side for Type III.

Useful Features of the Mixture Phase Diagram

Vapor/liquid phase boundary diagrams, such as those shown on Figures 11.1 through 11.5, are formed by the joining of two curves at the critical point. These are the

Type (I) Mixture
Normal Proportions

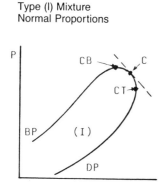

Type (II) Mixture
Lighter Component
Dominates

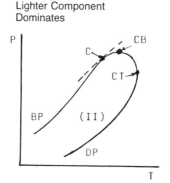

Type (III) Mixture
Heavier Component
Dominates

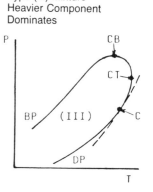

Figure 11.5. Relative locations of critical points for mixtures. (Dashed lines are loci of critical points: *C*—critical point; *CT*—cricondentherm point; *CB*—cricondenbar point.) (From Rowlinson, 1959.)

bubble point (0% vapor) and the dew point (100% vapor) curves, sometimes called "boundary curves." The area within these boundary curves is the "co-existing equilibrium vapor and liquid" region, i.e., the temperature-pressure combinations at which vapor and liquid will co-exist in equilibrium.

If lines of constant fraction vaporized are drawn within such a phase diagram, the result would be a family of curves that converge at the critical point at their upper ends and fan out in a gradual, uniform and conforming way at their lower ends. Boundary diagrams containing such lines of constant percent vapor are sometimes called *P-T-X* diagrams.

The boundary curves for the three mixture types shown on Figure 11.5 could be made into *P-T-X* diagrams by adding these percent vaporized lines. The top portions of these curves would contain some reversals.

Figure 11.6 contains *P-T* plots for ethane and *n*-heptane and for three mixtures of these two hydrocarbons, i.e., the vapor pressure curves for ethane and *n*-heptane plus the boundary curves for the three mixtures. These plots were made from experimental data obtained by Kay (14). The dashed curved line is the critical locus, drawn through the five critical points. The dashed straight line is the pseudo critical locus for this system, which was defined by Kay as the molar average of the critical temperatures and pressures of the mixture components.

Boundary curves, of the type shown in Figure 11.6, can be obtained accurately for binary mixtures of known composition by an experimental procedure that requires measuring only temperatures and pressures. These temperature and pressure readings are plotted to obtain boundary curves for several compositions.

From these charts, pressure versus composition cross-plots are made. Figure 11.7 shows pressure-composition cross-plots for five isotherms of the methane-ethane system, from Price and Kobayaski (22). From these *P-X*

plots, it is a simple matter to calculate $K = y/x$ values. This *K*-value method (experimental measurements plus graphical calculations) is limited to binary systems.

To illustrate the use of charts like Figure 11.7, find the *K*-values for methane and ethane at $-50°F$ and 700 psia. Read from Figure 11.7: m.f.s C_2 in vapor = 0.18; in liquid = 0.464. Complete calculations as follows:

	Ethane	Methane
Vapor	0.180	$1.0 - 0.180 = 0.82$
Liquid	0.464	$1.0 - 0.464 = 0.536$
$K = y/x$	$0.180/0.464 = 0.388$	$0.820/0.536 = 1.530$

This method can be used for binaries only.

Binary Interaction Coefficients

In Volume 1's Chapter 5, Equations 5.68 and 5.69 include the term k_{ij}, binary interaction coefficient, in combining the equation of state constants for components to obtain the values for mixtures by both Soave-RK and Peng-Robinson EOS's. The use of such binary interaction coefficients with equations of state is one of the reasons for the superiority of the Soave-RK (28) and the Peng-Robinson (21) EOS methods over the Chao-Seader and the Lee-Erbar-Edmister methods, which used an EOS for the vapor phase only and used empirical equations for the liquid phase properties.

Values of k_{ij} for use in VLE predictions are derived from *P-T-X-Y* measurements on a binary system by iterative calculations made to find a single k_{ij} value that gives the best agreement for calculated pressure, or another mixture property, with the experimental value for the same binary system. The experimental data used must be thermodynamically consistent and the derivation calculations must be made with the same EOS that the k_{ij} values are to be used with.

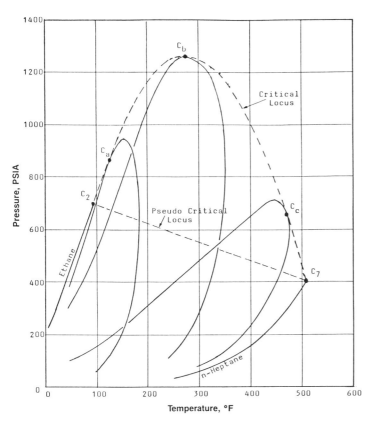

Figure 11.6. Critical loci for ethane, *n*-heptane, and three binaries. Wt. % C_2H_6: $C_2 = 100.0$; $C_a = 90.22$; $C_b = 50.25$; $C_7 = 0.0$. (From Kay, 1938.)

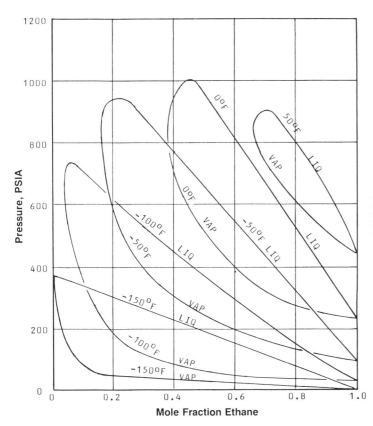

Figure 11.7. Pressure composition diagram for methane-ethane system. (From Price and Kobayaski, 1959.)

Ideally the optimum value of k_{ij} should minimize the errors in the prediction of all thermodynamic properties calculated from the EOS. In practice, one property like bubble-point pressure is used. The techniques for finding these binary interaction coefficients are described by Daubert et al. (6, 9), who defined and used an objective function of pressure, Q.

$$Q = \sum_1^N ((p_c - p_e)/p_e)^2 \qquad (11.4)$$

where N = number of points
  c = calculated pressure
  e = experimental pressure.

Paunovic et al. (20) proposed and used a fugacity objective function, as follows:

$$Q = \sum_1^N [\{(FV1 - FL1)/FV1\}^2$$

$$+ \{(FV2 - FL2)/FV2\}^2] \qquad (11.5)$$

where N = number of points
  $FV1$ = fugacity of component 1 in vapor
  $FL1$ = fugacity of component 1 in liquid
  $FV2$ = fugactiy of component 2 in vapor
  $FL2$ = fugacity of component 2 in liquid

With either of these functions, the procedure is to calculate Q for assumed values of k_{ij}, then plot Q against k_{ij}, and find the optimum k_{ij} at the minimum Q point on that plot. In evaluating Q by Equation 11.4, it is necessary to converge an equilibrium pressure calculation. Finding Q by Equation 11.5 is simpler in that no convergence is necessary; the fugacities are evaluated at the conditions of the experimental data.

For hydrocarbon systems containing non-hydrocarbons, such as CO_2, H_2S, CO, and N_2, the use of k_{ij} values is necessary. For hydrocarbons of the same kind, i.e., paraffin binaries or olefin binaries, k_{ij} values are not usually required. Consult References 12, 18, 19, 21, and 28 at the end of this chapter for more information.

The optimum value of k_{ij} will now be found for the carbon dioxide-normal hexane system, using the experimental data of Kalra et al. (12) and the Soave (28) EOS. P-T-X-Y data were complete and had been checked for thermodynamic consistency by the authors. An abridged tabulation, showing one-third of these data, is given in Table 11.1. It can be seen that the data are at temperatures from 99.5 to 399.3°F and pressures up to 1,931 psia.

Table 11.1
Experimental *P-T-X-Y* Data for Carbon Dioxide -*n*Heptane for Deriving Binary Interaction Coefficients (12)

Temperature °F	Pressure psia	Mol Fractions CO_2	
		Liquid	Vapor
99.5	30.5	0.0250	0.9490
	128.0	0.1000	0.9840
	526.0	0.4140	0.9930
	948.0	0.8350	0.9930
	1097.0	0.9490	0.9500
175.0	61.5	0.0310	0.8600
	230.0	0.1260	0.9560
	960.0	0.4980	0.9740
	1426.0	0.7150	0.9630
	1684.0	0.8478	0.9050
250.0	164.0	0.0730	0.8190
	450.0	0.1930	0.9210
	906.0	0.3760	0.9420
	1505.0	0.5890	0.9340
	1931.0	0.7680	0.8820
399.3	254.0	0.0420	0.3440
	720.0	0.2190	0.6810
	1228.0	0.4160	0.7680
	1439.0	0.5550	0.6380

Values of Q were computed at 24 evenly spaced values of k_{ij} ranging from 0.0 to 0.24, using Equation 11.5 and the Soave-RK EOS, and the resulting Q values were plotted against k_{ij} in Figure 11.8. Note that the curve through these points has a minimum Q value at $k_{ij} = 0.114$. This is the binary interaction coefficient for these data and the Soave-RK EOS.

Isenthalpic and Isentropic Processes

The preceding discussions were limited to flash processes in the two-phase region, wherein the variables considered were temperature, pressure, the fraction vaporized, and compositions of vapor and liquid. Other processes of interest are isenthalpic (constant enthalpy) and isentropic (constant entropy) processes, which are not restricted to either single- or two-phase conditions. Both isenthalpic and isentropic processes can occur without a phase change, i.e., the initial and/or final condition may be in the superheated vapor or the subcooled liquid states.

The isenthalpic process is sometimes called an adiabatic flash, even though the process may occur in the vapor or liquid single-phase region. Isenthalpic and isentropic processes are not really flashes in the true sense.

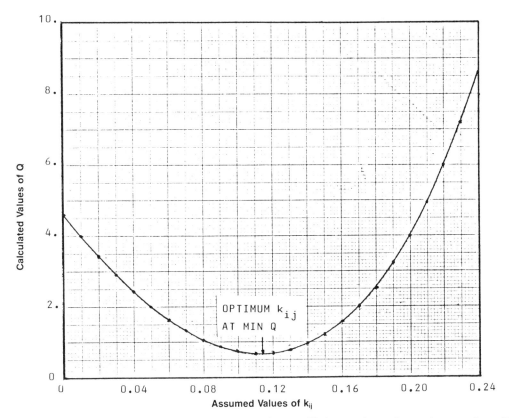

Figure 11.8. Objective function, Q, plotted against binary interaction coefficient to determine optimum value of k_{ij} for the carbon dioxide-normal heptane system.

Chapter 2 of Volume 1 points out that two thermodynamic properties must be specified to determine all other properties of a homogeneous mixture. In view of this, the adiabatic (or isentropic) flash is a process in which the enthalpy (or the entropy) is one of the two properties to be specified, along with the temperature (or pressure) as the other property. If the two-phase region is involved, as is frequently the case, then the flash calculations must accompany the isenthalpic or the isentropic process conditions.

Other Thermodynamic Processes

With thermodynamic algorithms for predicting properties of hydrocarbon mixtures and vapor-liquid phase equilibrium, calculation procedures can be developed for other thermodynamic single-stage processes of interest. These include gas compressor design, gas expander design, the flow of compressible fluids through pipes and orifices, and chemical reactors are examples of single-stage processes. Absorption, fractionation, and distillation are the multistage equilibrium processes of interest, but multistage processes are beyond the scope of this book.

Computer Program Development

The development of the computer programs for these vapor liquid equilibrium processes will be presented in Chapters 16 and 17.

References

1. American Petroleum Institute *Technical Data Book* (1980).
2. Akers, W. W., J. F. Burns, and W. R. Fairchild, *Ind. Eng. Chem.* 46 (12), 2531 (1954).
3. Asselineau, L., G. Bogdanic, and J. Vidal, "A Versatile Algorithm for Calculating Vapour-Liquid Equilibria," *Fluid Phase Equilibria,* 3(4), 273–290 (1979).
4. Baker, L. E. and K. D. Luks, "Critical Point and Saturation Pressure Calculations for Multipoint Systems," *J. Society of Petroleum Engineers,* 20(1), 15–24 (1980).
5. Etter, D. O. and W. B. Kay, "Critical Properties of Mixtures of Normal Paraffin Hydrocarbons," *J. Chem. Eng. Data* 6(3), 409–414 (1961).
6. Elliot, Jr., J. R. and T. E. Daubert, "Revised Procedures for Phase Equilibrium Calculations with Soave Equation of State," *IEC Proc. Des. & Dev.* 24(3), 743–748 (1985).
7. El-Twaty, A. I. and J. M. Prausnitz, "Correlation of K-factors for Mixtures of Hydrogen and Heavy Hydrocarbons," *Chem. Engr. Sci.,* 35(8) 1765–8 (1980).

8. Gonzalez, M. H. and A. L. Lee, "Dew and Bubble Points of Simulated Gases," *J. Chem. Eng. Data* 13(2), 172–176 (1968).

9. Graboski, M. S. and T. E. Daubert, "A Modified Soave Equation of State for Phase Equilibria Calculations. 1. Hydrocarbon Systems," *IEC Proc. Des. & Dev.*, 17(4), 443–448 (1978); "2. Systems Containing CO_2, H_2S, N_2, and CO," ibid 448–454 (1978); "3. Systems Containing Hydrogen," *IEC Proc. Des. & Dev.*, 18(2) 300–306 (1979).

10. Grieves, R. B. and G. Thodos, "The Cricondentherm and Cricondenbar Temperatures of Multicomponent Mixtures," *J. Soc. Petr. Eng.*, 3(12), 287–292 (1963); "The Cricondentherm and Cricondenbar Pressures of Multicomponent Hydrocarbon Mixtures," *J. Soc. Petr. Eng.*, 4(3), 240–246 (1964); "The Critical Temperatures and Pressures of Binary Systems—Hydrocarbons of All Types and Hydrogen," *AIChE J.*, 6(4), 561–566 (1960).

11. Heideman, R. A. and A. M. Khalil, "The Calculation of Critical Points," *AIChE J.* 20(5), 769–779, (1980).

12. Kalra, H., H. Kubota, D. B. Robinson, and H-J Ng, "Equilibrium Phase Properties of the Carbon Dioxide—n-Heptane System," *J. Chem. & Engr. Data*, 23(4), 317–321 (1978).

13. Katz, D. L. and A. Firoozabadi, "Predicting Phase Behavior of Condensate-Crude Oil Systems Using Methane Interaction Coefficients," *J. Pet. Tech.*, 20(11), 1649–55 (1978).

14. Kay, W. B., *Ind. Eng. Chem.* 30, 459 (1938).

15. Mathias, P. M. "A Versatile Phase Equilibrium Equation of State," *IEC Proc. Des. & Dev.*, 22(3), 385–391 (1983).

16. Michelsen, M. L., "Calculation of Phase Envelope and Critical Points for Multicomponent Mixtures," *Fluid Phase Equilibria* 4, 1 (1980).

17. Michelsen, M. L. and R. A. Heideman, "Calculation of Critical Points from Two-Constant Equations of State," *AIChE J.* 27(3), 521 (1981).

18. Moysan, J. M., M. J. Huron, H. I. Paradowski, and J. Vidal, "Prediction of the Solubility of Hydrogen in Hydrocarbon Solvents Through Cubic Equations of State," *Chem. Engr. Sci.* 38(7), 1085–92 (1983).

19. Oelrich, L., U. Plocker, J. M. Prausnitz, and H. Knapp, "Equation of State Methods for Computing Phase Equilibrium and Enthalpies," *Intn'l. Chem. Engr.*, 21(1), 1–16 (1981).

20. Paunovic, R., S. Jovanovic, and A. Mihajlov, "Rapid Computation of Binary Interaction Coefficients of an Equation of State for VLE Calculations," *Fluid Phase Equilibria*, 6(3-4). 141–148 (1981).

21. Peng, D.-Y. and D. B. Robinson, "A Rigorous Method for Predicting Critical Properties of Multicomponent Systems from an Equation of State," *AIChE J.*, 23(5) 137–144 (1977).

22. Price, A. R. and R. Kobayashi, *J. Chem. Eng. Data*, 4(1), 40–52 (1959).

23. Reid, R. C. and B. L. Beegle, "Critical Point Criteria in Legengre Transform Notation," *AIChE J.*, 23(5), 726–732 (1977).

24. Rowlinson, J. S., *Liquids and Liquid Mixtures*, Butterworth Scientific Publications, London, 360 pgs (1959).

25. Sage, B. H., W. N. Lacey, and J. G. Schaafsma, "Phase Equilibria in Hydrocarbon Systems," *Ind. Eng. Chem.* 26, 214–217 (1934).

26. Schwartzentruber, J., L. Ponce-Ramirez, and H. Renon, "Prediction of the Binary Parameters of a Cubic Equation of State from a Group Contribution Method," *IEC Proc. Des. Dev.* 25(3) 804–9 (1986).

27. Silverman, E. D. and G. Thodos, "Cricondentherm and Cricondenbars," *IEC Fundamentals*, 1, 299–303 (1962).

28. Soave, G., *Chem. Eng. Sci.*, 27, 1197 (1972).

29. Spencer, C. F., T. E. Daubert, and R. P. Danner, "A Critical Review of Correlations for the Critical Properties of Defined Mixtures," *AIChE J.*, 19(3) 522–527 (1973).

12

Graphical Phase Equilibria for Petroleum Fractions at Super-Atmospheric Pressures

Vapor-liquid phase equilibrium conditions are required in the design of most petroleum processing equipment, and there are three methods for obtaining information about them: experimental measurements, graphical correlations, and analytical calculations. This and the next chapter present charts previously developed and published to illustrate graphical procedures and to provide tools for making approximate calculations. Following that, Chapters 14 and 15 present analytical methods and computer applications.

The graphical methods of predicting phase equilibrium for petroleum fractions are based upon empirical correlations of experimental data on samples of the oils. Laboratory evaluation of crude oils includes many measurements—batch distillations to obtain boiling point and specific gravity assays; separation into the various products of gas, gasoline, naphtha, kerosene, diesel, etc.; analyses for sulfur, wax, and asphalts; analyses of viscosities of the middle and heavy oils; etc. All this information is useful in evaluating and pricing the crude oil. It is also useful to the refiner in designing the processing plants for manufacturing the petroleum products for the market.

For this book's objectives the boiling point and gravity assays are the only petroleum inspection data of interest. These assays are boiling point temperatures versus liquid volume percent, or volume fraction, distilled, and specific gravity versus liquid volume percent, or fraction

distilled. These distillations are made in batch laboratory stills that are charged with 100 to 1,000 ml, or more, oil, and are of two types. One method, which is designated as an ASTM (American Society for Testing Materials), is a differential distillation, i.e., without reflux so that the "components" of the oil are not collected pure in the order of their boiling points. The other method, which is designated as a TBP (true boiling point), is a refluxed distillation so that the "components" of the oil are distilled and collected nearly pure in order of their individual boiling points.

Data Sources

The sources of data correlations presented in this and the following chapter appear at the end of this chapter. Included are both experimental data correlations for atmospheric, vacuum, and high-pressure flash curves, and also the same kind of data for the true boiling point assays. Table 12.1 provides a key to these references, wherein two categories of literature are shown: (1) experimental data and (2) empirical correlations. The references giving data and correlations are divided according to pressure, i.e., atmospheric, super atmospheric and vacuum.

Two types of distillation experimental data were used in developing the correlations presented herein—analyti-

Table 12.1
**Key to Literature References on EFV Distillations
and ASTM and TBP Assays**

Data Type	Atmospheric Pressure	Super Atmospheric
Experimental		
EFV Data	2,3,5,6,8, 13,14,15,20, 23,25,26	1,5,6,13
TBP's (Feed)	11,17	
ASTM & TBP (V&L)	5,6,8	5,6
Empirical Correlations		
ASTM-TBP	2,5,6,17	
ASTM-EFV	4,5,6,18, 20,23,24,27	6,7,12
TBP-EFV	6,7,12	

cal distillations and equilibrium flash vaporization separations. The first are batch distillation assays run to define the mixture. The second are usually continuous flashes yielding vapor and liquid products that coexist in equilibrium. These data were from published sources and from the Richmond Laboratory of Chevron Research Company.

Analytical Distillations

Two laboratory experimental methods were used in developing the empirical correlations given in this and the following chapter. Examples of these two types of analytical distillation assays are:

ASTM Distillations
  D-86 for light petroleum products
  D-158 for gasolines through light gas oils
  D-1160 at 760 mm for middle oils
  D-1160 at 10 mm for heavy oils

TBP Distillations
  Podbielniak Hypercal
  Oldershaw column
  Spinning auger or band

The ASTM distillations are more rapid and cost less to run than the TBP distillations, while the latter are more accurate in defining the characteristics of the oil fractions. When available, the TBP assays are preferred as the basis for calculating properties, but they are often not available. ASTM assays are frequently the only analytical distillations available, so it is necessary to include ASTM's in the correlations and calculations.

TBP distillations usually require larger charge samples. The sample quantity depends upon the sizes of the still flash and the column, which are determined by the number and size of the TBP cuts that are to be collected for making specific gravity measurements. Typical numbers and sizes of these cuts may be ten 10% cuts or twenty 5% cuts. Sample requirements for analytical tests must be considered in selecting sizes of the equilibrium flash vaporization apparatus, as well as those of the batch distillation stills.

Equilibrium Flash Vaporization

Chapter 9 of Volume 1 describes equilibrium flash vaporization as a process in which the entire mixture separates into vapor and liquid phases that coexist in equilibrium, i.e., a reversible process. Chapter 11 of this volume presents thermodynamic methods of computing this process for a multicomponent system. This chapter looks at this process in a different way and develops empirical correlations relating the EFV (equilibrium flash vaporization) with the TBP and the ASTM assays. Such correlations do not have any theoretical justification. They do use the same parameters, however, i.e., temperature, pressure, and liquid volume fraction vaporized. These common parameters seem to have inspired engineers to develop and use these empirical correlations for several decades.

Equilibrium phase separations of petroleum mixtures are generally run on once-through flash vaporization units operated at a steady rate, slowly enough to reach equilibrium. Heat is supplied externally to the feed or internally during vaporization, or both. Other vapor-liquid equilibrium measuring devices, such as the vapor recirculating still and the stirred or rocked equilibrium cell or bomb, would be satisfactory for light oils but cannot be used for heavy oils because their longer holding times causes thermal decomposition or cracking and coking at temperatures greater than 700°F.

The EFV apparatus must be large enough to provide sufficiently large samples of the equilibrium vapor and liquid for use in making the desired analyses. On the other hand, the apparatus should be as small as possible to minimize equipment cost, time required to reach equilibrium, heat losses, and the quantities of oils handled.

VLE measurements can be made on oil fractions with a small sample in a PVT (pressure-volume-temperature) apparatus also. The smallest scale PVT apparatus is a small-bore, high-pressure glass capillary tubing, in which the sample is stirred by a smaller steel ball that is moved up and down by an external magnet. This is the apparatus that was used by Bahlke and Kay (2) in phase determinations for a gasoline and a naphtha. Properties of the coexisting equilibrium vapor and liquid were not determined, because it was impossible to take samples of the two phases.

This small-scale PVT apparatus can be operated at high pressures and temperatures, with good reproducibility, to make bubble- and dew-point observations. The fraction vaporized at intermediate temperatures, i.e., between the bubble and dew points, can be measured also in this type of apparatus, although these intermediate observations are of limited value, as the measured volumes of vapor and liquid are of unknown compositions, densities, and molecular weights. When samples of the equilibrium vapor and liquid can be taken and analyzed, the partially vaporized experiments on similar PVT apparatus can yield very useful data. Sampling is just not possible on most PVT apparatus, however.

Because of the requirement for samples and their analyses, vapor-liquid phase equilibria determinations for petroleum fractions are obtained in larger apparatus. The once-through EFV still is the most frequently used (5,6,8,14,15,19,21,22,24). As previously mentioned, the once-through EFV still may have external feed preheating, internal heating, or both. Two of the previous nine references (15,24) used internal heating in the EFV still, while all the others have feed preheating and only enough internal heat to compensate for heat losses.

As can be seen by the reference key in Table 12.1, most of the published experimental EFV data, up to 1961 when the first edition of this chapter was written, were for atmospheric pressure and vacuum conditions. Very few high-pressure data were available. Except for light oils, high pressures are not used in oil processing anyway. Another deficiency in the data available for correlation development is the lack of data on the properties (gravity and boiling point assays) on the equilibrium vapor and liquid products. With this latter information, it is not possible to convert flashed products from the observed quantities to liquid and vapor to weight (pounds or pound moles) quantities. Because of these and other deficiencies in the data, the empirical correlations developed from these available data will be approximations to be used only when observed EFV data are not available, or when it is not possible to make more rigorous calculations by the pseudo or hypothetical component method. The latter is presented in Chapters 14 and 15.

Empirical Correlations

Twenty-five years ago when this work was done, technical literature contained efforts to solve the petroleum phase behavior problem, with many correlating the three distillations, ASTM, TBP, and EFV at atmospheric pressure (6,12,18,23,24,27). None of these methods were sufficiently complete to permit making phase equilibria

calculations for all operating ranges of temperature and pressure encountered in petroleum refining. This and the following chapter present empirical correlations that can be used to predict phase relationships over a pressure range of 10 mm Hg to the focal point.

Two sets of charts were developed for predicting the equilibrium phase conditions for petroleum fractions. One set predicts the phase relationships for atmospheric and higher pressures, while the other predicts the phase relationships in the vacuum region. These relationships were developed from data on stocks of continuous boiling range. No stocks with discontinuous TBP-percent-off-vs.-temperature relationships were included in the development of the correlations.

The empirical correlations presented in this chapter for atmospheric and higher pressures are improvements over similar correlations given previously (7), while the correlations for vacuum conditions presented in the next chapter extend the applications of this empirical technique down to 10 mm. Figure 12.1 illustrates the interrelationship between atmospheric and 10-mm Hg distillations, with various conversion routes summarized in Table 12.2.

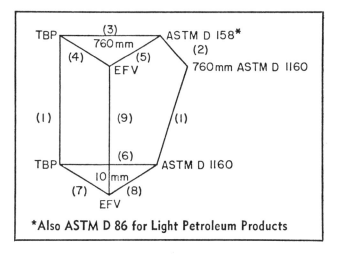

Figure 12.1. Interrelationships between empirical correlations relating ASTM-TBP-EFV at 760 and 10 mm Hg-pressure.

The average deviation of these charts from all available experimental data ranges from 4°F to 10°F. These charts are consistent and can be used satisfactorily for approximations in process design work. Detailed evaluations of these charts are given. This material was included in the first edition of this book because it offered a practical method of solving phase equilibria problems for petroleum fractions, which were solved by the more rigorous thermodynamic based K-values for mixtures of

Table 12.2
Interrelationships in Figure 12.1

Route	Figure	Remarks
1	13.1	Applicable for all stocks
2	13.2	Applicable for gas oils only
3	12.4 & 12.5	For light oils with N.B.P. < 800°F.
4	12.6 & 12.7	ditto
5	12.8 & 12.9	ditto
6	13.3	For heavy oils with N.B.P. > 800°F.
7	13.4 & 13.5	ditto
8	13.6 & 13.7	ditto
9	13.8	For both light & heavy oils

discrete components. This rigorous method is applied to petroleum fraction phase equilibrium problems in Chapters 14 and 15.

EFV Correlations

Empirical correlations for predicting equilibrium flash vaporization conditions (percent vapor at given temperature and pressure) are based on either of the analytical distillation curves, i.e., the ASTM or the TBP assay. All three distillations (ASTM, TBP, and EFV) may be plotted on the same coordinates, i.e., temperature versus liquid volume percent vaporized, when they are at the same pressure, which would usually be atmospheric (760 mm). Note that all three of these distillations can be run at other than atmospheric pressure. In Figure 12.1 all three distillations are shown at both 760 and 10 mm, and all could be run at those pressures. There is a standard procedure for the ASTM D-1160 at 10 mm, but no standards for the TBP or EFV at this low pressure. The EFV can also be run at super atmospheric pressures. Note that some of the routes, or paths, described in Figure 12.1 and Table 12.2 are vacuum processes covered in Chapter 13.

The ASTM and TBP temperature-vs.-volume-percent-off curves represent a series of points on batch distillation runs, in contrast to the EFV curve on the same coordinates, which represents a series of equilibrium flash vaporization separations, from 0% to 100% flashed. Another difference between these curves is that the EFV curves at different pressures must be obtained by experiments or by estimations from the curve at a reference pressure; whereas the ASTM and TBP distillation assays are usually run only at one pressure, namely atmospheric, except for special reduced pressure assays that are discussed in Chapter 13.

These three atmospheric pressure curves are shown in Figure 12.2 for a naphtha-kerosene blend. The lower

three curves that are labeled ASTM, TBP, and EFV 0 lb/in.² ga and cross one another are all for atmospheric pressure. In addition to these three 760-mm curves there are five other EFV curves at higher pressures, i.e., 20 to 300 lb/in.² ga. The higher pressure EFV curves might be obtained by experimental measurements or by a combination of experiments and cross plotting. Figure 12.2 is typical of plots obtained from similar data for other oil fractions.

The three atmospheric pressure distillation curves in Figure 12.2 show two characteristics that are common to such curves for all petroleum fractions, i.e., they always intersect, and the EFV curve is the flattest while the TBP curve is the steepest, suggesting an empirical correlation

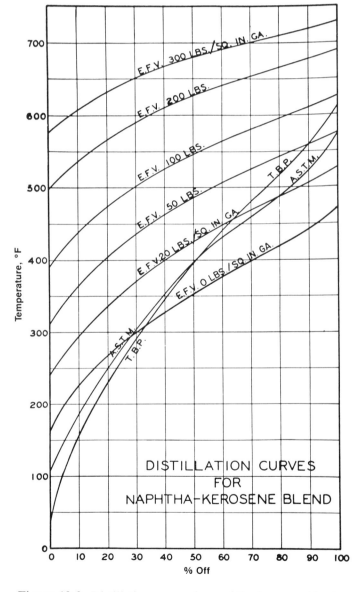

Figure 12.2. Distillation curves for naphtha-kerosene blend.

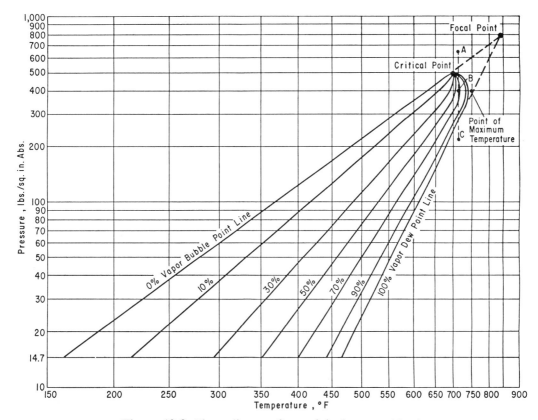

Figure 12.3. Phase diagram for naphtha-kerosene blend.

method. An atmospheric pressure EFV curve can be drawn from the ASTM or TBP with two empirical correlations, one to locate a key point on the EFV and another to give the EFV slopes, both from the corresponding values on the ASTM or TBP assay.

The vertical displacement of the higher pressure EFV curves on the temperature scale of Figure 12.2 suggests using similar coordinates to those of the vapor pressure charts in Chapter 7, i.e., log of vapor pressure versus reciprocal of absolute temperature plus a small constant, for constructing the *PTX* (pressure, temperature, percent vaporized) phase diagrams for petroleum fractions.

Figure 12.3 gives such a phase diagram for the naphtha-kerosene blend shown in Figure 12.2 via other coordinates. Figure 12.3 is typical of the phase diagrams that are obtained for petroleum fractions. On these coordinates, the lines of constant fraction vaporized are straight and intersect at a focal point, which is above the critical point. The EFV curves of Figure 12.2 are cross-plots of constant pressure paths across the phase diagram of Figure 12.3. Note that the bubble- and dew-point lines in Figure 12.3 fan out at lower pressures. Although these lines are straight on Figure 12.3, they should not be extrapolated to vacuum conditions.

Piromoov and Beiswenger (23) developed the first such empirical correlation in 1929, using experimental

"pipe still" or EFV distillation data on oils. In that correlation, the EFV curves were assumed to be straight lines on the temperature-vs.-percent-vaporized scales. This early work also assumed that the higher pressure EFV curve had the same slope as the atmospheric pressure curve. Both of these were oversimplified assumptions.

Edmister and Pollock (6) avoided these simplifications in 1948 in a widely used method with large-scale useable charts (7). That method was developed from experimental data on Mid-Continent crude oil fractions, with atmospheric TBP 50% points ranging from 150°F to 860°F. The present correlations, originally published in the first edition of this book in 1961, are similar to the earlier Edmister-Pollock charts, except that improvements were made in the high-temperature portion of the three-way correlations relating TBP, ASTM, and EFV 50% temperatures at atmospheric pressure. These improvements were based upon additional experimental data on high boiling fractions (3).

ASTM-TBP-EFV Relationships

A three-way correlation was developed between the EFV curve and the ASTM and TBP distillation assays, with all three at atmospheric pressure. This is the "keystone" of the empirical correlation method, as is evident by the interrelationships shown in Figure 12.1 and by the distillation curves shown in Figure 12.2.

The atmospheric pressure correlations given in Figures 12.4 through 12.9 are based upon using the 50% evaporated temperatures as the reference points on the ASTM, TBP, and EFV curves. Correlations of temperature differences for segments of the curves complete the representation of the distillation curves. The correlations for relating the ASTM and TBP 50% point temperatures and the temperature differences on these two assays permit constructing a TBP curve from an ASTM assay, or vice versa. Likewise, the correlations for estimating the

EFV 50% point temperatures and temperature differences from the corresponding ASTM or TBP values permit constructing the atmospheric EFV curve from the ASTM or TBP assay.

Figures 12.4 through 12.9 are 1961 versions of the 1948 Edmister-Pollock correlations (6). These revisions include: (1) more accurate ASTM-TBP 50% temperature for the entire range, (2) improvement in the high temperature regions of the ASTM-EFV and the TBP-EFV 50% temperature correlations, and (3) replotting the slope correlations as temperature differences for segments of the distillation curves to make the charts more convenient to use. Shortly after these improved correlations were published, they were evaluated by the

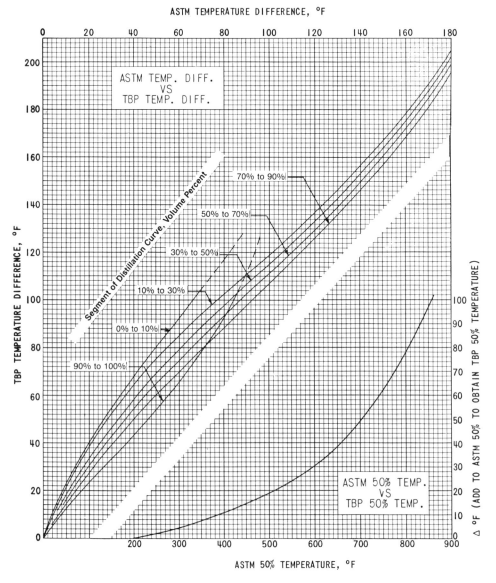

Figure 12.4. (lower right). ASTM 50% temperature vs. TBP 50% temperature.

Figure 12.5. (top left). ASTM temperature difference vs. TBP 50% temperature difference.

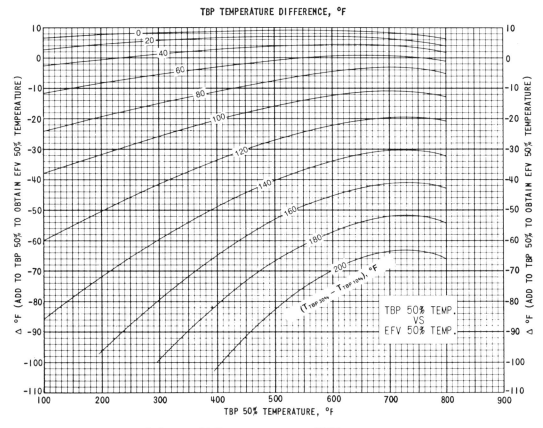

Figure 12.6. TBP 50% temperature vs. EFV percent temperature.

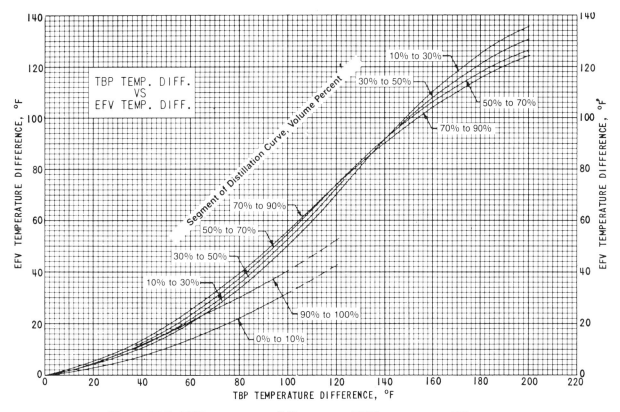

Figure 12.7. TBP temperature difference vs. EFV temperature difference.

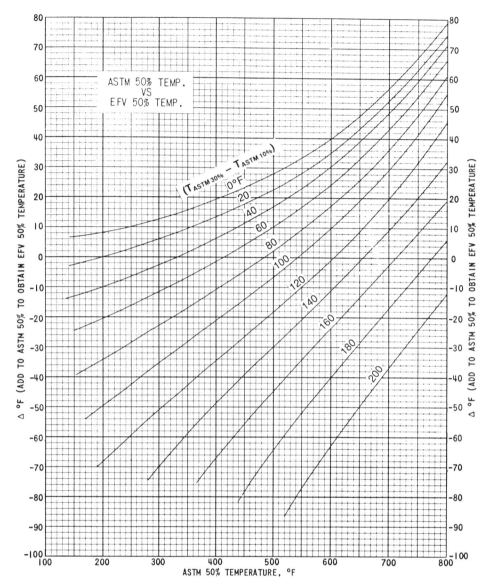

Figure 12.8. ASTM 50% temperature vs. EFV 50% temperature.

group that prepared the *API Technical Data Book* and several of these graphical correlations have since appeared in that valuable compilation.

The high temperature range of the atmospheric 50% point correlations and the low temperature range of the 10-mm Hg 50% point correlation (the latter is presented in Chapter 13) overlap each other by 125°F. In this region of overlap, these two correlations are consistent. Thus, for stocks that fall in this overlap region, either the atmospheric or the 10-mm Hg 50% point correlations may be used to obtain the same results.

Figures 12.4 and 12.5 may be used to convert ASTM assays to the equivalent TBP assays, and vice versa. As can be seen, the procedure is to start with the 50% points (Figure 12.4) and work up and down from that mid-point by the difference correlation (Figure 12.5). The

TBP > ASTM conversion should be more reliable than the ASTM > TBP conversion.

Figure 12.6, the TBP-EFV 50% point correlation and the comparable one of Edmister-Pollock (6) are similar, but with a revision in the lower than 650°F TBP 50% point region. This revision was made to give better agreement with experimental data. Above a TBP 50% point temperature of 650°F some compromises were made so that this portion of the correlation would agree better with vacuum data on heavy oils. Figure 12.7 is the temperature difference correlation that is used with Figure 12.6 in constructing an EFV curve from a TBP assay.

Figure 12.8, the ASTM-EFV 50% point correlation was developed to give a consistent three-way correlation of the 50% point temperatures of the EFV curve and the

ASTM and TBP assays. Figure 12.9 is the temperature difference correlation that is used with Figure 12.8 in constructing an EFV curve from an ASTM assay.

Detailed evaluations of the six charts, Figures 12.4 through 12.9, have been made and are shown in Tables 12.3, 12.4, and 12.5. These evaluations were made with experimental data of four sources (3,6,8,26). The experimental EFV data of Piromoov and Beiswenger (23) were not included because the temperatures in that work were measured with thermometers and no stem corrections seem to have been made, possibly resulting in reported EFV temperatures that are too low. Also, the reported ASTM distillations were run by the Saybolt 10% distillation method, which gives temperatures somewhat higher than the ASTM. The method of Maxwell (16) for predicting EFV curves from TBP or ASTM assays was primarily based upon the Piromoov-Beiswenger data, and as expected, predicted EFV temperatures that are 15°F to 20°F lower than those predicted by using Figures 12.4 through 12.9. A summary of the evaluations is given in Table 12.6.

Pressure Effect on EFV Curves

The effect of pressure on the EFV curves is estimated by constructing a phase diagram on $\log P$ versus $1/T$ co-ordinates, starting with the atmospheric EFV curve, previously predicted from the TBP or ASTM assays by use of Figures 12.6 and 12.7, or Figures 12.8 and 12.9. With the atmospheric EFV points determining the lower ends of the constant percent flashed lines, and the focal point at which these lines intersect, the phase diagram can be constructed. The method for locating the focal point previously proposed by Edmister and Pollock (6) used empirical relations between the critical and focal point temperatures and pressures. These empirical relationships were developed from experimental data on the critical point and phase diagram plots. Figures 12.10 and 12.11 are the critical point correlations. Figures 12.12 and 12.13 are the focal point correlations, in terms of the differences $(T_F - T_C)$ and $(P_F - P_C)$. These are also reprinted from previous publications (6,7).

For convenience, this focal point estimation method has been simplified to a three-parameter graphical correlation and incorporated into the phase diagram construction chart given in Figure 12.14. An example in the use of this chart is shown in Figure 12.15. By using Figures 12.4 through 12.15 phase conditions for petroleum fractions between atmospheric pressure and the focal point may be estimated. The phase diagram construction chart, given as Figure 12.14, has been designed for use with ASTM distillation assay data for predicting phase

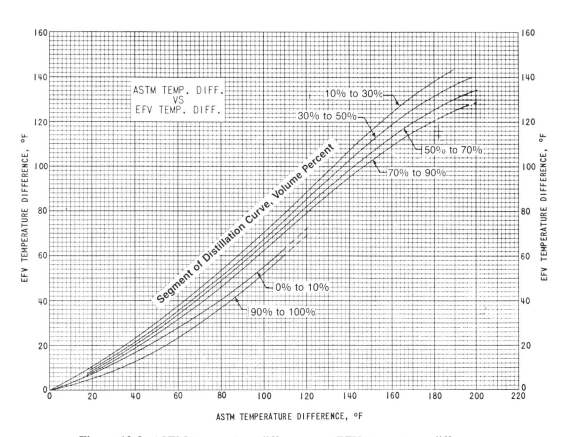

Figure 12.9. ASTM temperature difference vs. EFV temperature difference.

**Table 12.3
Evaluation of Atmos ASTM-TBP Correlation,
Figures 12.4 and 12.5**

Oil Stock (Reference)	Grav. °API	Volume Percent Distilled	Atm. ASTM Obs.	Atmos TBP Obs.	Atmos TBP Calc.	Diff. = Calc. − Obs.
Naphtha—	47.8	10	172	160	150	− 10
kerosene		30	298	289	292	+ 3
blend		50	393	396	403	+ 7
(6)		70	455	481	479	− 3
		90	516	559	550	− 9
Extraction	68.7	0	156	137	144	+ 7
naphtha		10	161	146	157	+ 11
(6)		30	163	153	162	+ 9
		50	164	155	164	+ 9
		70	168	166	172	+ 6
		90	180	190	189	− 1
		100	220	210	232	+ 22
Cleaner's	57.3	0	194	104	144	+ 40
naphtha		10	223	198	197	− 1
(6)		30	241	240	232	− 8
		50	256	259	258	− 1
		70	272	289	282	− 7
Virgin	63.9	0	115	62	48	− 14
light		10	158	122	120	− 2
naphtha		30	197	185	183	− 2
(6)		50	222	227	223	− 4
		70	248	257	261	+ 4
		90	292	325	315	− 10
Virgin	49.2	10	318	303	304	+ 1
heavy		30	328	327	326	− 1
naphtha		50	340	347	347	0
(6)		70	355	376	370	− 6
		90	380	416	404	− 12
Reformed	49.6	30	276	264	263	− 1
gasoline		50	312	318	317	− 1
(6)		70	341	351	358	+ 7
		90	381	410	408	− 2
Treated	49.7	10	170	148	137	− 11
pressure		30	266	277	253	− 24
distillate		50	340	367	347	− 20
(6)		70	392	429	413	− 16
		90	440	486	472	− 14
C.R.C.		10	140	90	106	+ 16
stock 1		30	170	155	158	+ 3
(3)		50	188	192	188	− 4
		70	202	210	210	0
		90	220	232	235	+ 3
C.R.C.		10	260	247	253	+ 6
stock 2		30	264	264	262	− 2
(3)		50	270	278	273	− 5
		70	275	286	281	− 5
		90	283	307	292	− 15

**Table 12.3
Continued**

Oil Stock (Reference)	Grav. °API	Volume Percent Distilled	Atm. ASTM Obs.	Atmos TBP Obs.	Atmos TBP Calc.	Diff. = Calc. − Obs.
C.R.C.		10	321	307	310	+ 3
stock 3		30	330	330	330	0
(3)		50	339	347	346	− 1
		70	348	370	360	− 10
		90	367	393	386	− 7
C.R.C.		10	408	397	400	+ 3
stock 4		30	420	425	425	0
(3)		50	430	447	443	− 4
		70	440	464	459	− 5
		90	456	490	481	− 9
C.R.C.		10	482	462	473	+ 9
stock 5		30	505	511	516	+ 5
(3)		50	518	540	539	− 1
		70	535	568	565	− 3
		90	560	603	599	− 4
C.R.C.		10	473	450	455	+ 5
stock 6		30	512	518	518	0
(3)		50	542	567	565	− 2
		70	567	602	601	− 1
		90	595	640	638	− 2
C.R.C.	37.0	10	436	409	416	+ 7
stock 7		30	493	494	498	+ 4
(3)		50	519	535	540	+ 5
		70	540	565	571	+ 6

conditions of petroleum fractions at super atmospheric pressure. As can be seen, the ASTM parameters used are the VABP (volumetric average boiling point), the 10 to 30% difference, and the 10 to 90% slope. Reference might be made to Figures 12.14 and 12.15 for clarification.

A calculation form in the upper left-hand corner of Figures 12.14 and 12.15 simplifies the estimation of the atmospheric EFV, the VABP, and the ASTM slopes. The three-parameter, right hand section of the chart is for locating the focal point. This focal point section was prepared from the original method, i.e., Figures 12.10 through 12.13. The three-parameter portion of Figure 12.14 is more convenient to use than the four original charts method, but probably not as reliable. The example in Figure 12.15 shows the application of this construction chart for a gasoline.

For heavy oils, the 90% evaporated temperature point on the ASTM will often be unavailable and must be estimated by extrapolation. For the heaviest oils, such as re-

(text continued on page 23)

Table 12.4
Evaluation of Atmos TBP-EFV Correlation, Figures 12.6 and 12.7

Oil Stock (Reference)	Grav. °API	Volume Percent Distilled	Atm. TBP Obs.	Atmos EFV Obs.	Atmos EFV Calc.	Diff. = Calc. – Obs.
Naphtha—	47.8	10	160	214	214	0
kerosene		30	289	295	295	0
blend		50	396	352	355	+ 3
(6)		70	481	399	397	− 2
		90	559	442	435	− 7
C.R.C.	37.0	10	409	475	477	+ 2
stock 7		30	494	501	514	+ 13
(3)		50	535	516	526	+ 10
		70	565	532	535	+ 3
C.R.C.	22.4	0	462	606	600	− 6
stock 8		10	557	630	629	− 1
(3)		30	617	652	649	− 3
		50	662	668	663	− 5
		70	705	684	677	− 7
		90	773	706	708	+ 2
C.R.C.	26.0	10	574	640	654	+ 14
stock 9		30	660	682	692	+ 10
(6)		50	720	718	714	− 4
		70	778	731	736	+ 5
		90	863	767	780	+ 13

Table 12.5
Evaluation of Atmos ASTM-EFV Correlation, Figures 12.8 and 12.9

Oil Stock (Reference)	Grav. °API	Volume Percent Distilled	Atm. TBP Obs.	Atmos EFV Obs.	Atmos EFV Calc.	Diff. = Calc. – Obs.
WCE-DHP		10	150	178	174	− 4
stock 1		30	200	208	204	− 4
(6)		50	235	224	222	− 2
		70	268	245	238	− 7
		90	325	273	268	− 5
WCE-DHP		0	203	258	260	+ 2
stock 2		10	255	278	283	+ 5
(6)		30	292	301	304	+ 3
		50	315	314	315	+ 1
		70	339	327	326	− 1
		90	400	354	358	+ 4
		100	475	380	391	+ 11
WCE-DHP		0	300	344	345	+ 1
stock 3		10	327	354	356	+ 2
(6)		30	358	371	374	+ 3
		50	375	380	382	+ 2
		70	393	388	390	+ 2
		90	415	400	399	− 1
		100	438	405	405	0

Table 12.5
Continued

Oil Stock (Reference)	Grav. °API	Volume Percent Distilled	Atm. ASTM Obs.	Atmos EFV Obs.	Atmos EFV Calc.	Diff. = Calc. – Obs.
WCE-DHP		0	230	298	304	+ 6
stock 4		10	285	322	329	+ 7
(6)		30	343	357	365	+ 8
		50	399	387	399	+ 11
		70	457	415	430	+ 15
		90	515	448	460	+ 12
		100	568	472	479	+ 7
WCE-DHP		0	388	448	451	+ 3
stock 5		10	424	457	466	+ 9
(6)		30	450	472	480	+ 8
		50	473	483	491	+ 8
		70	498	496	509	+ 13
		90	540	519	529	+ 10
		100	580	535	542	+ 7
WCE-DHP		0	415	474	481	+ 7
stock 6		10	456	484	498	+ 14
(6)		30	477	500	509	+ 9
		50	496	509	518	+ 9
WCE-DHP		0	432	484	491	+ 7
stock 7		10	462	498	503	+ 5
(6)		30	485	508	516	+ 8
		50	503	517	525	+ 8
		70	520	526	533	+ 7
		90	550	540	547	+ 7
		100	580	550	556	+ 6
WCE-DHP		0	410	480	488	+ 8
stock 8		10	454	500	507	+ 7
(6)		30	502	525	536	+ 11
		50	537	543	555	+ 12
		70	575	563	575	+ 12
		90	650	606	617	+ 11
		100	730	648	654	+ 6
WCE-DHP		0	507	570	584	+ 14
stock 9		10	535	585	595	+ 10
(6)		30	571	600	616	+ 16
		50	597	611	628	+ 17
		70	630	628	644	+ 16
		90	685	660	672	+ 12
		100	780	712	720	+ 8
WCE-DHP		0	536	607	620	+ 13
stock 10		10	580	621	639	+ 18
(6)		30	605	637	653	+ 16
		50	625	647	663	+ 16
		70	650	658	675	+ 17
		90	682	675	689	+ 14
		100	720	691	701	+ 10

(table continued on next page)

Table 12.5
Continued

Oil Stock (Reference)	Grav. °API	Volume Percent Distilled	Atm. ASTM Obs.	Atmos EFV Obs.	Atmos EFV Calc.	Diff. = Calc. – Obs.
WCE-DHP stock 11 (6)		0	570	644	664	+ 20
		10	612	660	682	+ 22
		30	645	678	701	+ 23
		50	670	692	714	+ 22
		70	700	707	728	+ 21
		90	756	739	757	+ 18
Pressure distillate (6)		10	165	213	192	– 21
		30	261	270	259	– 11
		50	330	315	301	– 14
		70	391	348	335	– 13
		90	463	382	375	– 7
		100	520	410	396	– 14
Pressure distillate (6)	50.8	0	80	122	147	+ 25
		10	165	191	191	0
		30	258	254	255	+ 1
		50	321	292	293	+ 1
		70	382	326	327	+ 1
		90	432	358	352	– 6
		100	500	384	380	– 4
Mid-Cont. fraction (6)		0	250	320	316	– 4
		10	284	334	326	– 8
		30	330	358	354	– 4
		50	380	384	382	– 2
		70	442	418	417	– 1
		90	525	467	465	– 2
		100	615	520	509	– 11
Mid-Cont. fraction (6)		0	280	352	364	+ 12
		10	327	380	385	+ 5
		30	408	424	439	+ 15
		50	490	472	491	+ 19
		70	584	527	551	+ 24
		90	706	603	632	+ 29
Treated pressure distillate (6)	52.3	0	65	127	126	– 1
		10	141	175	164	– 11
		30	244	237	237	0
		50	317	287	282	– 5
		70	378	325	316	– 9
		90	439	358	348	– 10
		100	493	390	368	– 22
Naphtha bottoms (6)	39.3	0	276	334	352	+ 18
		10	355	385	392	+ 7
		30	405	415	422	+ 7
		50	430	428	435	+ 7
		70	452	441	445	+ 4
		90	505	469	472	+ 3
		100	561	491	491	0

Table 12.5
Continued

Oil Stock (Reference)	Grav. °API	Volume Percent Distilled	Atm. ASTM Obs.	Atmos EFV Obs.	Atmos EFV Calc.	Diff. = Calc. – Obs.
Naphtha bottoms (6)	44.2	0	170	260	246	– 14
		10	258	300	292	– 8
		30	327	334	336	+ 2
		50	370	356	360	+ 4
		70	406	376	378	+ 2
		90	443	393	395	+ 2
		100	500	410	416	+ 6
Naphtha kerosene blend (6)	47.8	0	96	160	158	– 2
		10	172	214	196	– 18
		30	298	295	290	– 5
		50	393	352	353	+ 1
		70	455	399	388	– 11
		90	516	442	420	– 22
		100	575	464	443	– 21
Mid-Cont. gas oil (6)	35.5	0	475	533	551	+ 18
		10	512	551	566	+ 15
		30	543	573	584	+ 11
		50	570	589	598	+ 9
		70	604	607	615	+ 8
		90	663	636	646	+ 10
		100	730	665	673	+ 8
Mid-Cont. gas oil (6)	36.9	0	268	364	365	+ 1
		10	346	403	404	+ 1
		30	464	481	491	+ 10
		50	567	541	561	+ 20
		70	641	596	605	+ 9
		90	746	652	671	+ 19
		100	804	683	693	+ 10
Winkler gas oil-naphtha blend (6)	32.0	10	319	385	373	– 12
		30	432	457	455	– 2
		50	526	520	517	– 3
		70	612	571	571	0
		90	692	623	617	– 4
		100	747	650	637	– 13
Pipe Still pressure distillate (8)	50.3	10	165	203	185	– 18
		30	287	276	275	– 1
		50	365	332	324	– 8
		70	428	372	360	– 12
		90	500	403	400	– 3
Shell Still pressure distillate (8)	46.6	10	260	300	295	– 5
		30	367	360	371	+ 11
		50	431	400	409	+ 9
		70	468	430	428	– 2
		90	519	455	454	– 1

(table continued on next page)

Table 12.5
Continued

Oil Stock (Reference)	Grav. °API	Volume Percent Distilled	Atm. ASTM Obs.	Atmos EFV Obs.	Atmos EFV Calc.	Diff. = Calc. – Obs.
Texaco stock 1 (26)	37.2	10	494	535	530	– 5
		30	517	549	543	– 6
		50	534	563	551	– 12
		70	563	576	565	– 11
		90	620	589	595	+ 6
		100	703	615	634	+ 19
Texaco stock 2 (26)	55.0	10	235	262	260	– 2
		30	259	273	273	0
		50	281	283	284	+ 1
		70	304	295	294	– 1
		90	327	313	304	– 9
Texaco stock 3 (26)	48.0	10	264	315	300	– 15
		30	316	337	332	– 5
		50	357	359	354	– 5
		70	395	380	374	– 6
Texaco stock 4 (26)	43.0	10	382	409	413	+ 4
		30	394	415	419	+ 4
		50	410	422	427	+ 5
		70	426	428	434	+ 6
Texaco stock 5 (26)	43.0	10	293	342	343	+ 1
		30	371	405	395	– 10
		50	444	444	440	– 4
		70	516	481	482	+ 1
		90	589	518	523	+ 5
C.R.C. stock 7 (3)	37.0	0	330	437	424	– 7
		10	436	475	484	+ 9
		30	493	501	519	+ 18
		50	519	516	532	+ 16
		70	540	532	542	+ 10
		90	568	546	554	+ 8
C.R.C. stock 10 (3)	56.0	0	104	169	159	– 10
		10	172	206	192	– 14
		30	264	252	255	+ 3
		50	309	277	280	+ 3
		70	323	295	286	– 9
		90	337	310	292	– 19
		100	369	316	302	– 14

Table 12.6
Summary of Evaluations of the ASTM-TBP-EFV Correlation

Correlation (Figure No.)	Deviations °F Maximum	Deviations °F Average	Number of Experimental Points in Indicated Deviation Range 0–10°F	11–20°F	> 20°F
ASTM - TBP (12.4 & 12.5)	40	7	59	9	2
TBP - EFV (12.6 & 12.7)	14	6	17	3	0
ASTM - EFV (12.8 & 12.9)	29	9	126	55	11

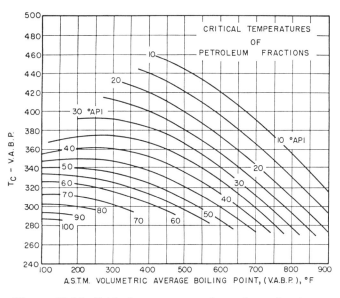

Figure 12.10. Critical temperature of petroleum fractions.

Example 1. Construct a phase diagram on Figure 12.14 for a stabilized gasoline with the following properties:

ASTM Distillation (Corrected for loss)

°API	IBP	10%	20%	30%	40%	50%	60%	70%	80%	90%	EP
61.6	63°F	117	156	186	213	239	262	287	312	342	400°F

Solution. The calculations made in obtaining the atmospheric pressure EFV curve are tabulated in the upper left-hand corner of Figure 12.15.

Step 1. The ASTM assay is tabulated in column 2.
Step 2. Temperature differences between indicated ASTM points are in column 3.

duced crudes, the 10-mm Hg ASTM D-1160 assay rather than the atmospheric ASTM may be available. Also the phase diagram of interest would be at vacuum conditions, requiring a different construction chart, which is discussed in Chapter 13.

The following is a solution of the example shown in Figure 12.15, including the detailed steps. The example is solved using the simpler three-parameter focal point correlation in Figure 12.14.

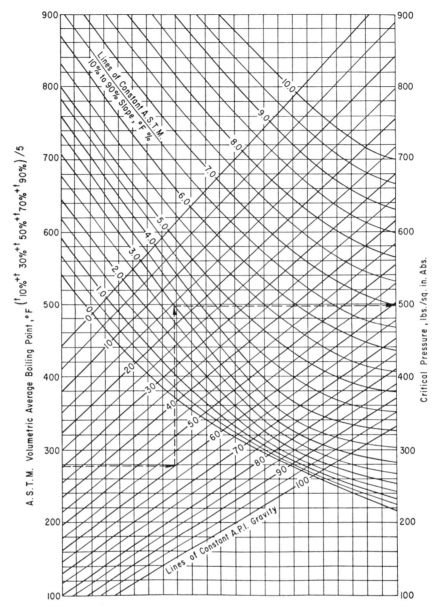

Figure 12.11. Critical pressure of petroleum fractions.

Step 3. From the ASTM 30-10% temperature difference of 69°F and the ASTM 50% point of
239°F, the EFV 50% is found from Figure 12.8
as 216°F and is entered in column 5.

Step 4. Temperature differences for the various segments on the EFV curve are found using the corresponding ASTM differences and Figure 12.9
and then tabulated in column 4.

Step 5. The remaining EFV 50% points in column 5 are
found by adding to or subtracting from the EFV
50% point the appropriate difference value in
column 4.

Step 6. The lower right corner of Figure 12.12 is entered with the ASTM VABP value and the focal

point is found by following the dashed lines and
the arrows.

Step 7. The EFV temperature points are entered on the
atmospheric pressure line.

Step 8. Straight lines are drawn from the atmospheric
EFV points to the focal point, thus completing
the phase diagram.

Figure 12.15 illustrates this example.

EFV Vapor and Liquid Properties

The previously described correlations are for predicting the *P-T-X* (pressure-temperature-percent vaporized)
relationships of petroleum fractions. In most design

(text continued on page 27)

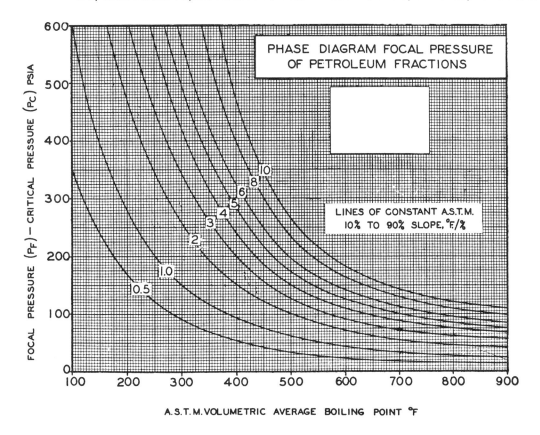

Figure 12.12. Phase diagram focal temperature of petroleum fractions.

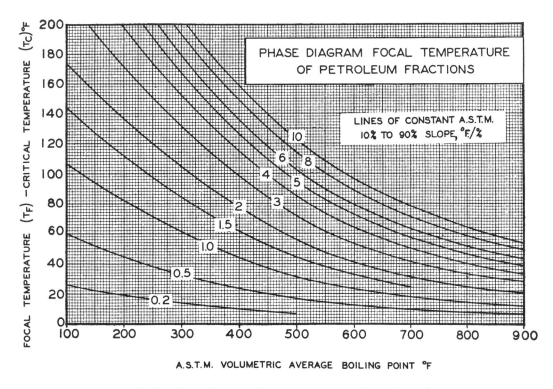

Figure 12.13. Phase diagram focal temperature of petroleum fractions.

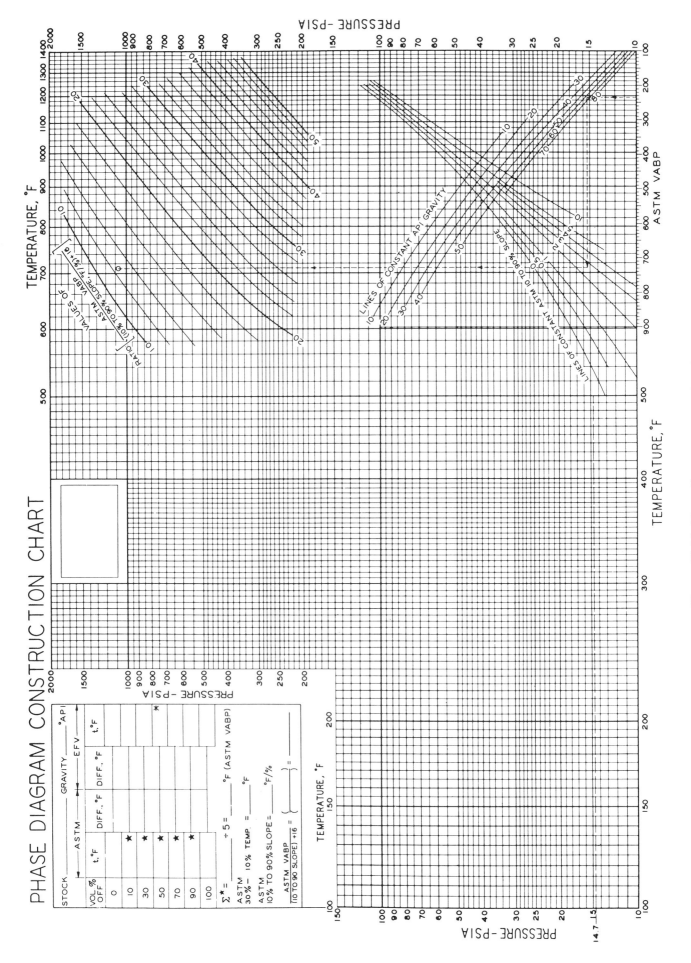

Figure 12.14. Phase diagram construction chart.

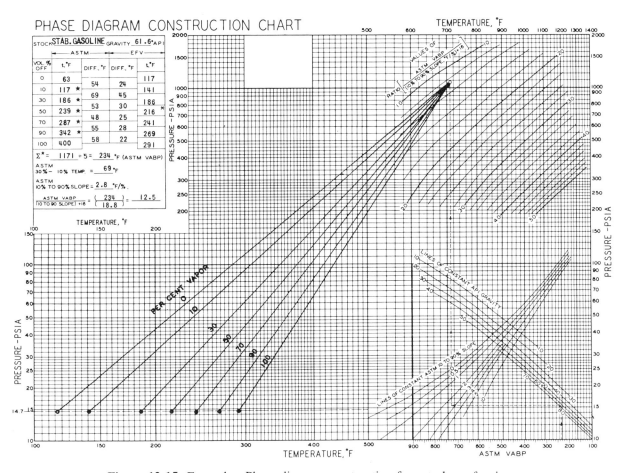

Figure 12.15. Example—Phase diagram construction for petroleum fractions.

problems the properties of the equilibrium vapor and liquid are required in addition to the *P-T-X* information.

From experimental data, three empirical correlations have been developed (12) for predicting the gravities and the ASTM assays of the equilibrium vapor and liquid from flash vaporization. These correlations are presented here as Figures 12.16, 12.17, and 12.18. These charts have no theoretical background and a rather tenuous empirical basis.

Figure 12.16 is used to estimate the API gravities of the vapor and liquid products from the EFV. The variables for this application are the ASTM 10-30% slope and the gravity of the charge plus the percent flashed. Vaporization pressure and temperature do not enter into this correlation as product gravities are assumed to be independent of the vaporization conditions.

Figures 12.17 and 12.18 are used for estimating the ASTM assays of the vapor and liquid products from the flashes. The readings from these charts are differences between the corresponding points on the product and the charge distillation assays. In these correlations the 10% and 70% points are used and the flashing pressure is a variable.

Example 2. For illustrating the use of Figures 12.17 and 12.18, product gravities and ASTM assays are estimated for the vapor and liquid from flashing a 43.2°API oil at 100 psig and 520°F. From a phase diagram it was found the 43.5% of the oil vaporized. The assays of the charge and the two equilibrium products are as follows:

ASTM Distillation Assay Points °F

Percent	Charge	Vapor	Liquid
10	248	233	308
30	322	272	370
50	374	307	411
70	424	347	452
90	486	399	504
°API	43.2	48.5	38.5

Step 1. From the ASTM assay of the charge—
ASTM 10-70 slope = (424 − 248)/60
= 2.9°F/%
ASTM 10-30 slope = (322 − 248)/20
= 3.7°F/%

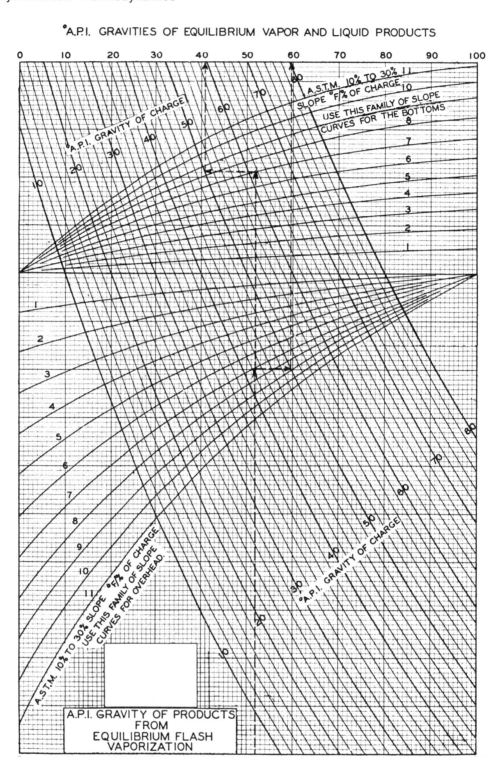

Figure 12.16. API gravity of products from equilibrium flash vaporization.

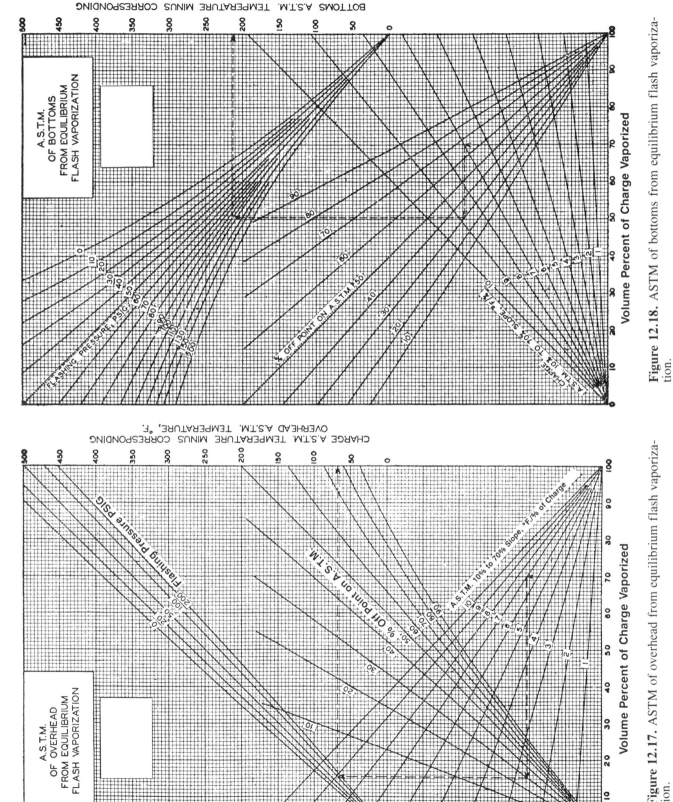

Figure 12.18. ASTM of bottoms from equilibrium flash vaporization.

Figure 12.17. ASTM of overhead from equilibrium flash vaporization.

Step 2. From Figure 12.16, find the gravities of the vapor and liquid by following the path indicated by dashed lines and arrows.
Step 3. From Figure 12.17, find the vapor ASTM assay points by following the path indicated.
Step 4. From Figure 12.18, find the liquid ASTM assay points by following the path indicated.

The results are shown in the previous tabulation.

References

1. Bahlke, W. H. and W. B. Kay, *Ind. Eng. Chem.*, 24, 291–301 (1932).
2. Beale and Docksey, *J. Inst. Pet. Tech.*, 21, 860 (1935).
3. Chevron Research Company, Richmond, CA.
4. Chu, J. C. and E. J. Staffel, *J. Inst. Pet. Tech.*, 41, 374 (March 1955).
5. Edmister, W. C., J. C. Reidel, and W. J. Merwin, *Trans. AIChE*, 39, 457–489 (1943).
6. Edmister, W. C. and D. H. Pollock, *Chem. Eng. Prog.*, 44, 905 (1948).
7. Edmister, W. C., *Pet. Ref.*, Oct. 1949, Nov. 1949, Dec. 1949.
8. Fancher, G. H., *Pet. Eng.*, 2, 176 (1931).
9. Filak, G. A., H. L. Sandlin, and G. L. Stockman, *Pet. Ref.*, 34, 4, 153 (1955).
10. Flanders, R. L. and G. M. Williams, Chevron Res. Co., Richmond, CA, private communication.
11. Geddes, R. L., *Ind. Eng. Chem.*, 33, 795 (1941).
12. Katz, D. L. and G. G. Brown, *Ind. Eng. Chem.*, 25, 1373–84 (1933).
13. Katz, D. L. and K. H. Hackmuth, *Ind. Eng. Chem.*, 29, 1072–77 (1937).
14. Leslie, E. H. and A. J. Good, *Ind. Eng. Chem.*, 19, 453–60 (1927).
15. Lockwood, J. A., R. L. LeTourneau, R. Matteson, and F. Sipos, *Anal. Chem.*, 23, 1398–1404 (1951).
16. Maxwell, J. B., *Data Book on Hydrocarbons*, pp. 222–229 D. van Nostrand Co., New York (1950).
17. Nelson, W. L. and M. Hansburg, *Oil & Gas J.*, pp. 45–47 (Aug. 3, 1939).
18. Nelson, W. L. and M. Souders, Jr., *Pet. Eng.*, 3, 1, 131 (1931).
19. Okamoto, K. K. and M. Van Winkle, *Pet. Ref.*, 28, 8, 113–120 (1949), 29, 1, 91–96 (1950).
20. Othmer, D. F., E. H. Ten Eyck, and S. Tolin, *Ind. Eng. Chem.*, 43, 1607–1613 (1951).
21. Packie, J. W., *Trans. AIChE*, 37, 51 (1941).
22. Paulsen, T. H., *Chem. Eng. Prog. Sym. Series*, 49, 6, 45 (1953).
23. Piromoov, R. S. and G. A. Beiswenger, *Proc. Amer. Petrol. Inst.*, 10, 2, 52–68 (1929).
24. Ragatz, E. G., E. R. McCartney, and R. E. Haylett, *Ind. Eng. Chem.*, 25, 975 (1935).
25. Smith, R. B., T. Dresser, H. F. Hopp, and T. H. Paulsen, *Ind. Eng. Chem.*, 43, 766–770 (1951).
26. Texaco, data by private communication.
27. Watson, K. M. and E. F. Nelson, *Ind. Eng. Chem.*, 25, 880 (1933).
28. Watson, K. M., E. F. Nelson and G. B. Murphy, *Ind. Eng. Chem.*, 27, 1460–1464 (1935).

13

Graphical Phase Equilibria for Petroleum Fractions at Sub-Atmospheric Pressures

Chapter 12 presented an empirical method for making vapor-liquid phase equilibria predictions for petroleum fractions at atmospheric and higher pressures. The phase diagram construction chart, Figure 12.14, was used in transforming atmospheric EFV curves to super-atmospheric pressures. The straightness of the constant-percent-vaporized lines between atmospheric and focal pressures suggest the possibility of extrapolating these lines to find the phase conditions at sub-atmospheric pressures.

Such an extrapolation is not reliable. Separate correlations are required for predicting phase equilibria conditions for heavy oils, such as reduced crudes and residua. The starting data for predicting the phase diagram for heavy oils will be a vacuum TBP or ASTM distillation assay, generally run at 10 mm Hg absolute pressure.

From the interrelationships shown in Figure 12.1, it is evident that there are two correlation routes that may be followed in going from a 10-mm assay to an EFV curve at some intermediate pressure, such as 50 mm Hg. Correcting the 10-mm TBP or ASTM up to 760 mm and then using the correlations of Chapter 12 to obtain the EFV and extrapolating the phase diagram down to 50 mm Hg is not recommended for heavy oils.

The alternate method wherein the 10-mm EFV curve is first estimated and then transformed up to 50 mm Hg is the better method. For heavy gas oils calculations by either procedure should give the same EFV curves, but EFV curves for higher boiling oils than heavy gas oil should be predicted by vacuum correlations. Likewise, EFV curves for lower boiling oils than heavy gas oil should be predicted by the high-pressure correlation.

Empirical correlations for predicting phase relationships in the vacuum region are presented in the chapter for high boiling petroleum fractions. These are based on experimental data from various sources.

Conversion of ASTM Distillation Assays

An empirical relationship between the 10-mm ASTM D-1160 and the ASTM D-158 assays has been developed from experimental data on gas oils. This method transposes the 10-mm ASTM D-1160 to a pseudo-ASTM D-1160 at atmospheric pressure by the use of a vapor pressure chart, Figure 13.1. This pseudo-ASTM D-1160 is then converted to an atmospheric ASTM D-158 by using volume-percent-distilled corrections plus stem corrections. This is done by the graphical correlation given on Figure 13.2.

Figure 13.2 is limited to gas oils. Evaluation of Figures 13.1 and 13.2 was made with 104 experimental points for 35 gas oils on the distillation curves. The results are given in Table 13.1, a summary of which follows:

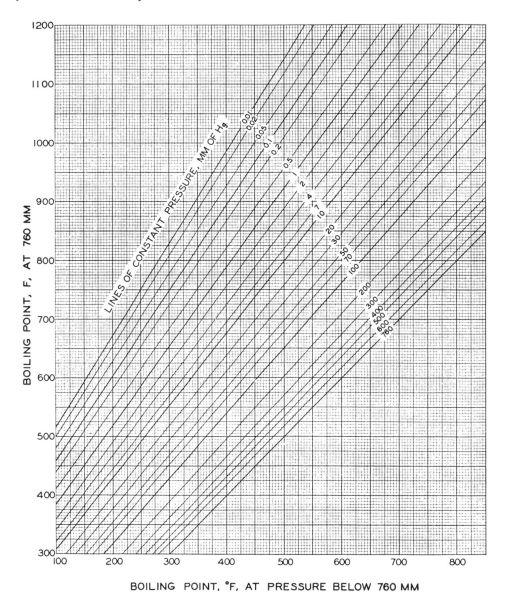

Figure 13.1. High-temperature boiling point conversion chart for narrow cut petroleum fractions and pure hydrocarbons.

Deviations (all points)

Maximum	62°F
Average	10°F

Number of Points

In deviation range:	0–10°F	68
	11–20°F	26
	> 20°F	10

The following example illustrates the use of Figures 13.1 and 13.2.

Example 1. Determine the ASTM D-158 distillation assay for a gas oil that has the following ASTM D-1160 at 10 mm Hg:

% Evap	0	10	30	50	70	90
°F	174	337	393	434	483	542

Solution.

Step 1. Transpose 10-mm D-1160 temperatures to atmospheric pressure using vapor pressure chart Figure 13.1 to obtain D-1160 at 760 mm.

Step 2. Determine Δ°F correction from Figure 13.2.

Step 3. Add Δ°F to the 760-mm D-1160 to obtain the ASTM D-158 assay temperatures.

The results are tabulated as follows:

Percent Off	Step 1		Step 2	Step 3
	ASTM D-1160		$\Delta\,^\circ F$	ASTM D-158
	10 mm	760 mm		
0	174	398	68	466
10	337	598	− 6	592
30	393	667	− 28	639
50	434	717	− 42	675
70	483	777	− 57	720
90	542	850	(− 73)	(777)

Phase Equilibria for Sub-Atmospheric Pressures

When using ASTM D-1160 "Reduced Pressure Distillation of Petroleum Products" as a control method for heavy, high-boiling petroleum products, it is necessary to develop correlations for predicting EFV and TBP assays from the simple ASTM D-1160 analytical distillation. The ASTM D-1160 procedure does not specify any particular operation pressure. However, most refinery operations specify a pressure of 10 mm Hg for the procedure, and therefore the bulk of the available ASTM D-1160 data are at 10 mm.

An empirical method for predicting the sub-atmospheric EFV curves of reduced crudes was published in 1953 by Paulsen (9), but this method is not entirely satisfactory because it is tedious and requires the 5-mm Hg Engler distillation assay. Also, three-way correlations of atmospheric pressure curves are not applicable to heavy oils or residua. These stocks must be distilled under reduced pressures to avoid thermal decomposition as much of the oil boils above 700°F, the temperature at which cracking occurs. Vacuum distillation data are necessary for developing correlations of ASTM-TBP-EFV correlations.

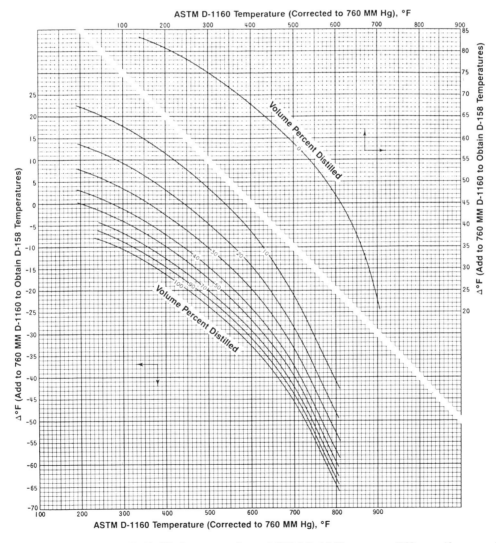

Figure 13.2. ASTM D-158 distillation curve from ASTM D-1160 curve at 760 mm (for gas oils).

Table 13.1
Evaluation of Figures 13.1 and 13.2 for Predicting 760-mm ASTM D-1160 Distillation Assays from 10-mm ASTM D-1160 Assays (4)

Stock	Vol %	Pseudo D-1160 via 10-mm Data and Figure 13.1	ASTM D-1160 Obs.	Calc via Figure 13.2	Diff. = Calc-Obs.
Gas oil 1	10	690	670	671	+ 1
Gas oil 2	10	582	560	578	+18
	30	682	648	652	+ 4
Gas oil 3	10	562	548	560	+12
	30	657	621	630	+ 9
	50	727	682	682	0
Gas oil 4	10	571	556	568	+12
	30	656	641	630	−11
	50	771	695	717	+22
Gas oil 5	10	629	610	619	+ 9
	30	655	640	629	−11
	50	685	662	648	−16
	70	739	690	690	0
Gas oil 6	10	542	563	542	−21
	30	690	655	658	+ 3
Gas oil 7	10	514	519	516	− 3
	30	634	621	610	−11
	50	766	699	714	+15
Gas oil 8	10	646	622	634	+12
	30	667	648	639	− 9
	50	692	669	654	−15
	70	731	689	683	− 6
Gas oil 9	10	483	499	488	−11
	30	634	621	610	−11
	50	766	699	714	+15
Gas oil 10	10	347	353	362	+ 9
	30	382	380	381	+ 1
	50	403	411	393	−18
	70	492	469	463	+ 8
Gas oil 11	10	569	568	566	− 2
	30	636	614	612	− 2
	50	674	648	638	−10
	70	720	682	674	− 8
Gas oil 12	10	583	528	578	+50
	50	643	634	611	−23
Gas oil 13	10	711	687	688	+ 1
Gas oil 14	10	496	492	500	+ 8
	50	527	512	507	− 5
	90	572	550	544	− 6
Gas oil 15	10	616	606	608	+ 2
	50	675	651	639	−12
	90	739	698	688	−10

Table 13.1
Continued

Stock	Vol %	Pseudo D-1160 via 10-mm Data and Figure 13.1	ASTM D-1160 Obs.	Calc via Figure 13.2	Diff. = Calc-Obs.
Gas oil 16	10	511	518	513	− 5
	30	550	534	535	+ 1
	50	569	548	545	− 3
	70	598	567	569	+ 2
	90	624	592	590	− 2
Gas oil 17	10	579	616	575	−41
	30	665	664	637	−27
	50	738	705	691	−14
	70	795	740	735	− 5
Gas oil 18	10	484	485	489	+ 4
	30	546	533	532	− 1
	50	611	582	583	+ 1
	70	671	639	633	− 6
	90	776	712	718	+ 6
Gas oil 19	10	522	532	524	− 8
	30	630	603	607	+ 4
	50	705	663	665	+ 2
Gas oil 20	10	583	577	578	+ 1
	30	670	654	642	−12
	50	725	691	681	−10
Gas oil 21	10	524	576	526	−50
	50	671	628	636	+ 8
	90	811	684	746	+62
Gas oil 22	10	662	646	647	+ 1
Gas oil 23	10	465	487	472	−15
	50	640	614	609	− 5
	90	820	731	753	+ 2
Gas oil 24	10	700	666	679	+11
Gas oil 25	10	573	563	570	+ 7
	50	745	694	697	+ 3
Gas oil 26	10	455	466	463	− 3
	50	511	513	493	−20
	90	720	668	675	+ 7
Gas oil 27	10	465	468	472	+ 4
	50	514	512	496	−16
Gas oil 28	10	588	595	583	−12
	50	721	691	678	−13
Gas oil 29	10	517	531	519	−12
	50	665	644	631	−13
	90	772	720	715	− 5
Gas oil 30	10	583	582	580	− 2
	50	725	694	681	−13
Gas oil 31	10	599	589	593	+ 4
	30	668	647	640	− 7
	50	718	685	675	−10
	70	778	719	721	+ 2

(table continued on next page)

Table 13.1
Continued

Stock	Vol %	Pseudo D-1160 via 10-mm Data and Figure 13.1	ASTM D-1160 Obs.	Calc via Figure 13.2	Diff. = Calc-Obs.
Gas oil 32	10	444	452	452	0
	30	500	500	590	− 10
	50	566	540	543	+ 3
	70	610	579	580	+ 1
	90	671	627	630	+ 3
Gas oil 33	10	625	617	615	− 2
	30	702	671	668	− 3
	50	754	698	704	+ 6
	70	810	726	747	+21
Gas oil 34	10	573	567	570	+ 3
	30	646	620	621	+ 1
	50	695	657	656	− 1
	70	730	693	683	− 10
	90	811	745	746	+ 1

Only a few sets of adequate experimental distillation were available for heavy stocks at low pressures when this work was being done. To augment these meager data, rather complete distillations (ASTM D-1160, TBP, and EFV) were obtained on three heavy oils at sub-atmospheric pressures. These experimental data are presented in Table 13.2. These data plus previously published data formed the basis for the correlations presented in this chapter.

The three-way correlation relating the ASTM D-1160, TBP, and EFV, all at 10 mm Hg, are shown as Figures 13.3 through 13.7. On these charts, the 50% temperatures and the temperature differences for various segments of the distillation curves are correlated. The low temperature range of the 10-mm Hg 50% correlations and the high temperature range of the atmospheric 50% correlations overlap each other by about 125°F. In this region of overlap these two sets of correlations are consistent. Thus, for stocks that fall in this overlap region, either the atmospheric pressure or the 10-mm Hg pressure 50% correlations may be used to obtain the same results. These charts were developed from data on five heavy stocks, including a heavy gas oil and four residua.

These correlations have been compared with available experimental data on heavy oils. Details of the evaluation are given in Table 13.3. The distribution of the deviations and the maximum deviations are as follows:

Table 13.2
Experimental ASTM-TBP-EFV Data for Residua Under Vacuum (2)

Stock	% off	ASTM D-1160			TBP		EFV	
		5 mm	10 mm	100 mm	10 mm	10 mm	25 mm	100 mm
Resid 1	0	118	182	275	170	275		400
	10	270	302	422	303	338		465
	20	328	350	478	357	377		510
	30	372	395	525	405	414		546
	40	420	450	577	456	452		585
	50	475	503	631	501	494		627
	60	530	560	690	552	538		670
	70	593	625		620*	588	630	
	80	667*	698*			655	700	
Resid 2	0	258	269	350	260	450	500	592
	10	448	466	580	464	511	562	655
	20	498	523	640	532	552	605	698
	30	542	572	683	575	591	647	
	40	582	610		615*	630	691	
	50	632	660		660*	670	733	
Resid 3	0	260	302	410	278	395		532
	10	390	422	550	412	455		592
	20	455	484	606	480	500		637
	30	490	528	643	524	537		678
	40	532	563		562	574		
	50	586	615		615*	620		
	60	650	680			682		

* Estimated

Table 13.3
Evaluation of Empirical Correlations (Figures 13.3–13.7) for Estimating 10-mm TBP and EFV (2)

Stock	% off	Observed 10 mm Hg Distillations °F ASTM D-1160	TBP	EFV	Calc. 10-mm Assays TBP Fig. 13.3	EFV Fig. 13.4/ 13.5	EFV Fig. 13.6/ 13.7	TBP (calc -obs) Fig. 13.3	EFV (calc -obs) Fig. 13.4/ 13.5	EFV (calc -obs) Fig. 13.6/ 13.7
Resid 1	0	182	170	275	160		268	−10		− 7
	10	302	303	338	293	338	329	−10	0	− 9
	30	395	405	414	391	414	402	−14	0	−12
	50	503	501	494	503	484	489	+ 2	−10	− 5
	70	625	620*	588	625	578	586	+ 5	−10	− 2
Resid 2	0	269	260	450						
	10	466	464	511	456	512	505	− 8	+ 1	− 6
	30	572	575	591	566	600	591	− 9	+ 9	0
	50	660	660*	670	660	660	660	0	−10	−10
Resid 3	0	302	278	395	279		393	+ 1		− 2
	10	422	412	455	412	457	454	0	+ 2	− 1
	30	528	524	537	522	544	540	− 2	− 7	+ 2
	50	615	615*	620	615	610	610	0	−10	−10
Gas oil 31	0	174		333						
	10	337	317	354	322	348	356	+ 5	− 6	+ 2
	30	393	388	396	386	395	396	− 2	− 1	0
	50	434	435	422	434	422	424	− 1	0	+ 2
	70	483	483	445	483	450	452	0	+ 5	+ 7
	90	542	552	481	542	495	488	−10	+14	+ 7
Gas oil 35	0		350	440		464			+24	
	10		439	488		486			− 2	
	30		489	524		515			− 9	
	50		531	543		538			− 5	
	70		579*	568		566			− 2	
Resid 4	10		298	324		327			+ 3	
	30		416	405		420			+15	
	50		517	492		496			+ 4	
Resid 5	10		280	314		311			− 3	
	30		378	390		384			− 6	
	50		478	466		459			− 7	
Resid 6	10		356	397		393			− 4	
	30		491	502		501			− 1	
	50		616	601		601			0	

* Estimated

	ASTM D-1160 to TBP	TBP to EFV	ASTM D-1160 to EFV
Deviations			
Maximum	14°F	24°F	12°F
Average	5°F	6°F	5°F
Number of points in Deviation range:			
0 to 10°F	16	26	16
11 to 20°F	1	2	1
> 20°F	0	1	0

Example 2. The application of Figures 13.1, 13.2, and 13.3 will be illustrated by finding the atmospheric TBP curve for a heavy gas oil from the following information:

10-mm ASTM D-1160 Assay

Volume % Off	IBP	10	30	50	70	90
Temperature °F	397	454	496	531	579	640

Solution. Following are the step-by-step calculations and a table of the results.

Step 1. The 10-mm ASTM D-1160 assay is tabulated in columns 1 and 2.

Step 2. Differences between adjacent points of the D-1160 are found and entered in column 3.

Step 3. The 10-mm TBP 50% point is found by Figure 13.3 to be 531°F, equal to the 10-mm ASTM D-1160 50% point. It is entered in the fifth column at the proper point and identified by an asterisk.

Step 4. Using the ASTM D-1160 differences in column 3 and Figure 13.3, the 10-mm TBP differences are found and tabulated in column 4.

Step 5. Points on the 10-mm TBP curve are obtained by subtracting or adding the differences found in Step 4 to the 531°F TBP 50% point.

Step 6. Transpose the 10-mm TBP points to 760-mm TBP points via Figure 13.1.

10 mm ASTM D-1160			10-mm TBP		760-mm TBP
Vol % Off	Step 1 T°F	Step 2 ΔT°F	Step 4 ΔT°F	Step 5 T°F	Step 6 T°F
0	397			350	615
		57	89		
10	454			439	724
		42	50		
30	496			489	786
		35	42		
50	531			531*	837
		48	48		
70	579			579	896
		61	61		
90	640			640	971

* At 10 mm, ASTM D-1160 50% Pt. = TBP 50% Pt. = 531°F (Step 3)

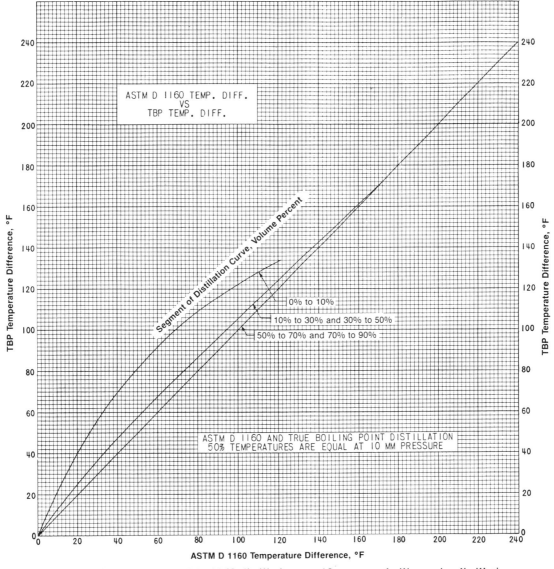

Figure 13.3. 10 mm ASTM D-1160 distillation vs. 10 mm true boiling point distillation.

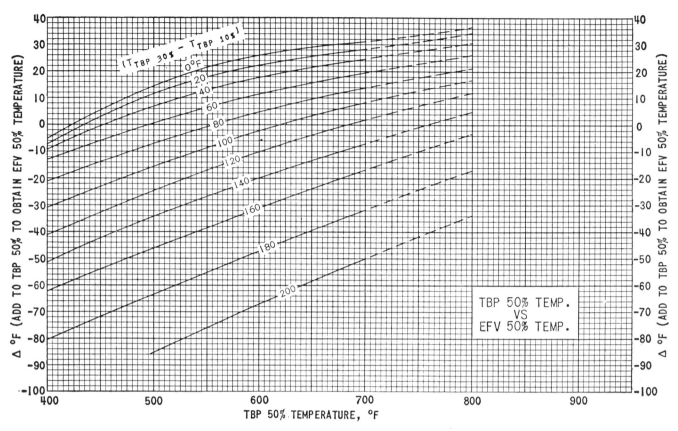

Figure 13.4. 10 mm TBP 50% temperature vs. 10 mm EFV 50% temperature (top right).

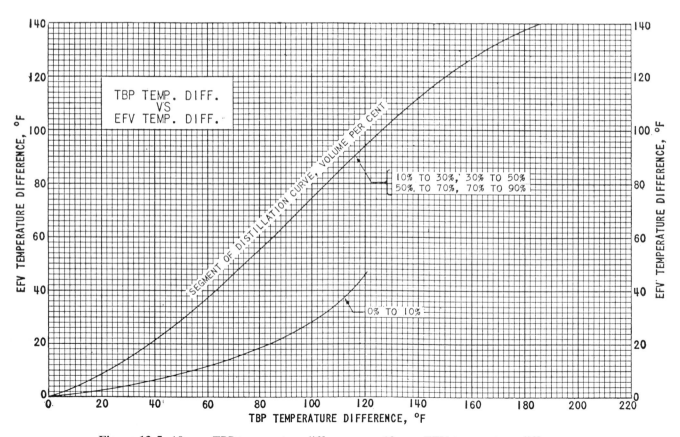

Figure 13.5. 10 mm TBP temperature difference vs. 10 mm EFV temperature differences.

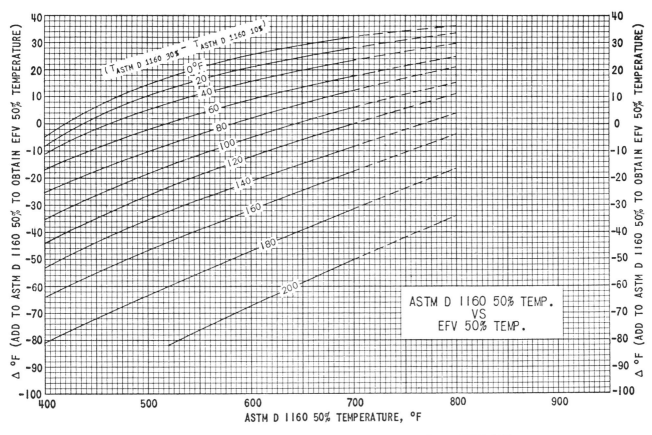

Figure 13.6. 10 mm ASTM D-1160 50% temperature vs. 10 mm EFV 50% temperature.

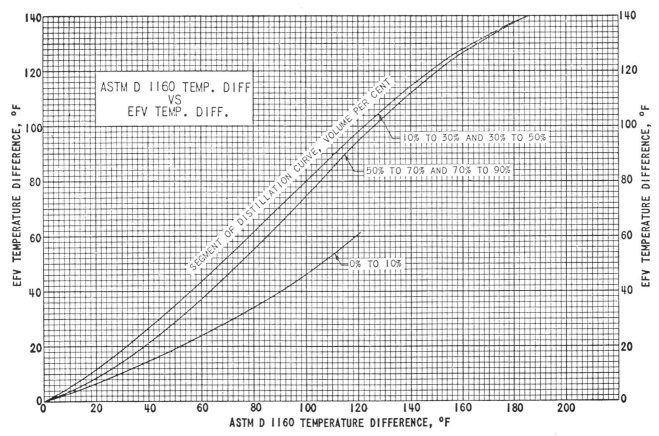

Figure 13.7. 10 mm ASTM D-1160 temperature difference vs. 10 mm EFV temperature differences.

Pressure Effect on 30% and 50% Points

A correlation was developed for predicting the effect of pressure on the 30% and 50% points of the EFV curve in the vacuum region and is given here as Figure 13.8. This correlation was developed from experimental EFV data covering a pressure range of 5 mm to 760 mm Hg and included 16 petroleum fractions ranging from a 52°API naphtha to an 11°API residuum. Figure 13.8 has been evaluated for predicting the 30% and 50% points and for transposing the EFV curve in the vacuum region. The details of this evaluation are given in Tables 13.4 and 13.5. The distribution of deviations plus the maximum and average deviations are summarized in the first two lines of the following:

By use of the temperature-pressure relationship given in Figure 13.8 for the EFV 50% point and the assumption that the slope of the EFV curve is constant in the sub-atmospheric pressure region, the EFV curve can be satisfactorily transposed from one pressure to another. This assumption of constant EFV slope is justified by the excellent agreement of this method of transposing the EFV curves with the experimental data. Transposition of the EFV curves by this method has been evaluated over a pressure range of 10 mm to 760 mm Hg. The detailed results are shown in Tables 13.6, 13.7 and 13.8, and are summarized in the the third line of the previous summary.

	30 % Pt. Calc.	50 % Pt. Calc.	EFV Transposing Calculation
Deviations			
Maximum	23°F	25°F	25°F
Average	5°F	4°F	4°F
Number of points in Deviation range:			
0 to 10°F	83	90	319
11 to 20°F	8	4	15
> 20°F	1	1	4

Table 13.4
Evaluation of Figure 13.8 for Transposing EFV 30% Points (°F) in Vacuum Region

T°F for:		10 mm Hg 30% EFV Obs.	25 mm Hg			50 mm Hg			100 mm Hg			200 mm Hg			400 mm Hg			760 mm Hg		
			30% EFV		Calc -Obs	30% EFV		Calc -Obs	30% EFV		Calc -Obs	30% EFV		Calc -Obs	30% EFV		Calc -Obs	30% EFV		Calc -Obs
Source	Stock		Obs.	Calc	Diff	Obs.	Calc	Diff	Obs.	Calc	Diff	Obs.	Calc	Diff	Obs.	Calc	Diff	Obs.	Calc	Diff
Othmer	Oil 1								179	177	− 2	214	213	− 1	257*	250	− 7	298	290	− 8
et.al.	Oil 2								193	195	+ 2	228	232	+ 4	268*	270	+ 2	309	313	+ 4
(8)	Oil 3	408	453	458	+ 5	498	498	0	538*	540	+ 2	585	588	+ 3	640*	640	0	692*	696	+ 4
	Oil 8	457	506*	507	+ 1	546	547	+ 1	592*	592	0	642	639	− 3	696*	690	− 6	754*	747	− 7
C.R.C.	Rsd 1	414	465	464	− 1	505*	504	0	547	546	− 1	595*	594	− 1	650*	646	− 4	702*	701	− 1
Lab.	Rsd 2	591	647	641	− 6	688*	683	− 5	737	729	− 8	786*	774	−12	838*	822	−16	890*	882	− 8
(2)	Rsd 3	537	588*	588	0	631*	630	− 1	678	676	− 2	728*	721	− 7	780*	770	−10	834*	827	− 7
	G.O.35	525	572*	575	+ 3	611	615	+ 4	653*	661	+ 8	698*	707	+ 9	751*	756	+ 5	800*	813	+13
	G.O.36	370*	417*	420	+ 3	455	459	+ 4	498*	500	+ 2	543*	547	+ 4	595*	598	+ 3	648*	654	+ 6
	Rsd 5	390*	440*	440	0	481	480	− 1	524*	520	− 4	573*	568	− 5	628*	620	− 8	685*	677	− 8
Okamoto	Oil 1	182	219*	224	+ 5	253	257	+ 4	284	291	+ 7	322	333	+11	366	381	+15	405*	428	+23
& Van	Oil 2	254	209*	299	0	335	338	+ 3	376	374	− 2	413	418	+ 5	471	471	0	518*	522	+ 4
Winkle	Oil 3	192	238*	235	− 3	272	270	− 2	312	303	− 9	355	346	− 9	406	395	−11	453*	443	−10
(6,7)	Oil 4	247	289*	292	+ 3	325	330	+ 5	366	365	− 1	410	410	0	465	462	− 3	515*	513	− 2
	Oil 5	175	215	216	+ 1	248	248	0	287	283	− 4	328	325	− 3	375	362	− 3	422*	418	− 4
	Oil 6	224	270*	268	− 2	302	304	+ 2	347	339	− 8	390	383	− 7	450	434	−16	495*	484	−11
Abs. Averages of Differences			=	2.36			2.29			3.88			5.25			6.81			7.50	

* Estimate from plots

Table 13.5
Evaluation of Figure 13.8 for Transposing EFV 50% Points (°F) in Vacuum Region

| T°F for: | | 10 mm Hg 50% EFV | 25 mm Hg | | | 50 mm Hg | | | 100 mm Hg | | | 200 mm Hg | | | 400 mm Hg | | | 760 mm Hg | | |
| | | | 50% EFV | | Calc-Obs | 50% EFV | | Calc-Obs | 50% EFV | | Calc-Obs | 50% EFV | | Calc-Obs | 50% EFV | | Calc-Obs | 50% EFV | | Calc-Obs |
Source	Stock	Obs.	Obs.	Calc	Diff	Obs.	Calc	Diff	Obs.	Calc	Diff	Obs.	Calc	Diff	Obs.	Calc	Diff	Obs.	Calc	Diff
Othmer	Oil 1					156	156	0	190	186	− 4	224	224	0	267*	258	− 9	309	301	− 8
et.al.	Oil 2	105*	140*	140	0	172	165	− 7	201	197	− 4	236	235	− 1	275*	273	− 2	316	315	− 1
(8)	Oil 3	451	500	500	0	540*	540	0	583*	585	+ 2	629	632	+ 3	683*	683	0	738*	741	+ 3
	Oil 8	458	509*	508	− 1	548	548	0	592*	592	0	644	641	− 3	692*	692	0	750*	748	− 2
C.R.C.	Rsd 1	494	540	544	+ 4	583*	585	+ 2	623	630	+ 7	677*	677	0	725*	726	+ 1	780*	785	+ 5
Lab.	Rsd 2	670	720*	719	− 1	762*	762	0	805	806	+ 1	847*	848	+ 1	895*	895	0	952*	952	0
(2)	Rsd 3	620	672*	670	− 2	712*	712	0	762	757	− 5	807*	801	− 6	860*	848	−12	910*	906	− 4
	G.O. 35	543	592*	593	+ 1	631	635	+ 4	680*	680	0	725*	725	0	778*	775	− 3	828*	832	+ 4
	G.O. 36	382*	431*	431	0	472	470	− 2	514*	511	− 3	560*	558	− 2	612*	611	− 1	668*	667	− 1
	Rsd 5	466*	516*	516	0	557	556	− 1	600*	601	+ 1	650*	648	− 2	700*	699	− 1	758*	756	− 2
Okamoto	Oil 1	199	238*	233	− 5	268	276	+ 8	301	310	+ 9	340	352	+12	382	402	+20	427*	452	+25
& Van	Oil 2	275	318*	321	+ 3	355	361	+ 6	395	397	+ 2	430	442	+12	486	495	+ 9	538*	547	+ 9
Winkle	Oil 3	208	250*	252	+ 2	288	287	− 1	328	321	− 7	370	364	− 6	422	415	− 7	474*	465	− 9
(6,7)	Oil 4	269	313*	315	+ 2	350	355	+ 5	390	391	+ 1	434	435	+ 1	488	488	0	531*	540	+ 9
	Oil 5	192	232*	235	+ 3	265	267	+ 2	303	302	− 1	341	344	+ 3	388	394	+ 6	433*	442	+ 9
	Oil 6	254	300*	301	+ 1	333	338	+ 5	375	375	0	422	419	− 3	480	472	− 8	522*	523	+ 1
Abs. Averages of Differences		=			1.67			2.69			2.94			2.94			4.94			5.50

* Estimate from plots

Following are two examples that illustrate the use of Figure 13.8.

Example 3. Find the 10-mm Hg EFV curve for a petroleum fraction that has the following 760-mm EFV.

Vol % Off	0	10	30	50	70	90	100
Temp. °F	440	492	518	538	558	590	615

Solution.

Step 1. The 760-mm EFV temperatures are tabulated in column 2 of the following tabulation.

Step 2. Differences between adjacent 760-mm temperatures are in column 3.

Step 3. The 10-mm EFV 50% temperature is found to be 268°F by using Figure 13.8 and the 760-mm EFV temperature of 538°F.

Step 4. Assuming a constant slope for the vacuum EFV curves, other points on the 10-mm EFV curve are found by subtracting and adding the difference found in Step 2 to the 10-mm EFV 50% temperature.

	Step 1	Step 2	Step 3
Vol %
Off	760 mm EFV
T°F	ΔT°F	10 mm EFV
T°F
0	440		170
		52	
10	492		222
		26	
30	518		248
		20	
50	538		268*
		20	
70	558		288
		32	
90	590		320
		25	
100	615		345

* From Figure 13.8

Example 4. A petroleum fraction has a 25-mm EFV 50% point of 250°F. Find the 50% temperature on the 300-mm Hg EFV curve.

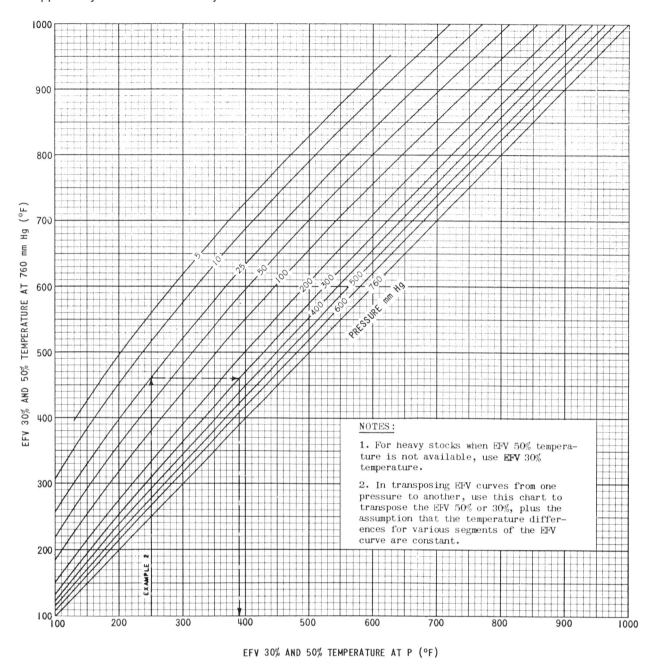

NOTES:

1. For heavy stocks when EFV 50% temperature is not available, use EFV 30% temperature.

2. In transposing EFV curves from one pressure to another, use this chart to transpose the EFV 50% or 30%, plus the assumption that the temperature differences for various segments of the EFV curve are constant.

EFV 30% AND 50% TEMPERATURE AT P (°F)

Figure 13.8. Effect of pressure on the 30% and 50% point temperatures on the vacuum EFV.

Solution. Enter Figure 13.8 at 250°F on the abscissa and follow the dashed lines and arrows to locate the 50% temperature on the 300 mm EFV curve as 390°F.

Vacuum Phase Diagram Construction

Phase diagrams for petroleum fractions at super-atmospheric pressures can be constructed satisfactorily by us-ing the focal point method shown on Figures 12.14 and 12.15. This method cannot be extended into the vacuum region without introducing considerable error. However, experimental EFV data show that lines of constant-vol-ume-percent-vaporized are also straight lines on a log P versus $1/T$ coordinates plot for the vacuum region; this phenomena is a feature of the super-atmospheric phase diagram.

Figure 13.9 illustrates phase diagram construction for pressures below atmospheric. The following example shows how the calculations are made and the diagram drawn. Note that the calculation steps are similar to the steps followed in Example 2.

Example 5. Construct a phase diagram in the 10- to 760-mm vacuum region for a residuum having the following assay:

10-mm True Boiling Point Distillation						
Vol % Off	0	10	30	50	70	90
Temp. °F	178	300	402	507	625	748

Solution.

Step 1. The 10-mm TBP distillation points are tabulated in columns 1 and 2 of the following tabulation.

Table 13.6

Evaluation of Method for Transposing EFV Curves at Sub-Atmospheric Pressure (8)

Stock	Press mm Hg	IBP Obs	IBP Calc	IBP Calc -Obs Diff	10% Obs	10% Calc	10% Calc -Obs Diff	30% Obs	30% Calc	30% Calc -Obs Diff	50% Obs	50% Calc	50% Calc -Obs Diff	70% Obs	70% Calc	70% Calc -Obs Diff	90% Obs	90% Calc	90% Calc -Obs Diff
Oil 1	50	121	121	0	134	133	− 1	148	146	− 2	156	156	0	164	165	+ 1	175	176	+ 1
	100	152	151	− 1	166	163	− 3	179	176	− 3	190	186	− 4	198	195	− 3	207	206	− 1
	200	191	189	− 2	202	201	− 1	214	214	0	224	224	0	233	233	0	246	244	− 2
	760	275	272	− 3	286	284	− 3	298	291	− 7	309	301	− 8	317	310	− 7	328	321	− 7
Oil 2	50	148	136	−12	156	146	−10	165	157	− 8	172	165	− 7	178	172	− 6	183	180	− 3
	100	175	168	− 7	182	178	− 4	193	189	− 4	201	197	− 4	208	204	− 4	216	212	− 4
	200	208	206	− 2	218	216	− 2	227	227	0	236	235	− 1	243	242	− 1	251	250	− 1
	760	283	286	+ 3	295	296	+ 1	309	307	− 2	316	315	− 1	325	322	− 3	336	330	− 6
Oil 3	10	287	284	− 3	350	349	− 1	408	405	− 3	451	451	0	495	498	+ 3			
	25	338	334	− 4	401	398	− 3	453	454	+ 1	500	500	0	549	548	− 1			
	200	473	466	− 7	533	530	− 3	584	586	+ 2	629	632	+ 3						
Oil 8	10	447	442	− 5	454	450	− 4	457	456	− 1	458	458	0	463	464	+ 1	490	490	0
	50	528	530	+ 2	537	540	+ 3	546	546	0	548	548	0	551	554	+ 3	576	580	+ 4
	200				632	634	+ 2	642	639	− 3	644	641	− 3	651	647	− 4			

Table 13.7

Evaluation of Method for Transposing EFV Curves at Sub-Atmospheric Pressure (2)

Stock	Press mm Hg	IBP Obs	IBP Calc	IBP Calc -Obs Diff	10% Obs	10% Calc	10% Calc -Obs Diff	30% Obs	30% Calc	30% Calc -Obs Diff	50% Obs	50% Calc	50% Calc -Obs Diff	70% Obs	70% Calc	70% Calc -Obs Diff	90% Obs	90% Calc	90% Calc -Obs Diff
Rsd 1	10	275	274	− 1	338	343	+ 5	414	417	+ 3	494	494	0	588	585	− 3			
	25	320	324	+ 4	392	393	+ 1	464	467	+ 2	540	544	+ 4	630	635	+ 5			
	100	400	410	+10	473	479	+ 6	547	553	+ 6	623	630	+ 7	712	721	+ 9			
Rsd 2	10	450	452	+ 2	511	513	+ 2	591	595	+ 4	670	670	0	765	773	+ 8			
Rsd 3	10	395	391	− 4	455	451	− 4	537	536	− 1	620	620	0						
	100	533	529	− 4	592	589	− 3	678	673	− 5	762	757	− 5						
G.O. 35	10	440	437	− 3	488	487	− 1	524	523	− 1	543	543	0	568	566	− 2			
	50	523	529	+ 6	576	579	+ 3	611	615	+ 4	631	635	+ 4	652	658	+ 6			
G.O. 36	50	410	407	− 3	434	429	− 5	456	453	− 3	472	470	− 2	488	485	− 3	510	505	− 4
	250	515	512	− 3	535	534	− 1	560	558	− 2	578	575	− 3	591	590	− 1	609	610	+ 1

Table 13.8
Evaluation of Method for Transposing EFV Curves at Sub-Atmospheric Pressure (6,7)

T°F for:		IBP			10%			30%			50%			70%			90%			FBP		
Oil	Press mm Hg	Obs	Calc	Calc -Obs Diff	Obs	Calc	Calc -Obs Diff	Obs	Calc	Calc -Obs Diff	Obs	Calc	Calc -Obs Diff	Obs	Calc	Calc -Obs Diff	Obs	Calc	Calc -Obs Diff	Obs	Calc	Calc -Obs Diff
No. 1	10	130	126	− 4	154	154	0	182	183	+ 1	199	199	0	216	217	+ 1	241	242	+ 1	260	259	− 1
	50	197	203	+ 6	223	231	+ 8	253	260	+ 7	268	276	+ 8	276	294	+ 7	312	319	+ 7	329	336	+ 7
	100	227	237	+10	257	265	+ 8	284	294	+10	301	310	+ 9	320	328	+ 8	348	353	+ 5	364	370	+ 6
	200	265	279	+14	293	279	+14	322	336	+14	340	352	+12	360	370	+10	385	395	+10	399	412	+13
	400	304	329	+25	337	357	+20	366	386	+20	382	402	+20	398	420	+22	422	445	+23	439	462	+23
No. 2	10	184	179	− 5	222	227	+ 5	254	257	+ 3	275	275	0	298	297	− 1	331	330	− 1	355	357	+ 2
	50	263	265	+ 2	305	309	+ 4	335	343	+ 8	355	361	+ 6	377	383	+ 6	410	416	+ 6	438	443	+ 5
	100	302	301	− 1	346	349	+ 3	376	379	+ 3	395	397	+ 2	417	419	+ 2	450	452	+ 2	476	479	+ 3
	200	342	346	+ 4	385	394	+ 9	413	424	+11	430	442	+12	453	464	+11	488	497	+ 9	516	524	+ 8
	400	390	399	+ 9	442	447	+ 5	471	477	+ 6	486	495	+ 9	505	517	+12	538	550	+12	568	577	+ 9
No. 3	10	163	158	− 5	174	172	− 2	192	192	0	208	208	0	237	232	− 5	277	272	− 5	301	305	+ 4
	50	240	237	− 3	253	251	− 2	272	271	− 1	288	287	− 1	314	311	− 3	352	351	− 1	383	384	+ 1
	100	278	271	− 7	292	285	− 7	312	305	− 7	328	321	− 7	352	345	− 7	390	385	− 5	424	418	− 6
	200	320	314	− 6	334	328	− 6	355	348	− 7	370	364	− 6	396	388	− 8	437	428	− 9	471	461	−10
	400	369	365	− 4	385	379	− 6	406	399	− 7	422	415	− 7	445	439	− 6	488	479	− 9	522	512	−10
No. 4	10	170	163	− 7	210	207	− 3	247	246	− 1	269	269	0	293	293	0	326	326	0	349	352	+ 3
	50	247	249	+ 2	285	293	+ 8	325	332	+ 7	350	355	+ 5	375	380	+ 5	408	413	+ 5	433	439	+ 6
	100	281	285	+ 4	328	329	+ 1	366	368	+ 2	390	391	+ 1	416	416	0	448	449	+ 1	475	475	0
	200	324	329	+ 5	372	373	+ 1	410	412	+ 2	434	435	+ 1	458	460	+ 2	491	493	+ 2	520	519	− 1
	400	377	382	+ 5	424	426	+ 2	465	465	0	488	488	0	511	513	+ 2	543	546	+ 3	568	572	+ 4
No. 5	10	147	143	− 4	160	158	− 2	175	177	+ 2	192	192	0	211	209	− 2	234	231	− 3	248	248	0
	50	216	218	+ 2	230	233	+ 3	248	252	+ 4	265	267	+ 2	283	284	+ 1	307	306	− 1	322	323	+ 1
	100	254	253	− 1	267	268	+ 1	287	287	0	303	302	− 1	321	319	− 2	341	341	0	359	358	− 1
	200	292	295	+ 3	308	310	+ 2	328	329	+ 1	341	344	+ 3	357	361	+ 4	380	383	+ 3	398	400	+ 2
	400	337	345	+ 8	355	360	+ 5	375	379	+ 4	388	394	+ 6	402	411	+ 9	422	433	+11	440	450	+10
No. 6	10	152	153	+ 1	185	185	0	224	224	0	254	254	0	286	283	− 3	321	319	− 2	345	346	+ 1
	50	234	237	+ 3	265	269	+ 4	302	308	+ 6	333	338	+ 5	364	367	+ 3	398	403	+ 5	425	430	+ 5
	100	279	274	− 5	307	306	− 1	347	345	− 2	375	375	0	403	404	+ 1	440	440	0	468	467	− 1
	200	319	318	− 1	352	350	− 2	390	389	− 1	422	419	− 3	451	448	− 3	487	484	− 3	518	511	− 7
	400	370	368	− 2	407	403	− 4	450	442	− 8	480	472	− 8	507	501	− 6	545	537	− 8	572	564	− 8

Step 2. Temperature differences for adjacent points of the TBP curve are found and entered in column 3.

Step 3. From the 10-mm TBP 30-10% difference of 102°F and the TBP 50% point of 507°F, the 10-mm EFV 50% point is found from Figure 13.4 to be 492°F, which is entered at the proper point in column 5.

Step 4. Using the temperature differences found in Step 2 (tabulated in column 3) and Figure 13.5, the 10-mm EFV temperature differences are found and entered in column 4.

Step 5. The EFV temperature differences in column 4 are added to or subtracted from the 10-mm EFV 50% temperature found in Step 3 to get the re-maining point on the 10-mm EFV curve. These are in column 5.

Step 6. The 10-mm EFV 50% temperature is transposed to 760 mm by using Figure 13.8. Since the EFV temperature differences in the vacuum region are unaffected by the pressure, determine the 760-mm EFV curve by subtracting or adding the differences found in Step 4 to the 760 mm EFV 50% temperature point. These values are given in column 6.

Step 7. From the points at 10 and 760 mm, the phase diagram is then constructed by drawing straight lines of constant percent vaporized, as shown in Figure 13.9, where the coordinates are log P versus $1/T$.

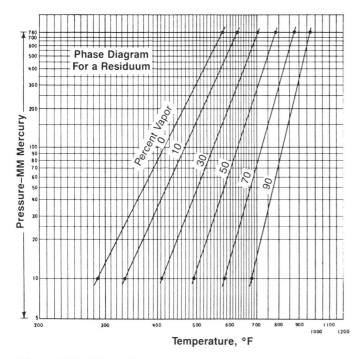

Figure 13.9. Phase diagram for a residuum in Example 5.

10-mm TBP			10-mm EFV		760-mm EFV
Step 1		Step 2	Step 4	Step 5	Step 6
Vol % Off	T°F	ΔT°F	ΔT°F	T°F	T°F
0	178			287	577
		122	48		
10	300			335	635
		102	77		
30	402			412	702
		105	80		
50	507			492*	782**
		118	92		
70	625			584	874
		123	97		
90	748			681	971

  * 10-mm EFV 50% temperature found from Figure 13.4.
** 760-mm EFV 50% temperature found from Figure 13.8.

Consistency Between Atmospheric and Vacuum EFV Correlations

The three-way correlation relating the atmospheric ASTM, TBP, and EFV curves, presented as Figures 12.4 through 12.9, are applicable to oil fractions with normal boiling points up to 800°F. For heavier stocks, similar relationships for the 10-mm ASTM D-1160, TBP, and EFV curves, given here as Figures 13.3 through 13.7,

are applicable. Figure 13.8 relates the EFV 30% and 50% temperatures at various sub-atmospheric pressures, while the TBP and ASTM curves at 10 and 760 mm are related by Figures 13.1 and 13.2.

Figure 12.1 illustrates the interrelationships. There are nine correlations that are consistent and should not be extrapolated. The atmospheric three-way correlations and the 10-mm Hg three-way correlations overlap each by about 125°F. In this region these two correlations are consistent. Thus, for stocks that fall in this overlap region, either the atmospheric or the 10-mm Hg correlations may be used to predict the EFV curve.

Plant data on vacuum flashing residua have been published (3) with laboratory assays on the reduced crude charge stocks. An attempt was made to evaluate these correlations with these data. The results are inconclusive, however, because of uncertainties about the laboratory distillation assay data and the corrected flash zone pressure in the plant.

Assuming that the laboratory distillation is a TBP, that Figures 13.4 and 13.5 apply, and that the pressures given were corrected rather than observed pressures, this calculation method gave volume percent flashed results that were an average of 8% high, varying from 2% to 13% high. This indicated that the correlations presented here predict vacuum flash zone temperatures that are from 10 to 60°F low. This is significant and requires further checking.

References

1. Beale and Docksey, *J. Inst. Pet. Tech.*, 22, 860 (1935).
2. Chevron Research Company, Richmond, CA.
3. Filak, G. A., H. L. Sandlin, and C. J. Stockman, *Pet. Ref.*, 34, 4, 153 (Apr. 1955).
4. Flanders, R. L. and G. M. Williams, Chevron Res. Co., Richmond, CA.
5. Nelson, W. L. and M. Souders, Jr., *Pet. Eng.*, 3, 1, 131 (1931).
6. Okamoto, K. K. and M. Van Winkle, *Pet. Ref.*, 28, 8, 113–120 (1949), 29, 1, 91–96 (1950).
7. Okamoto, K. K. and M. Van Winkle, *Ind. Eng. Chem.*, 45, 429 (1951).
8. Othmer, D. F., E. H. Ten Eyck, and S. Tolin, *Ind. Eng. Chem.*, 43, 1607–1613 (1951).
9. Paulsen, T. H., Chem. Eng. Prog. Sym. Series, 49, 6, 45 (1953).
10. Smith, R. B., T. Dresser, H. F. Hopp, and T. H. Paulsen, *Ind. Eng. Chem.*, 43, 766–770 (1951).
11. Watson, K. M., E. F. Nelson, and G. B. Murphy, *Ind. Eng. Chem.*, 27, 1460–1464 (1935).
12. White, R. R. and G. G. Brown, *Ind. Eng. Chem.*, 34, 1162 (1942).

14

Integral Technique for Petroleum Vapor-Liquid Equilibrium Calculations

Graphical methods, such as those presented in Chapters 12 and 13 for predicting vapor-liquid phase behavior of petroleum fractions, could be put into equation form for computer applications by curve fitting the graphs and preparing suitable programs for applying the equations. This has not yet been done and would not be justified. Calculations by such a procedure would be inconsistent with the type of vapor-liquid equilibrium calculations described in Chapter 11, which describes a method that is universally used for mixtures of discrete components because it is more accurate and powerful than a procedure based on empirical curve fits to charts.

Applying the same VLE calculation procedure to petroleum fractions and to mixtures of discrete components, such as hydrocarbons, permits making similar calculations for mixtures of petroleum and natural gas in exactly the same way. Likewise, gaseous mixtures containing a "C6 + " cut are handled in the same manner. This chapter summarizes previous work. Chapter 15 covers methods of dividing petroleum fractions and "C6 + " cuts into acceptable pseudo or hypothetical components into a format that can be handled conveniently on a computer.

Mixture Types

This section addresses multicomponent mixtures of three types of components, identified as: discrete, pseudo, and hypothetical components. As used here, these names have the following meanings. A *discrete* component is a pure component with known chemical structure and molecular properties. A *pseudo* component is similar to a pure component in that it has a zero boiling range representing a single point on the TBP assay of a petroleum fraction, for which the structure is unknown and characterizing properties must be estimated. A *hypothetical* component is a small slice of a petroleum fraction, having a boiling range of 5 to 50°F, being a narrow boiling mixture of close boiling components, for which the chemical structure is unknown so characterizing properties must be estimated.

Petroleum fractions are essentially continua mixtures of an infinite number of pseudo components, the amount of each pseudo component being infinitesimally small. Light hydrocarbon mixtures, on the other hand, are mixtures of a finite number of discrete components, that can be named and identified by formulae and other properties. As gases often occur with oil fractions in production and manufacturing processes, mixtures containing both discrete and pseudo or hypothetical components appear frequently in hydrocarbon process engineering. Calculation methods must be versatile enough for such systems.

Calculation Methods

The empirical method is limited to petroleum fractions and cannot be applied to phase equilibria predictions for mixtures containing discrete components. This leads immediately to the conclusion that the so-called theoretical method, described in Chapter 11, in which thermodynamics is applied to the process and equilibrium distribution ratios ($K_i = y_i/x_i$) are computed in an iterative procedure, is the correct method to use. The next step is to devise the best procedure for representing petroleum fractions by a finite number of pseudo or hypothetical components. After that, the physical and thermal properties of these components must be estimated. Then, and only then, can the thermodynamic properties and the K-values be predicted so that VLE calculation can be made.

A continua mixture, i.e., a petroleum fraction, may be represented as a multicomponent mixture of a finite number of hypothetical components of equal, or unequal, size cuts made on the TBP assay. For example, there might be 20 cuts of 5-volume-percent width, each cut being identified by its mean boiling point. Another way would be to divide the oil fraction into narrow cuts of equal boiling width but of varying quantities. Thus, the manual breakdown might be made in terms of either coordinate of the TBP assay, i.e., the volume-percent abscissa or the boiling-point ordinate. For irregularly shaped TBP curves, the mean boiling points of these cuts may be taken at irregular intervals or the sizes may vary, depending upon the shape of the TBP curve. In addition to the mean boiling points of these cuts, specific gravities will be needed to complete the definition of the hypothetical components. If a specific-gravity-versus-percent-off assay is not available then a procedure for estimating the gravities of the cuts from the gravity of the entire fraction must be used. Each of the narrow cuts, or components, is characterized by its average boiling point and average specific gravity.

Dividing a petroleum fraction into a finite number of narrow cuts, whether by equal percentage widths, equal boiling widths, or by irregular intervals, is equivalent to representing a smooth TBP curve by a series of plateaus. The more plateaus, or hypothetical components, the more accurate is the representation and the more tedious are the calculations made in terms of these components. The ultimate in precision would be an infinite component breakdown, if this could be done in a convenient and practical way. Such a method has been proposed for making distillation calculations for petroleum fractions (1,2,3,4,5,6,7). Although the principle of this "integral" method is simple, previous applications were made by techniques that were either awkward or approximate. A numerical and graphical application of this integral technique will be presented here to illustrate the procedure.

The principles of the integral method will be illustrated by four types of VLE calculations for petroleum fractions: equilibrium flash vaporization, bubble and dew points, fractional distillation, and batch differential vaporization. Simplifications have been made in applying the integral method in these demonstrations. Conventional phase equilibrium calculations are made for infinitesimal increments of the mixture, the results plotted,

and the area under the curve is found to check VLE conditions and to obtain the total phase distribution for the mixture.

Previous Work

Katz and Brown (7) noted in 1933 that the same relationships used in making equilibrium vaporization calculations for simple mixtures, such as light hydrocarbons, could be applied in a differential form to complex mixtures, such as petroleum fractions in which the composition is expressed in the form of a true boiling-point curve. Integrals rather than summations of the component contributions to equilibrium vapor and liquid were shown in the discussion of this application, but summations of the contributions of the segments were used in the illustrations.

A few years later, White and Brown (11) obtained experimental phase equilibria data on petroleum fractions (naphtha, furnace oil, and their mixtures) at temperatures of 300 to 820°F and pressures of 50 to 700 psia. From the results of the equilibrium flash vaporizations, the vapor-liquid distribution ratios, i.e., K-values, were derived for hypothetical components, i.e., narrow cuts of 50°F width. Quantities of these 50°F cuts were read from the TBP curves. These liquid volume quantities were converted to lb-moles, using specific gravities and molecular weights, and then the K-values were calculated for the arbitrary components appearing in the coexisting equilibrium vapor and liquid. These K-values were plotted against temperature and pressure for average boiling points of 325 to 725°F at 50°F intervals.

A more fundamental method of analyzing these equilibrium vaporization for petroleum fractions would have been to work with infinitesimal cuts or increments of one degree or less. Applying the procedure used by White and Brown to such small arbitrary components would have magnified the data processing considerably. Experimental vapor-liquid equilibrium data for petroleum fractions can be put into K-value form by using the integral technique presented in this and the next chapter.

In 1947, Harbert (6) proposed using differential curves instead of dividing the petroleum fractions into mixtures of narrow boiling cuts or "components," or by making vaporization calculations by empirical methods based on laboratory assays. In this previous work the TBP curve was used to define the complex mixtures of petroleum fractions. Harbert put the TBP curve into differential form, so it could be used in making vapor-liquid equilibria calculations, by multiplying dM/dT_B vs. T_B, where M represents moles and T_B represents boiling points. This differential is related to the reciprocal slope of the TBP curve by the following:

$$(dM/dT_B) = (dV/dT_B)(dM/dV)$$
$$= (dV/dT_B)(D/MW) \qquad (14.1)$$

where V = volume fraction distilled on TBP assay
        D = liquid density of increment distilled
      MW = molecular weight of increment distilled

Reciprocal slopes of the TBP curve, i.e., dV/dT_B, were plotted against boiling point temperature and then converted to a plot of dM/dT_B versus T_B by using Equation 14.1. Harbert used such a plot, with K-values, in making vaporization calculations, plotting K versus T_B on the same chart with the dM/dT_B versus T_B curve. The area under the dM/dT_B versus T_B curve is moles and the shape of the curve depends upon mixture character and not upon the conditions of vaporization. For different conditions, another auxiliary K-value scale is required. Bubble and dew point and flash calculations were made by using these differential curves.

The integral technique presented here is based upon the same basic principles as Harbert's work (6), and the procedure is more convenient in applications. These will be described by means of demonstrations on four examples plus a brief summary of the fundamental concepts of the method.

Integral Technique

Petroleum may be regarded as continua mixtures of an infinite number of hydrocarbons, each appearing in an infinitesimal amount. On the true boiling point distillation curve for the "mixture," each of these "components" will be a point, as contrasted to the plateaus that represent the finite amounts of components on the TBP curves of light hydrocarbon mixtures.

In distillation calculations for "finite" mixtures, the vapor-liquid distribution, and the properties of the equilibrium phases are found by the summations of the distributions and properties of the components. For continua mixtures of an indefinite number of components, the distillation calculations follow the same basic principles with the "summation" becoming an "integration." This distinction is illustrated by the following bubble-point relationships:

$$\sum K_i x_i = 1.0 \text{ for "finite" mixture} \qquad (14.2)$$

$$\int_0^1 K dm_F = 1.0 \text{ for "continua" mixture} \qquad (14.3)$$

For a mixture of finite components, the values of $K_i x_i$ are summed for all components. When the K-values are taken at the correct bubble-point temperature, this sum is

unity. For the continuum, an integration is performed by determining the area under a curve of K versus m_F. The points for this curve are located by finding the values of K at the system temperature and pressure for each of several points on the TBP curve. At the bubble point, the area under the K versus m_F curve equals unity.

Points on the distillation curve of the bubble-point vapor can be found in the integral procedure by finding the area under the curve up to the value of m_F in question. The ratio of this area to the total area under the K versus m_F curve gives the value of m_V. The same analogy between the summations and integral procedures holds for the dew-point and other distillation calculations.

Component Ratios. The integral technique can be used in equilibrium flash vaporization calculations or fractional distillation calculations and may be applied to increments or components of any amounts. In this way separation calculations can be made for continua by computing the ratios for several points, or infinitesimal components, and then finding the totals by integration. In distillation calculations it is often more convenient to use component ratios for designating the distribution of the infinitesimal components to the products, as will be illustrated later.

Data Requirements. The TBP points referred to above in defining the infinitesimally small "components" must be points on a molar TBP curve, i.e., a plot of boiling point against mole percent, or mole fraction, of the original mixture distilled. Calculations based on such molar TBP curves result in vapor and liquid products that are defined by similar curves.

Liquid volume quantities and temperatures are the measurements made in the laboratory distillations and the results reported are $T°F$ and volume-percent or volume-fraction-off. Therefore, the observed liquid volume fractions must be converted to mole-fraction-off. This conversion can be made by the graphs given as Figures 14.1 and 14.2. Instantaneous moles per 100 gallons are given in Figure 14.1 as a function of normal boiling and gravity or the W & N characterization factor. Delta percents for converting the TBP curve from volumetric basis to a molar basis are given in Figure 14.2 as a function of TBP slope and the Watson characterization factor, K, and the percent-off point on the volumetric TBP curve.

More recent methods became available (10) after the previously cited were proposed for converting the standard ASTM distillation curves to molar true boiling point curves.

Vapor-liquid K-values for the "components" of this mixture are required also. These may be evaluated from vapor pressures, fugacities, or empirical correlations.

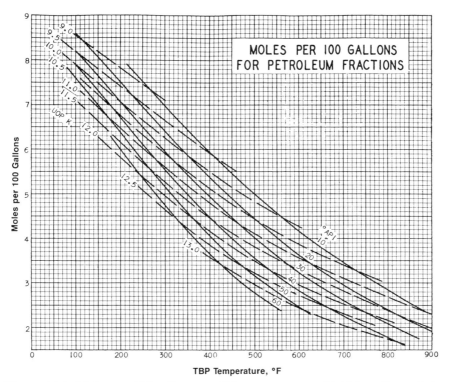

Figure 14.1. Moles per 100 gallons vs. TBP temperature and °API or Watson K for petroleum fractions.

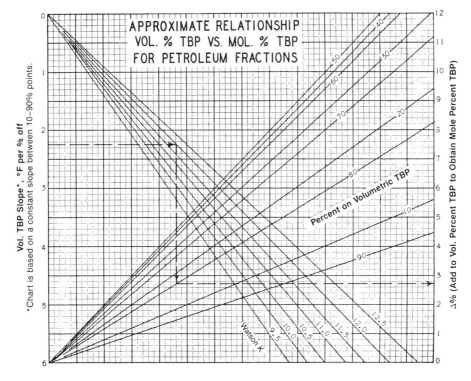

Figure 14.2. Approximate relationship volume percent TBP vs. mole percent TBP for petroleum fraction.

Equilibrium Flash Vaporization

Flash calculations for a continua mixture by the integral technique are basically the same as the flash calculations for a finite component mixture. Both involve assuming a V/F ratio, finding the K-values, then calculating the component distributions to the vapor and liquid phases, summing up to get total V and L, and finally checking the assumed V/F ratio. Different configurations of the flash process and the mathematical expressions for the calculations were given in Chapter 9. For present purposes it is convenient to arrange these flash expressions as follows:

$$v/f = 1/(1 + L/KV) \qquad (14.4)$$

and

$$l/f = 1/(1 + KV/L) \qquad (14.5)$$

Since these expressions give component distribution fractions, they can be applied to the components of finite mixtures or to the infinitesimal components of continua, equally well. In the integral flash calculations the values of v/f and l/f (note that $l/f = 1 - v/f$) are found for several points along the molar TBP curve and plotted against m_F. The area under this curve gives the vapor-liquid split for the mixture. If this checks for the assumed split, equilibrium is satisfied. If not, another vapor-liquid split must be assumed and the calculations repeated until convergence is obtained.

The integral method differs from the finite component method in that calculations are made for several selected points on the molar TBP curve and the totals are found by integrating under a curve instead of by a summation.

Example 1. Calculate the equilibrium vapor and liquid from flashing a 39.6°API petroleum fraction at 390°F and 22 psia. The oil has the following volumetric TBP assay: 0% 200°F, 10% 247°F, 30% 291°F, 50% 334°F, 70% 387°F, 90% 490°F, and 100% 600°F.

Solution. Figures 14.3 and 14.4 illustrate the solution of this example. Calculation steps are as follows:

Step 1. Convert the volumetric TBP to a molar TBP and plot. These calculations are made by using Figure 14.1 and the numerical values are shown in Table 14.1. Figure 14.1 is applied by first finding the intersection of the 50% TBP temperature of 334°F and the 39.6°API gravity, which point is approximately on the 11.2 Watson K line. Then the "moles per 100 gallons" values are read at the intersections of this 11.2 Watson K line and the average boiling temperatures for

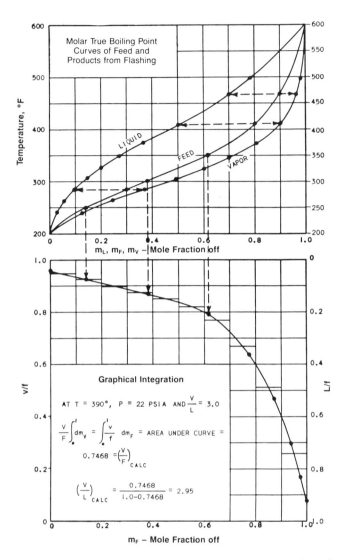

Figure 14.3. Molar true boiling point curves of feed and products from flashing—Example 1.

Figure 14.4. Graphical integration for flashing—Example 1.

Table 14.1
Conversion of Volumetric TBP to Molar TBP Curve for Example 1 Using Figure 14.1

Liq. Vol. %	Gallons	t, °F	Ave. t°F	Moles/100 Gal.	Moles	Mole Fraction	Moles off
0		200					0
	10		223	6.15	0.615	0.1215	
10		247					0.1215
	20		269	5.70	1.140	0.2254	
30		291					0.3469
	20		312	5.33	1.066	0.2107	
50		334					0.5576
	20		360	4.96	0.992	0.1961	
70		387					0.7537
	20		438	4.38	0.876	0.1732	
90		490					0.9269
	10		545	3.70	0.370	0.0731	
100		600					1.0000
Σ					5.059	1.0000	

each segment. It should be noted that the Watson K determined by using the TBP 50% temperature is the same characterization factor that would be obtained from the equivalent ASTM 50% temperature. In this way, the moles-off were found. These were plotted against the temperatures (third column of Table 14.1) in Figure 14.3.

Step 1 (alternate). Table 14.2 shows the conversion of the volumetric TBP to the molar TBP by the use of Figure 14.2. As can be seen, these results agree with those in Table 14.1. They can also be plotted to get a molar TBP curve for the feed.

Step 2. Assume L/V ratio and find K values at the flash conditions of 390°F and 22 psia for each 50°F "component." The K values given in Table 14.3 for this solution are vapor pressure/total pressure ratios. Values of the component ratios v/f and l/f are calculated, using Equations 14.4 and 14.5. These values are in Table 14.3.

Step 3. The v/f and l/f values are plotted in Figure 14.4.

Step 4. The area under the curve in Figure 14.4 is found. See the insert on Figure 14.4 and the numerical calculations in Table 14.4. The area of 2.95 is to be compared with the initial assumption of 3.0 for V/L, as given in the heading of Table 14.3. For these purposes, this is satisfactory.

Step 5. At the 0.1 abscissa divisions on Figure 14.3, points are located for the assays of the vapor and liquid products. These points are found relative to the assay curve of the feed in the manner shown by the horizontal dashed lines and arrows, using the prorated accumulated areas for the liquid and vapor from Step 4. The resulting vapor and liquid TBP curves versus mol-fraction-off are then drawn through these points as shown in Figure 14.3 for this flash.

Bubble and Dew Points

In principle, bubble- and dew-point calculations by the integral method are the same as calculations for finite component mixture by the conventional summation method. Equation 14.3 is the relationship used in making the integral bubble-point calculations. Values of K at an assumed temperature for several points on the molar TBP curve are plotted against m_F and the area under the curve found. An area of unity indicates a correct temperature assumption.

The integral relationship for the dew-point calculation is

$$\int_0^1 dm_F/K = 1.0 \qquad (14.6)$$

Table 14.2
Conversion of Volumetric TBP to Molar TBP Curve for Example 1 Using Figure 14.2

Liquid Volume Percent Off	TBP Temp., °F	△ Percent	mF Moles Off
0	200	0	0
10	247	2.2	0.122
30	291	4.8	0.348
50	334	5.5	0.555
70	387	4.4	0.7440
90	490	1.8	0.9180
100	600	0	1.00

Table 14.3
Calculation of Separation Functions for Example 1 at $(V/L)_{\text{assumed}} = 3.0$

	SEPARATION FUNCTIONS AT (V/L) ASSUMED = 3.0			
t, °F	K, 390°F, 22 psia	KV/L	1/f	v/f
200	7.40	22.2	0.0431	0.9569
250	4.00	12.0	0.0769	0.9231
300	2.20	6.6	0.1316	0.8684
350	1.30	3.90	0.204	0.796
400	0.59	1.77	0.361	0.639
450	0.30	0.90	0.526	0.474
500	0.14	0.42	0.704	0.296
550	0.064	0.192	0.838	0.162
600	0.027	0.081	0.925	0.075

Table 14.4
Integration Calculations—Example 1

	INTEGRATION CALCULATIONS					
	Mean Ordinate		Liquid Areas		Vapor Areas	
mF	1/f	v/f	Incr.	Acc./Total	Incr.	Acc./Total
0				0		0
	0.054	0.946	0.0054		0.0946	
0.1				0.02133		0.1267
	0.0775	0.9225	0.00775		0.09225	
0.2				0.05193		0.2502
	0.1025	0.8975	0.01025		0.08975	
0.3				0.09242		0.3704
	0.1275	0.8725	0.01275		0.08725	
0.4				0.1428		0.4872
	0.1510	0.8490	0.01510		0.08490	
0.5				0.2024		0.6009
	0.1810	0.8190	0.01810		0.08190	
0.6				0.2739		0.7105
	0.2380	0.7620	0.02380		0.07620	
0.7				0.3679		0.8126
	0.3380	0.6620	0.03380		0.06620	
0.8				0.5014		0.9012
	0.5000	0.5000	0.05000		0.05000	
0.9				0.6988		0.9682
	0.7625	0.2375	0.07625		0.02375	
1.0				1.0000		1.0000
			0.2532		0.7468	

Values of $1/K$ at an assumed temperature for several points on the molar TBP curve are plotted against m_F and the area under the curve is found. An area of unity indicates a correct temperature assumption.

Bubble-point calculations are sensitive to the front-end K values, while dew-point calculations are sensitive to the high boiling-point end K values. For these reasons, K values are found and plotted at more frequent intervals at the low boiling end for the bubble-point calculations and at the high boiling-point end for the dew-point calculations.

Example 2. Find the bubble and dew points for the petroleum fraction of Example 1, the pressure being 22 psia, using vapor pressure K values. Estimate the TBP assays of the bubble-point vapor and the dew-point liquid.

Solution. The calculations are given in Tables 14.5 and 14.6 and Figures 14.5 and 14.6. Calculation steps are:

Step 1. Temperature is assumed and K values are found for several pseudo components (as defined by the temperature points on the molar TBP curve) and a plot of K, or $1/K$, versus m_F prepared. The abscissa m_F is the mol-fraction-off on the TBP temperature point defining the pseudo component of interest. Both curves, i.e., K vs m_F and $1/K$ vs m_F, are shown in Figure 14.6.

Step 2. For the bubble point, the area under the K vs m_F curve must equal unity. For the dew point, the area under the $1/K$ vs m_F curve must be unity. These areas may be found by using a planimeter or by counting squares on graph paper, or by tabulation of incremental areas, as shown in Tables 14.5 and 14.6.

Step 3. The TBP curves of the bubble-point vapor and the dew-point liquid can now be constructed by plotting values of m_V and m_L from Tables 14.5 and 14.6 against temperature. These are shown in Figure 14.5.

Fractional Distillation

The integral technique can be applied to multistage distillation calculations, as well as to single-stage flash vaporization. Calculations are made to find the distribution of the infinitesimal "components" of the feed to the top and bottom products of the fractionator, in the same way that such calculations are made for mixtures of finite components.

For the present discussion, ratio type distribution equations for multicomponent, multistage fractionation will be used. The following equations are in terms of the absorption and stripping factors. For the enriching section:

$$l_n/d = A_0 A_1 A_2 A_3 \ldots A_n + A_1 A_3 A_3 \ldots A_n + A_2 A_3 \ldots A_n + \ldots + A_n \quad (14.7)$$

For the exhausting section:

Table 14.5
Bubble Point Temperature Calculation at 327°F

t°F	m_F	$\triangle m_F$	K 22 psia	Mean K	Area	Values of m_V = Acc. Areas for Bubble Point Vapor TBP
200	0		3.85			0
		0.060		3.24	0.1944	
225	0.060		2.63			0.1944
		0.080		2.30	0.1840	
250	0.14		1.98			0.3784
		0.120		1.72	0.2064	
275	0.26		1.45			0.5848
		0.120		1.24	0.1488	
300	0.38		1.02			0.7336
		0.24		0.755	0.1812	
350	0.62		0.49			0.9148
		0.17		0.356	0.0652	
400	0.79		0.222			0.9800
		0.09		0.161	0.0145	
450	0.88		0.100			0.9945
		0.055		0.0728	0.0040	
500	0.935		0.0455			0.9985
		0.038		0.0312	0.0012	
550	0.973		0.0168			0.9997
		0.027		0.0116	0.0003	
600	1.000		0.0064			1.0000

Table 14.6
Dew-Point Temperature Calculation at 435°F

t°F	m_F	$\triangle m_F$	$\frac{1}{K}$ 22 psia	Mean $\frac{1}{K}$	Area	Values of m_L = Acc. Areas for Dew Point Liquid TBP
200	0		0.0936			0
		0.14		0.1283	0.0180	
250	0.14		0.1630			0.0180
		0.24		0.2225	0.0534	
300	0.38		0.2820			0.0714
		0.24		0.3880	0.0931	
350	0.62		0.494			0.1645
		0.17		0.7035	0.1196	
400	0.79		0.913			0.2841
		0.05		1.085	0.0542	
425	0.84		1.257			0.3383
		0.04		1.508	0.0603	
450	0.88		1.760			0.3986
		0.03		2.130	0.0639	
475	0.91		2.500			0.4625
		0.025		3.024	0.0756	
500	0.935		3.548			0.5381
		0.020		4.302	0.0860	
525	0.955		5.057			0.6241
		0.018		6.146	0.1106	
550	0.973		7.236			0.7347
		0.017		8.323	0.1415	
575	0.990		9.410			0.8762
		0.010		12.38	0.1238	
600	1.000		15.35			1.0000

$$v_m/b = S_0 S_1 S_2 S_3 \ldots S_m + S_1 S_2 S_3 \ldots S_m + S_2 S_3 \ldots S_m + \ldots + S_m \quad (14.8)$$

For meshing:

$$d/f = (v_m/b + v_f/f)/(l_n/d + v_m/b + 1) \quad (14.9)$$

$$b/f = (l_n/d + l_f/f)/(l_n/d + v_m/b + 1) \quad (14.10)$$

where v_f/f and l_f/f are component distribution ratios for feed flashing, subscripts designate plate locations from terminals (top and bottom) toward the feed, and where $A = L/KV = 1/S$ and $A_0 =$ Reflux ratio for total condenser.

These component distribution ratio equations represent one of several ways that such calculations might be

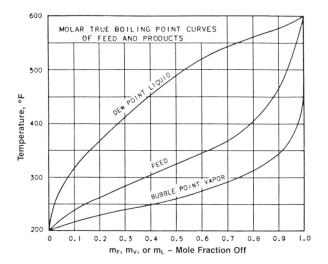

Figure 14.5. Molar true boiling point curves of feed and products for bubble and dew point calculations—Example 2.

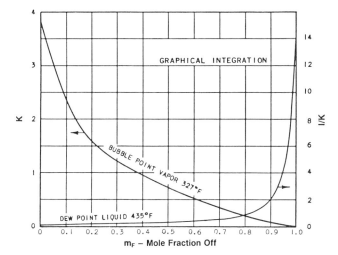

Figure 14.6. Graphical integration for bubble and dew point calculations—Example 2.

made. These equations are not part of the integral technique. The component distribution ratios might be found by a short-cut method, if desired.

After separation functions are evaluated and component distribution ratios found, by these or other equations, the values of d/f and b/f are plotted against m_F. The area under this curve gives D/F and B/F. Areas up to different values of m_F are used to determine points on molar product TBP curves. An interesting feature of the integral method, as applied to fractional distillation, is that distribution ratios may be estimated for several closely spaced points near the cut point and at a few widely spaced points on the balance of the feed TBP curve.

Example 3. A gasoline rerun column, consisting of a reboiler and one theoretical stage below the feed entry

point plus two theoretical stages and a total condenser above, will be calculated to illustrate the use of the integral technique in fractional distillation. In this problem, the feed stock is defined by a true boiling point distillation curve, and the reflux and boilup are fixed. The problem is to calculate the products.

This problem is from Thiele and Geddes (9) and the same problem used previously (5) to illustrate the integral method. The present solution is simpler than previous solutions. The temperatures and quantities of vapor and liquid given by Thiele and Geddes for each stage are used in the following solution.

Solution. This calculation is identical to the flash calculation except in the evaluation of the separation functions, in which the effects of the number of equilibrium stages and the reflux are included. The separation functions are evaluated for the pseudo components in the vicinity of the cut point, because of the sharpness of the desired separation. These separation functions are based upon an assumed temperature gradient and assumed liquid and vapor traffics from stage to stage. Thus, the calculations give the products expected from the assumed operation. Figures 14.7 and 14.8 plus Table 14.7 show the calculations and the results.

Calculations giving the d/f values for the components are plotted at the proper value of m_F and the area under the curve is found. This equals D/F, from which D is found and the desired overall stock balance checked. Product TBP curves are plotted with points obtained from the areas up to the m_F values. Following are the step-by-step procedures that go with the graphs and numerical parts of this solution.

Step 1. Plot the molar TBP curve for the feed, deriving points for this curve from the volumetric TBP and Figure 14.1 or 14.2, if necessary. See Figure 14.7 for the resulting plot.

Step 2. Assume the number of equilibrium stages, temperatures, and the liquid and vapor traffics, all given by Thiele-Geddes in this case.

Step 3. Select several "component" points and evaluate the separation functions. This is done in Table 14.7.

Step 4. Plot d/f values from Step 3 against m_F in Figure 14.8.

Step 5. Integrate, finding area under the curve in Figure 14.8. This is equal to D/F, giving $D = 620$ and checking the material balance.

Step 6. Find the areas under the curve in Figure 14.8 up to different values of m_F and from these estimate the molar TBP curves of the products. See Table 14.8 for the calculations and Figure 14.7 for the curves.

In most problems, the temperatures and quantities of vapor and liquid will not be known in the beginning. These unknowns must be assumed before these calculations can be made. Assumptions may then be checked by computing the quantities and assays of the intrastage va-

por and liquid streams, using the product quantities and the temperature versus m_F curves from the previous trial.

Batch Distillation

Simple batch distillation is a differential process because vapors are removed as they are formed. Solution of batch distillation problems for a mixture like this is defined by its molar TBP curve, and is easily done by the integral form of the Rayleigh equation and a trial and error integral method. The integrated Rayleigh equation is

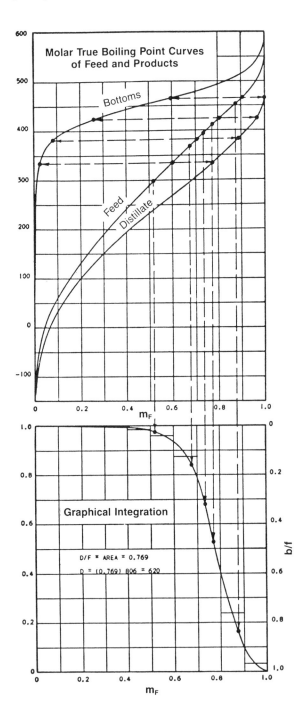

Figure 14.7. Molar true boiling point curves of feed and products of fractional distillation—Example 3.

Figure 14.8. Graphical integration for fractional distillation calculations—Example 3.

Table 14.7
Separation Functions—Example 3

SEPARATION FUNCTIONS					
Equivalent H—C. Atm BP °F	C_9 298	C_{11} 360	C_{12} 393	C_{13} 415	C_{15} 453
Reboiler: t = 435°F, L = 196, V = 1079, V/L = 5.5					
K	5.18	2.25	1.697	1.316	0.789
S_0 = KV/L	28.5	12.4	9.33	7.24	4.34
Plate 1: t = 365°F, L = 1275, V = 818, V/L = 0.642					
K	2.4	0.934	0.667	0.486	0.279
S_1 = KV/L	1.54	0.5996	0.428	0.312	0.179
Plate 2: t = 355, L = 208, V = 822, L/V = 0.253					
K	2.14	0.816	0.575	0.414	0.237
A_2 = L/KV	0.1182	0.310	0.440	0.611	1.066
Plate 3: t = 322, L = 212, V = 822, L/V = 0.258					
K	1.38	0.492	0.337	0.238	0.130
A_3 = L/KV	0.187	0.524	0.765	1.083	1.981
Enriching with reflux from total condenser: R = 0.3475					
$A_3 (1 + R)$	0.252	0.706	1.031	1.459	2.669
$A_2 [1 + A_3 (1 + R)]$ = ln/d	0.148	0.5288	0.8936	1.503	3.912
Stripping					
$S_0 + 1$	29.5	13.4	10.33	8.24	5.34
$S_1 (S_0 + 1)$ = v_m/b	45.6	8.03	4.43	2.57	0.955
Feed Plate Mesh					
ln/d + v_m/b + 1	46.75	9.56	6.32	5.07	5.86
$\dfrac{d}{f} = \dfrac{v_m/b}{\ln/d + v_m/b + 1}$	0.975	0.839	0.701	0.487	0.163

Table 14.8
Integration Calculations—Example 3

	INTEGRATION CALCULATIONS						
	DISTILLATE				BOTTOMS		
m_F	Mean Ordinate	Incr. Areas	Pro-rated Incr.	Acc. Pro-rated	Incr. Areas	Pro-rated Incr.	Acc. Pro-rated
0				0			0
	0.9999	0.09999	0.1306		0.00001	0.0001	
0.1				0.1306			0.0001
	0.999	0.0999	0.1303		0.0001	0.0004	
0.2				0.2609			0.0005
	0.998	0.0998	0.1293		0.0002	0.0009	
0.3				0.3902			0.0014
	0.995	0.0995	0.1290		0.0005	0.0022	
0.4				0.5193			0.0036
	0.986	0.0986	0.1282		0.0014	0.0060	
0.5				0.6475			0.0096
	0.960	0.0960	0.1248		0.0040	0.0173	
0.6				0.7723			0.0269
	0.875	0.0875	0.1138		0.0125	0.0541	
0.7				0.8861			0.0810
	0.600	0.060	0.0780		0.0400	0.1730	
0.8				0.9641			0.2540
	0.240	0.024	0.0312		0.0760	0.3290	
0.9				0.9953			0.5830
	0.036	0.0036	0.0047		0.0964	0.4170	
1.0				1.0000			1.000
		0.76889	1.000		0.23111	1.00	

$$(X''/X')_i = (X''/X')\alpha_n \qquad (14.11)$$

where $\alpha = K_i/K_n$

The ratio X''/X' for each component is the ratio of moles in the liquid after distillation to the moles in the liquid before distillation. The relative volatilities should be constant over the temperature range of the process, but an average value of α can be used.

The method of solution for a constant batch distillation is to assume a final temperature, fix the X''/X' ratio for the point on the molar TBP curve, and then calculate the X''/X' ratio for other points on the TBP curve. New mole fraction values in the remaining liquid at other points on molar TBP of the feed are found by:

$$\int_0^{\text{old m.f.}} (X''/X')dx \div \int_0^1 (X''/X')dx \qquad (14.12)$$

The assumed temperature is correct for the fixed X''/X' when

$$\int_0^1 K\,dx = 1.0, \text{ where } x \text{ is the mole fraction in the remaining liquid.}$$

If $K = p^o/p$, then $\int_0^1 p^o\,dx = p$ is the "correct" criterion.

Example 4. Find the volume of fuel lost by evaporation as an airplane rises from sea level to 52,000 feet, assuming that the fuel remains at 100°F and is vented to the atmosphere. The fuel has a 0.745 specific gravity, a Reid Vapor Pressure (RVP) of 2.86 lb, and the following ASTM D-86:

IBB	5%	10%	15%	20%	25%	30%	40%	50%
113	180	192	202	210	219	225	240	257

IBB	60%	70%	80%	90%	95%	EP	L/R*
113	280	310	350	410	457	480°F	1%/1%

* Loss and Residue in laboratory distillation

Solution. The evaporation of fuel from the tank is a batch or differential distillation at the reduced pressure of 52,000 feet altitude. The assumption that the tank remains at 100°F requires heat input to the fuel tank equal to the latent heat of vaporization plus convective heat losses, if any.

Step 1. Plot the above ASTM D-86 temperatures against liquid-volume-percent-off, correcting the latter for the 1% loss.

Step 2. Convert the ASTM D-86 curve to a TBP curve via Figure 12.5.

Step 3. Convert the volumetric TBP curve from Step 2 to a molar TBP curve via Figure 14.1 and then plot, as on Figure 14.9.

Step 4. Calculate the vapor pressure of the original fuel at 100°F by finding the area under a curve drawn through several vapor pressure points plotted against the corresponding mole fraction distilled on the TBP assay. This calculated vapor pressure is compared with the RVP to check the two conversions, i.e., ASTM to TBP and vol-% TBP to mol-fraction TBP. See Figure 14.10.

Step 5. Find relative volatilities of hydrocarbons corresponding to various points along the TBP curve, using *n*-octane as the reference component.

Step 6. Assume the ratio of any one component in the liquid before and after the distillation. Compute similar ratios for other components from relative volatilities using the relationship:

$$(X''/X')_{Cnn} = (X''/X')^\alpha_{C8}$$

Plot these ratios against mole-percent-off. See Figure 14.11.

Step 7. Check the assumption in Step 6 by plotting vapor pressure versus the new mole-percent (lower curve in Figure 14.10), finding the new mole-percent values from the proata areas under the curve in Figure 14.11. If the resulting vapor pressure equals the atmospheric pressure at the

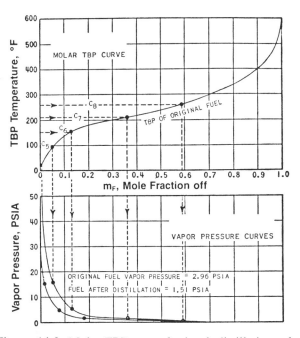

Figure 14.9. Molar TBP curve for batch distillation calculations—Example 4.

Figure 14.10. Vapor pressure curves for batch distillation calculations—Example 4.

high altitude, then the assumption of Step 6 was correct. If not, Step 6 must be repeated with another assumption.

Step 8. Calculate liquid volume of fuel lost by converting the molar loss curve in Figure 14.11 to a liquid volume loss curve and find the area under the curve. Plot this curve and the gallons per mol of original fuel on the same graph. These two plots are on Figure 14.12. Find the area under each curve. The fuel loss is the ratio of the two areas, or 11%, as shown on Figure 14.12.

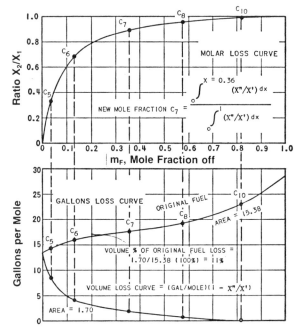

Figure 14.11. Molar loss curve for batch distillation—Example 4.

Figure 14.12. Gallons loss curve for batch distillation—Example 4.

Notation

L = moles of liquid mixture
V = moles of vapor mixture
F = moles of feed mixture
D = moles of distillate mixture
B = moles of bottoms mixture
l = moles of any infinitesimal increment* of the liquid
v = moles of any infinitesimal increment* of the vapor
f = moles of any infinitesimal increment* of the feed
d = moles of any infinitesimal increment* of the distillate
b = moles of any infinitesimal increment* of the bottoms

$K = y/x$ = vapor-liquid equilibria ratio
$\alpha = K_2/K_1$ = relative volatility
y = mole fraction of any infinitesimal increment* of the vapor
x = mole fraction of any infinitesimal increment* of the liquid
X = moles of component in liquid
$A = L/KV$ = absorption factor at indicated point or stage
$S = KV/L$ = stripping factor at indicated point or stage
m_F = mole fraction off on TBP assay of feed
m_L = mole fraction off on TBP assay of equilibrium liquid
m_V = mole fraction off on TBP assay of equilibrium vapor
M = moles of mixture
P = pressure
p^o = vapor pressure
T_B = boiling point
R = reflux ratio

Subscripts

0 refers to reboiler or condenser
1,2,3, etc. indicate equilibrium stages numbered from the condenser or reboiler toward the feed
n refers to bottom stage in enriching section
m refers to the top stage in the stripping section
f refers to the feed
i and n designate any component

References

1. Bowman, J. R., *Ind. Eng. Chem.*, 41, 2004 (1949); 43, 2622, (1951).
2. Bowman, J. R. and W. C. Edmister, *Ind. Eng. Chem.*, 43, 2625 (1951).
3. Edmister, W. C. and J. R. Bowman, Chem. Eng. Prog. Symp. Series, 48, 2, 112, (1952); 48, 3, 46 (1952).
4. Edmister, W. C. and D. H. Buchanan, Chem. Eng. Prog. Symp. Series, 49, 6, 69 (1953).
5. Edmister, W. C., *Ind. Eng. Chem.*, 47, 1685 (1955).
6. Harbert, W. D., *Ind. Eng. Chem.*, 39, 118 (1947); *Pet. Ref.*, 26, 132 (1947).
7. Katz, D. L. and G. G. Brown, *Ind. Eng. Chem.*, 25, 1373 (1933).
8. Smith, R. L. and K. M. Watson, *Ind. Eng. Chem.*, 29, 1408 (1937).
9. Thiele, E. W. and R. L. Geddes, *Ind. Eng. Chem.*, 25, 289 (1933).
10. Van Winkle, M. *Distillation*, McGraw-Hill, New York (1967), p. 127.
11. White, R. R. and G. G. Brown, *Ind. Eng. Chem.*, 34, 1162 (1933).

* An indefinite number of these infinitesimal increments, i.e., components, make up each petroleum fraction.

15

Pseudo Components of Petroleum For Computer VLE Calculations

Chapter 14 presented the integral method of making vapor-liquid equilibria calculation for petroleum fractions and illustrated it with four examples. Although theoretically sound, the numerical and graphical techniques used in these applications are not consistent with the conventional methods used for mixtures of discrete components. Also, it would be awkward to put these numerical and graphical procedures onto the computer. Accordingly, the integral method is modified in this chapter to be consistent with the procedure described in Chapter 11 for multicomponent mixtures of discrete substances, making it practical for computer applications.

Modified Integral Solution

The basis for this modification is the employment of a numerical integration technique, suggested in 1971 by Taylor and Edmister (5) for use in solving petroleum and natural gas processes. The following description of that improvement in the graphical integration technique is from that reference.

The material balance of an equilibrium flash vaporization for a component represented by a point on the molar TBP curve for the feed is

$$F\ dm_F = V\ dm_V + L\ dm_L \tag{15.1}$$

Since m_F is the total mole fraction distilled up to the temperature T on the TBP curve of the feed, the differential

dm_F represents the mole fraction in the feed of the component having a true boiling point of T. Analogous meanings are associated with dm_L and dm_V. The equilibrium relationship for this component is

$$dm_V = K\ dm_L \tag{15.2}$$

Hence, by combining Equations 15.1 and 15.2

$$dm_L = (F\ dm_F)/(L + KV) \tag{15.3}$$

Integration of Equation 15.3 over the range of all components in the feed gives

$$\int_0^1 dm_L = 1 = \int_0^1 (1/(L + KV))F\ dm_F \tag{15.4}$$

In one form of the flash problem, the flash pressure and the quantities of liquid and vapor products are specified and the flash temperature and the characteristics of the liquid and vapor products must be determined. The solution is obtained by iterative calculations. Flash temperatures are assumed until Equation 15.4 is satisfied.

Equation 15.4 was solved by graphical integration in Example 1 of Chapter 14. It can also be solved by numerical integration, via the well-known Simpson's Rule (3) as proposed and illustrated by Taylor and Edmister (5) for making petroleum VLE calculations. In this method the curve being integrated must be divided into an even number of equally sized divisions n on the m_F scale, giving $n + 1$ temperature points on the TBP curve. Thus, the interval width h on the m_F abscissa scale is the same for all divisions, while the temperature intervals will be irregular. The total intervals of integration includes all of the "components" in the feed.

For example, 20 divisions along the m_F scale gives a total of 21 pseudo components that are defined by the boiling points and specific gravities at these points. With 10 divisions, the total pseudo components would be 11. Any even number of divisions may be used, although 10 or 20 are the usual numbers chosen. The quantity m_F takes on fractional values from 0 to unity. This corresponds to components with true boiling points ranging from the initial to the final boiling point of the mixture.

When this interval of integration is broken into n sub-intervals for the application of Simpson's Rule, the integral is approximated by a summation, as follows:

$$\int_0^1 \frac{dm_f}{L + KV} = \frac{\frac{1}{3}h}{(L + KV)_{t_1}}$$

$$+ \sum_{i=2}^{n} \frac{(\frac{1}{3}h)\{3 + (-1)^i\}}{(L + KV)_{t_i}}$$

$$+ \frac{\frac{1}{3}h}{(L + KV)_{t_{n+1}}} \tag{15.5}$$

where h = interval width, fraction of m_F
    n = number of divisions (when $n = 10$, $h = 0.10$; when $n = 20$, $h = 0.05$)
    t_1 = initial temperature on TBP assay
    t_i = TBP assay temperature at point i
    t_{n+1} = final temperature on TBP assay

The notation $(L + KV)|_{t_i}$ indicates that the distribution coefficient K is evaluated for that component whose TBP temperature is t_i. For example, K_{t_i} is the value of K for the component whose TBP is the initial boiling point of the mixture. If 10 sub-intervals are used for approximating the integral, K_{t_2} is the value of K for the component whose TBP corresponds to a value of 0.1 for m_F on the TBP curve of the feed.

The expression corresponding to Equation 15.4 for a finite mixture is

$$\sum_{i=1}^{c} \frac{Fx_{F_i}}{L + K_iV} = 1.0 \tag{15.6}$$

Quantities $(\frac{1}{3}h)$, $(\frac{1}{3}h)(3 + (-1)^i)$, and $(\frac{1}{3}h)$ appearing in Equation 15.5 might be considered the composi-

tions of a pseudo-feed containing $n - 1$ components. The integral can then be expressed as

$$\int_0^1 \frac{dm_F}{L + KV} = \sum_{i=1}^{n+1} \frac{x_{F_i}}{L + K_iV} \tag{15.7}$$

That is $xF_1 = x_{F_{n+1}} = \frac{1}{3} h$ and
    $x_{F_i} = (\frac{1}{3} h)(3 + (-1)^i)$ for $2 \le i \le n$ (15.8)

where x_{F_i} = mole fraction of pseudo-component i in feed.

In this manner the integral technique can be used in standard programs written for finite component mixtures, such as the programs described in Chapters 16 and 17.

The various points on the TBP curve that characterize the liquid product of a flash process are obtained by integrating Equation 15.3 from the initial boiling point where $m_F = 0$ to the value of m_F corresponding to the TBP of the point in question. That is

$$m_{L|t} = \int_0^{m_{F|t}} \frac{F}{L + KV} dm_F \tag{15.9}$$

The subscript $L|t$ refers to "liquid" at assay temperature "t" and the subscript $F|t$ refers to "feed" at assay temperature "t." The molar TBP curve of the vapor product is found by a similar calculation

$$m_{V|t} = \int_0^{m_{F|t}} \frac{KF}{L + KV} dm_F \tag{15.10}$$

Analogous meanings are associated with the subscripts $V|t$ and $F|t$ as those given for $L|t$ and $F|t$.

The procedure using a pseudo-feed composition in the evaluation of the integral in Equation 15.4 cannot be used for Equations 15.9 and 15.10. Simpson's Rule must be reapplied in the appropriate manner over each of the desired intervals of the integration.

Example 1. For illustrating integration by Simpson's Rule method, comparing with the chord-area method, a flash calculation for a petroleum kerosene will be solved numerically. This illustration starts with a molar TBP of the naphtha feed. Only the final trial V/L ratio of 3.0 is shown and the temperature and pressure conditions plus the K values are given. After finding a converged solution, the molar assays of the flash vapor and liquid are calculated. The calculations and results are given in Tables 15.1 and 15.2.

Table 15.1
Alternate Numerical Integrations in Flash Calculations for a Petroleum Kerosene

			Chord-Area Method with 20 increments		Simpson's Rule Method with 21 pseudo components	
	Molar TBP	K Values				
m	BP°F	K @ T&P	$F/(L + KV)$	Δm_L	xf_i Eq. 15.8	$xf_i/(1 + KV/L)$
0.00	200	7.4	0.172414		0.016667	0.000718
				0.0097551		
0.05	225	5.8	0.217391		0.066667	0.003623
				0.0126690		
0.10	240	4.8	0.259740		0.033333	0.002164
				0.0141858		
0.15	252	4.0	0.307692		0.066667	0.005128
				0.0166289		
0.20	261	3.4	0.357143		0.033333	0.002976
				0.0189286		
0.25	273	3.0	0.400000		0.066667	0.006667
				0.0206383		
0.30	280	2.8	0.425532		0.033333	0.003546
				0.0220019		
0.35	280	2.6	0.454545		0.066667	0.007576
				0.0245215		
0.40	300	2.2	0.526316		0.033333	0.004386
				0.0274436		
0.45	310	2.0	0.571429		0.066667	0.009524
				0.0299107		
0.50	320	1.8	0.625000		0.033333	0.005208
				0.0328664		
0.55	331	1.6	0.689655		0.066667	0.011494
				0.0374043		
0.60	347	1.35	0.792079		0.033333	0.006601
				0.0422739		
0.65	358	1.15	0.898874		0.066667	0.014981
				0.0474719		
0.70	370	1.00	1.000000		0.033333	0.008333
				0.0572580		
0.75	388	0.70	1.290333		0.066667	0.021505
				0.0742749		
0.80	410	0.46	1.680672		0.033333	0.014006
				0.0918522		
0.85	440	0.34	1.980198		0.066667	0.033003
				0.1086760		
0.90	470	0.23	2.366864		0.033333	0.019724
				0.1389926		
0.95	525	0.095	3.112840		0.066667	0.051881
				0.1703280		
1.00	600	0.027	3.700278		0.016667	0.015418
			Totals	0.9960816		0.248462
			vs	1.0000000		0.250000

For the chord-area method solution in Table 15.1, the area for each increment is the product of the average value of $(F/(L + KV))$ for the increment and the width of the increment or delta m_F, which was 0.05 in this case. This product is in the fifth column and is midway between the initial and final points of the increment involved. The sum of these areas will be equal to the value of unity for a converged solution.

For the Simpson's Rule solution, 21 pseudo components were created by applying Equation 15.8. These xf_i

are given in column six of Table 15.1. In column seven are values of $xl_i = xf_i/(1 + KV/L)$. It will be noted that $(L + KV) = (1 + KV/L)L$, which is the reason that column seven total is equal to L at convergence, while the total of column five is only unity.

In Table 15.2 are the calculations to obtain mole-fraction-off points for the assays of the flash products. Column one gives the boiling points, which make one coordinate of each assay, i.e., for feed, liquid, and vapor. Column five gives the mole-fraction-off for the feed assay, while columns six and seven give these values for liquid and vapor products. These are found from the pseudo component compositions, xl_i and $xv_i = xf_i - xl_i$, by

$$m_L = \frac{\sum_1^i xl_i}{\sum_1^{k+1} xl_i} \text{ and } m_V = \frac{\sum_1^i xv_i}{\sum_1^{k+1} xv_i}$$

The summations in the numerators are from the first pseudo component to each point in turn. The summa-

tions in the denominator are for all of the pseudo components.

An examination of the data in columns 5, 6, and 7 of Table 15.2 shows that the abscissa intervals on the product assays vary. These data could be plotted and new temperatures read from the curves at even values of m_L and m_V, if these were needed.

Example 2. This example illustrates the calculation steps required for converting the molar TBP assays to volumetric TBP assays for the vapor and liquid products from flashing the petroleum kerosene. These calculations are made for each of the 21 pseudo components, which are defined by their atmospheric boiling points, and are given in Tables 15.3 and 15.4. The calculations are made for the feed as well because the starting assay for Example 1 was a molar TBP assay, rather than the usual volumetric TBP assay.

Note that the intervals in the "volume-fraction-off" columns in Table 15.4 are irregular. These volumetric assays could be plotted with boiling point temperature as the ordinate and volume-fraction-off as the abscissa.

Table 15.2
Calculations of Points on Molar TBP Assays of Vapor and Liquid Products
from Flashing Petroleum Kerosene

Comp. BP°F	Pseudo-Component Compositions			Molar TBP Assay Abscissas		
	xf_i	xl_i	xv_i	m_F	m_L	m_V
200	0.016667	0.000718	0.015949	0.00	0.00	0.00
225	0.066667	0.003623	0.063044	0.05	0.00903	0.05416
240	0.033333	0.002164	0.031169	0.10	0.02107	0.11657
252	0.066667	0.005128	0.061539	0.15	0.03624	0.17896
261	0.033333	0.002976	0.030357	0.20	0.05309	0.24081
273	0.066667	0.006667	0.060000	0.25	0.07315	0.30162
280	0.033333	0.003546	0.029787	0.30	0.09440	0.36204
290	0.066667	0.007576	0.059091	0.35	0.11753	0.42186
300	0.033333	0.004386	0.028947	0.40	0.14242	0.48111
310	0.066667	0.009524	0.057143	0.45	0.17135	0.53905
320	0.033333	0.005208	0.028125	0.50	0.20200	0.59643
331	0.066667	0.011494	0.055173	0.55	0.23674	0.65249
347	0.033333	0.006601	0.026732	0.60	0.27438	0.70762
358	0.066667	0.014981	0.051686	0.65	0.31928	0.76039
370	0.033333	0.008333	0.025000	0.70	0.36777	0.81200
388	0.066667	0.021505	0.045162	0.75	0.42984	0.85922
410	0.033333	0.014006	0.019327	0.80	0.50371	0.90262
440	0.066667	0.033003	0.033664	0.85	0.60149	0.93828
470	0.033333	0.019724	0.013609	0.90	0.71107	0.97010
525	0.066667	0.051881	0.014786	0.95	0.86000	0.98921
600	0.016667	0.015418	0.001249	1.00	1.00000	1.00000
Sums	1.000000	0.248462	0.751538			

Table 15.3
Volumetric Equivalents of Pseudo Component Quantities for Feed and Products from Flashing Petroleum Kerosene

BP°F	Mol/100G	Feed M/100	Feed Equiv Gal	Liquid M/100	Liquid Equiv Gal	Vapor M/100	Vapor Equiv Gal
200	5.60	1.6667	0.297625	0.0718	0.012820	1.5949	0.28480
225	5.45	6.6667	1.223242	0.3623	0.066480	6.3044	1.15677
240	5.30	3.3333	0.628930	0.2164	0.040830	3.1169	0.58809
252	5.20	6.6667	1.282050	0.5128	0.098615	6.1339	1.18344
261	5.10	3.3333	0.653590	0.2976	0.058353	3.0357	0.59524
273	5.00	6.6667	1.333334	0.6667	0.133334	6.0000	1.20000
280	4.95	3.3333	0.673390	0.3546	0.071640	2.9787	0.60176
290	4.90	6.6667	1.360551	0.7576	0.154612	5.9091	1.20594
300	4.85	3.3333	0.686280	0.4386	0.090433	2.8947	0.59685
310	4.75	6.6667	1.403520	0.9524	0.200505	5.7143	1.20301
320	4.70	3.3333	0.709210	0.5208	0.110809	2.8125	0.59840
331	4.60	6.6667	1.449283	1.1494	0.249870	5.5173	1.19941
347	4.50	3.3333	0.740740	0.6601	0.146689	2.6732	0.59404
358	4.45	6.6667	1.498135	1.4981	0.336652	5.1686	1.16150
370	4.40	3.3333	0.757575	0.8333	0.143832	2.5000	0.56818
388	4.30	6.6667	1.550395	2.1505	0.500116	4.5162	1.05028
410	4.00	3.3333	0.833333	1.4006	0.350150	1.9327	0.48318
440	3.85	6.6667	1.731600	3.3003	0.857220	3.3664	0.87439
470	3.70	3.3333	0.900890	1.9724	0.533080	1.3609	0.36781
525	3.30	6.6667	1.801819	5.1881	1.572152	1.4786	0.44806
600	2.80	1.6667	0.595300	1.5418	0.550640	0.1249	0.04461

Table 15.4
Calculation of Volume-Fractions-Off for TBP Assays of Feed to and Products from Flashing Petroleum Kerosene

BP°F	Feed Cumul. Gal.	Feed V.F. Off	Liquid Cumul. Gal.	Liquid V.F. Off	Vapor Cumul. Gal.	Vapor V.F. Off
200	0.00000	0.000000	0.00000	0.000000	0.00000	0.000000
225	0.76043	0.035978	0.03965	0.006609	0.72079	0.045500
240	1.37206	0.069190	0.09580	0.015968	1.59322	0.100589
252	2.32755	0.110128	0.16552	0.027589	2.47898	0.156491
261	3.29536	0.155920	0.24401	0.040668	3.36832	0.212633
273	4.28883	0.202926	0.33985	0.056646	4.26594	0.269297
280	5.29219	0.250399	0.44234	0.073728	5.16682	0.326168
290	6.30916	0.298518	0.55546	0.092583	6.07067	0.383225
300	7.33258	0.346941	0.67799	0.113005	6.97207	0.445756
310	8.37748	0.396380	0.82346	0.137252	7.87200	0.496938
320	9.43385	0.446362	0.97911	0.163196	8.77270	0.553797
331	10.51310	0.497427	1.15945	0.193255	9.67161	0.610543
347	11.60811	0.549237	1.35773	0.226290	10.56833	0.667150
358	12.72755	0.602203	1.59940	0.266585	11.44610	0.722560
370	13.85540	0.655568	1.83964	0.306628	12.31094	0.777160
388	15.00938	0.710168	2.16162	0.360294	13.12017	0.828241
410	16.20125	0.766561	2.58675	0.431154	13.88685	0.876640
440	17.48369	0.827240	3.19044	0.531743	14.56564	0.919490
470	18.79994	0.889521	3.88559	0.647642	15.18674	0.958698
525	20.15107	0.953447	4.03820	0.823088	15.59467	0.984450
600	21.13496	1.000000	5.99960	1.000000	15.88410	1.000000

From these plots, new temperature values could be found for even values for fraction or percent off, i.e., 10%, 20%, etc.

Computer Methods

The calculations shown in Tables 15.1, 15.2, 15.3, and 15.4 illustrate the numerical methods of solving problems like those in Examples 1 and 2 and help develop computer programs for doing the same kind of calculation faster and more accurately. Computer methods for dividing a petroleum oil into small cuts or slices and then characterizing these pseudo components requires analytical equations and procedures for each step of the calculations. Many of these equations and procedures are already available in the technical literature (1,2,3,4). This available material will be supplemented by other procedures for this application.

Note that the starting assay in Example 1 was a molar TBP, which cannot be obtained experimentally in the laboratory and would not be the starting point of a real-life problem. Normally, we would start with a volumetric TBP or a volumetric ASTM assay, which can be obtained in the laboratory. Starting with a volumetric TBP assay requires a conversion to molar step. Starting with the ASTM volumetric requires two extra steps: "volumetric ASTM to volumetric TBP" and "volumetric TBP to molar TBP."

ASTM-TBP Conversion

ASTM assays usually involve atmospheric-boiling-point-temperatures versus liquid-volume-percent-off and contain seven points consisting of the temperatures at the IBP (initial boiling point), the 10% to 90% (at even 10% intervals) distilled points, and the FBP (final boiling point). As stated in Chapter 12, there are three such ASTM assays of interest here, i.e., the ASTM D-86, the ASTM D-158, and the ASTM D-1160. For present purposes, the designation "ASTM Assay" refers to any, or all, three. Likewise, the designation "TBP Assay" refers to any true boiling point distillation made in a batch laboratory column at a high reflux rate. A graphical method for interconverting the ASTM and TBP assays was given in Figures 12.4 and 12.5. An analytical method of making these conversions is needed for computer applications.

Such a program was written in 1975 by Lion and Edmister (4). The program is named Subroutine ASSAY and is given in Table 15.5. Like Figures 12.4 and 12.5, ASSAY relates the 50% point temperatures of the ASTM and the TBP and also the slopes of different segments, 0–10, 10–20, etc., of the two assays. As written, ASSAY is for converting the ASTM to the TBP only. The

same empirical equations can also be applied to the (TBP-ASTM) conversion, if needed. This would be a useful exercise for the student.

The COMMON/DDD is for the transfer of information in and out of this and some other subroutines. In a different application this data transfer might be via the argument of the call statement. The DATA and DIMENSION statements are for this program only. A correction to the ASTM temperatures for cracking will be made by DO-loop 3, if necessary, this being signaled by a calling argument of KEY = 1 and by temperatures greater than 475°F, both being required. Temperatures from an ASTM D-1160 assay that has been run at 10 mm Hg, for example, and the temperatures corrected to equivalent 760 mm values, might not require a cracking correction. This situation should be known to the user and signaled by KEY = 2.

DO-loop 10 computes the slopes of the ASTM 0-10, 10-30, 30-50, 50-70, 70-90, and 90-100% segments in °F/vol%. DO-loop 10 also finds the values of the TBP-ASTM temperature differences at the corresponding assay locations. Next, the 50% point temperature on the TBP curve is found from the 50% point on the ASTM Assay, designated ASTM(4) and TBP(4) in the program. DO-loop 20 calculates the remaining six assay points.

Conversion of TBP Assays

In the previous section, a program was presented for converting the ASTM assay an equivalent volumetric TBP assay, similar to the type obtained from a laboratory distillation. There is also a weight TBP assay, i.e., boiling temperature versus weight-percent-off instead of liquid volume percent off. Such weight-basis assays are obtained from analyses made on the gas chromatograph. Both volumetric and weight assays must be converted to a molar basis before breaking the oil down into 11 or 21 pseudo components because vapor-liquid equilibrium calculations are made with quantities expressed in moles.

Thus, two TBP assay conversion steps are of interest, volumetric-to-weight and weight-to-molar. The first step requires gravities and the second step requires molecular weights. As molecular weight is a function of boiling point and gravity for narrow boiling cuts of petroleum, the gravity assay and the boiling point assay are both needed. In most crude oil evaluations, both gravity and boiling point assays are obtained in the laboratory. In these cases all of the data needed to represent the oil fraction by pseudo components are available.

The volumetric-to-weight conversion involves multiplying each volume increment, usually 5 or 10%, by its gravity, summing for all increments and then dividing through by the total, obtaining weight fractions. These

Table 15.5
FORTRAN Program for Conversion of ASTM to TBP Distillation Assay

```
      SUBROUTINE    ASSAY(KEY)
      COMMON/DDD/NSIMP,NPOINT,XAPI,ASTM(7),XTBP(21),GRAV(21),XOF(21),
             TBP(21),XMOL(21),XN(21),AMW,XGRAV,FEED(5),TITLE(20)
      DIMENSION CV(6),A(7,5),TAUX97
      DATA A/3.47343E-2,3.00864E-2,2.177006E-2,1.284808E-2.-6.8517E-3,
      1.18498E-2,1.1061821E + 1,2.54839,2.11575,1.80539,1.62486,1.54216,
      1.30941,1.798E-2,-0.35301,-0.32401,-0.21701,-0.17628,-0.17765,
      0.10068,3.16E-5,4.0851E-2,3.1598E-2,1.8244E-2,1.5496E-2,
      1.877E-2,1.2669E-2,-5.5E-8,-1.6997E-3,-8.526E-4,-2.912E-4,
      2.712E-4,-751805E-4,-3.36E-4,9.51E-11/
      DATA CV/10.0,4*20.0,10.0/
      GO TO (1,2),KEY
  1   DO 3 I = 1,7
      IF(ASTM(I) − 475.) 3,3,7
  7   ASTM(I) = ASTM(I) + EXP(2.302585*(0.00473*ASTM(I) − 1.587))
  3   CONTINUE
  2   DO 10 I = 1,6
      S = (ASTM(I + 1) − ASTM(I))/CV(I)
 10   TAUX(I) = (((A(I,5)*S + A(I,4))*S + A(I,3))*S + A(I,2))*S + A(I,1)
      S = ASTM(4)
      TAUX(7) = (((A(7,5)*S + A(7,4))*S + A(7,3))*S + A(7,2))*S + A(7,1)
      TBP(4) = ASTM(4) + TAUX(7)
      I = 3
      DO 20 J = 5,7
      TBP(I) = TBP(I + 1) − TAUX(I)*CV(I)
      TBP(J) = TBP(J − 1) + TAUX(J − 1)*CV(J − 1)
 20   I = I − 1
      RETURN
      END
```

derived weight fractions will vary in size from the IBP to the FBP. Mole fractions are obtained by dividing the weight fraction of each increment by its molecular weight, summing for all increments and then dividing through by the sum, obtaining mole fractions, which also vary in size. In practice it is more convenient to combine these two conversion factors for making the volumetric to molar conversion, i.e., multiply by SG/WM, the specific gravity/molecular weight ratio.

After making the conversion of the volume or weight-percent-off scale to a mole-fraction-off scale, the resulting temperature vs. mole fraction values must be curve fitted and temperatures interpolated at even values of mole fraction, i.e., 0.05, 0.10, 0.15, . . ., 0.90, 0.95. The resulting molar TBP is then ready for division into pseudo components.

When the laboratory analysis includes a specific-gravity-versus-volume-percent-off assay, the same procedure is applied to the gravity assay to obtain a gravity-versus-mole-fraction-off curve. Boiling point and specific gravity are the properties of the pseudo components that are used in predicting other properties, such as critical conditions, etc.

Sometimes a gravity is not available and it is necessary to proceed with an overall gravity of the entire petroleum fraction. This is usually the case with ASTM distillation assays. In these cases, it is necessary to approximate the gravity assay, a procedure that will be shown later.

Lion and Edmister (4) described a program for volumetric-to-molar-TBP conversions, using gravities and true boiling points from the volumetric assays and molecular weight. Molecular weight can be obtained from a recent equation given in the APITDB Procedure 2B2.1 (1), which was given in Chapter 7 of Volume 1 as:

$$M = 204.38 \frac{\exp(0.00218\ TT)\exp(-3.07\ SG)}{TT^{0.118}\ SG^{1.88}} \quad (7.2)$$

where M = molecular weight of petroleum fraction
        TT = mean boiling point of petroleum fraction, °R
        SG = specific gravity, 60°F/60°F.

A computer program, based on Equation 7.2, can be prepared for making the conversion of liquid volume fraction abscissa values to equivalent mole fraction values, for which a new abscissa scale can be used for both the true boiling point assay and the gravity assay. The calculations made are similar to those in Tables 15.3 and 15.4, where the conversion was from molar to volumetric, rather than from volumetric to molar.

The new "mole-fraction-off" abscissa scales of the TBP and Grav assays have divisions that are not at even intervals, or in round numbers. In order to get even intervals and round numbers, i.e., 0.1, 0.2, 0.3, etc., it is necessary to curve fit the TBP-temperature-versus-mole-fraction-off coordinates and find, via the equation, new TBP values that correspond to the even abscissa values. This can also be done for the specific gravity versus mole fraction off coordinates.

Pseudo Component Breakdown

From the molar TBP assay, obtained by the previous procedures, an oil fraction can be divided into pseudo components, by Simpson's Rule, as given in Equation 15.8. A small computer program, SIMPSO, for making these feed breakdown calculations has been prepared. In this program, the number of increments in the molar TBP assay of the feed may be ten 10% increments, or twenty 5% increments and the 11 or 21 pseudo components, thus obtained, will be identified by the true boiling point temperatures of the molar TBP assay. With these boiling point temperatures and the specific gravities, the component breakdown is ready for properties calculations.

Petroleum Component Properties

The following expressions are recommended for calculating properties of the pseudo components of petroleum. Equation 7.2 was given for molecular weights. The following equations, which are Equations 3, 4 and 5 of Kesler and Lee (2) and were given in Chapter 7 of Volume 1, are recommended for critical temperature, critical pressure, and acentric factor:

$$T_c = 341.7 + 811.0\,SG + (0.4244 + 0.1174\,SG)\,TT + (0.4669 - 3.2623\,SG)10^5/TT \quad (7.16)$$

$$\ln P_c = 8.3634 - 0.0566/SG - (0.24244 + 2.2898/SG + 0.11857/SG^2)\,10^{-3}\,TT + (1.4685 + 3.648/SG + 0.47227/SG^2)\,10^{-7}\,TT^2 - (0.42019 + 1.6977/SG^2)\,10^{-10}\,TT^3 \quad (7.17)$$

$$WC = -7.904 + 0.1352\,K - 0.00746\,K^2 + 8.359\,BR + (1.408 - 0.01063\,K)/BR \text{ (for } BR > 0.8) \quad (7.26)$$

where
T_c = critical temperature, °R
P_c = critical pressure, psia
TT = normal boiling point, °R
$BR = TT/T_c$ = reduced normal boiling point
WC = acentric factor

When $BR < 0.0$ or $BR = 0.8$, another WC equation is recommended instead of Equation 7.26, namely a rearrangement of Equation 7.25. Equation 7.34 for the ideal gas state heat capacities of petroleum fractions is also given in Chapter 7 of Volume 1. Additional material on enthalpies appears in Chapter 18.

Examples of Pseudo Component Breakdown

The results from two computer breakdown calculations for typical oils are given in Tables 15.6 and 15.7.

Example 1. The first breakdown calculation is for a petroleum naphtha that is defined by an ASTM distillation assay and API gravity. These starting data and the calculated properties are given in Table 15.6. The seven-point assay information, given in the top part of the table, includes the derived TBP distillation assay points and also the corresponding specific gravities, both at the same seven points. Working from the TBP and SG assay information, the feed was cut into 11 pseudo components and the values of the normal boiling points and specific gravities were calculated. This pseudo component information is given in the lower part of Table 15.6.

Example 2. Similar pseudo component information is given in Table 15.7 for Sumatran Light Crude Oil, which was defined by 21-point TBP and SG assays. These crude oil breakdown data will be used in the next illustration that shows the preparation of an input data file.

Example 3. Assay and pseudo component calculations are illustrated by this example on Sumatran Light Crude Oil, for which Tables 15.7 and 15.8 give the computer results. The original assays were 21-point volumetric true-boiling-point and gravity assays. With these data, the oil was broken down into 21 pseudo components, creating a data file for later VLE calculations.

Pseudo component mole fractions, normal boiling points, and specific gravities were first calculated. These are in Tables 15.7 and 15.8, the latter being part of an input data summary. Then the thermodynamic property and flash calculations were made using the same programs that are used for the mixtures of discrete components. These are given in Table 15.8. The use of the same thermodynamic methods for pure and pseudo components is made possible by using empirical equations for

Table 15.6
Break Down of Petroleum Naphtha into 11 Pseudo Components from
a 7-Point ASTM Distillation Assay and 66 API Gravity

Assay Properties of the Petroleum Naphtha							
Point	IBP	10%	30%	50%	70%	90%	EP
ASTM, °F	130.0	190.0	240.0	280.0	320.0	360.0	400.0
TBP, °F	50.2	142.6	217.7	275.8	329.4	379.5	422.9
Spec. Grav.	.6372	.6736	.7005	.7200	.7370	.7523	.7651

Pseudo Component Breakdown and Properties							
Pseudo	COMP1	COMP2	COMP3	COMP4	COMP5	COMP6	COMP7
Mol. Fr.	.0333	.1333	.0667	.1333	.0667	.1333	.0667
NBP, F.	50.2	119.2	165.7	197.7	223.9	252.7	281.3
Sp. Gr.	.6372	.6645	.6823	.6938	.7026	.7123	.7217
Pseudo	COMP8	COMP9	COMP10	COMP11			
Mol. Fr.	.1333	.0667	.1333	.0333			
NBP, F.	310.8	338.0	368.6	422.9			
Sp. Gr.	.7312	.7398	.7491	.7651			

Table 15.7
Break Down of Sumatran Light Crude Oil into 21 Pseudo Components
from TBP Distillation and Specific Gravity Assays of 21 Points

Assays of the Sumatran Light Crude							
Point	IBP	5%	10%	15%	20%	25%	30%
TBP, °F	0.0	195.0	265.0	320.0	370.0	418.0	460.0
Spec. Grav.	.6112	.7057	.7377	.7559	.7796	.7936	.8114
Point	35%	40%	45%	50%	55%	60%	65%
TBP, °F	504.0	540.0	579.0	615.0	653.0	685.0	721.0
Spec. Grav.	.8198	.8227	.8260	.8314	.8398	.8473	.8498
Point	70%	75%	80%	85%	90%	95%	EP
TBP, °F	759.0	796.0	831.0	869.0	910.0	950.0	995.0
Spec. Grav.	.8524	.8571	.8618	.8681	.8751	.8861	.8888

Properties of the Pseudo Components							
Pseudo	COMP1	COMP2	COMP3	COMP4	COMP5	COMP6	COMP7
Mol. Fr.	.0167	.0667	.0333	.0667	.0333	.0667	.0333
NBP, F.	0.0	105.0	188.4	228.9	268.4	303.7	337.0
Sp. Gr.	.6112	.6624	.6955	.7260	.7380	.7476	.7653
Pseudo	COMP8	COMP9	COMP10	COMP11	COMP12	COMP13	COMP14
Mol. Fr.	.0667	.0333	.0667	.0333	.0667	.0333	.0667
NBP, F.	372.0	407.8	441.9	476.4	512.8	548.7	586.4
Sp. Gr.	.7803	.7905	.8031	.8154	.8210	.8232	.8272
Pseudo	COMP15	COMP16	COMP17	COMP18	COMP19	COMP20	COMP21
Mol. Fr.	.0333	.0667	.0333	.0667	.0333	.0667	.0167
NBP, F.	630.8	674.5	715.0	770.1	829.3	902.3	993.0
Sp. Gr.	.8340	.8442	.8495	.8535	.8613	.8739	.8896

Table 15.8
Sumatran Light Crude Oil (Top 75%) Pseudo Component Breakdown
Part of Data Input File for VLE Calculations

Comp. ID	Feed Mols/hr	Boil Pt °F	Specific Gravity	Component Name
1	20.235	0.00	0.6112	SUMATRAN LIGHT OIL COMP 01
2	80.941	105.01	0.6624	SUMATRAN LIGHT OIL COMP 02
3	40.471	188.41	0.6955	SUMATRAN LIGHT OIL COMP 03
4	80.941	228.89	0.7260	SUMATRAN LIGHT OIL COMP 04
5	40.471	268.39	0.7380	SUMATRAN LIGHT OIL COMP 05
6	80.941	303.67	0.7476	SUMATRAN LIGHT OIL COMP 06
7	40.471	337.02	0.7653	SUMATRAN LIGHT OIL COMP 07
8	80.941	371.95	0.7803	SUMATRAN LIGHT OIL COMP 08
9	40.471	407.83	0.7905	SUMATRAN LIGHT OIL COMP 09
10	80.941	441.86	0.8031	SUMATRAN LIHGT OIL COMP 10
11	40.471	476.41	0.8154	SUMATRAN LIGHT OIL COMP 11
12	80.941	512.75	0.8210	SUMATRAN LIGHT OIL COMP 12
13	40.471	548.69	0.8232	SUMATRAN LIGHT OIL COMP 13
14	80.941	586.38	0.8272	SUMATRAN LIGHT OIL COMP 14
15	40.471	630.75	0.8340	SUMATRAN LIGHT OIL COMP 15
16	80.941	674.50	0.8442	SUMATRAN LIGHT OIL COMP 16
17	40.471	715.00	0.8495	SUMATRAN LIGHT OIL COMP 17
18	80.941	770.06	0.8535	SUMATRAN LIGHT OIL COMP 18
19	40.471	829.25	0.8613	SUMATRAN LIGHT OIL COMP 19
20	80.941	902.25	0.8739	SUMATRAN LIGHT OIL COMP 20
21	20.235	993.00	0.8896	SUMATRAN LIGHT OIL COMP 21

physical properties, such as critical constants, molecular weight, etc., as functions of the NBP and SG of the pseudo components.

References

1. Amer. Petrol. Inst., Div. of Ref'g., Technical Data Book (1980).

2. Kesler, M. G. and B. I. Lee, *Hydro. Proc.* p. 153 March (1976).

3. Lapidus, L., *Digital Computation for Chemical Engineers*, p. 49, McGraw-Hill Book Company, New York (1962).

4. Lion, A. R. and W. C. Edmister, *Hydro. Proc.* p. 119, August (1975).

5. Taylor, D. L. and W. C. Edmister, *AIChE J*, 17, 6, 1324 (1971).

16

Vapor-Liquid Equilibria Predictions by Computer

During the past three decades computers have been used in various ways for solving hydrocarbon vapor-liquid equilibria problems. Included in the early computer applications were methods based on curve fits of graphical correlations to obtain empirical equations for calculating the K-values.

Empirical Curve Fit Methods

One method applied this procedure to the K_i vs. pressure (with lines of constant temperature) charts for different hydrocarbons and convergence pressures, obtaining empirical equations relating the K-value for each hydrocarbon to the pressure, temperature, and convergence pressure. Another applied this curve fitting procedure to the generalized fugacity coefficient charts of Edmister and Ruby (7), from which equations for the fugacity coefficients of the liquid and vapor phases as functions of the temperature, pressure, and boiling-point ratios (i.e., ratios of the individual component boiling points to the molar average boiling point of the mixture) of the components in liquid and vapor phases were obtained.

With computer programs solving the empirical equations for the K-values, these two methods were applied to process calculations. However, these, and other such methods, were soon replaced by more accurate and versatile methods, using equations of state (EOS) and other more fundamental concepts.

Equation Methods

In 1940, Benedict et. al. (2) developed an eight-constant equation of state for light hydrocarbons and their mixtures in liquid and/or vapor phases and with it prepared a book of fugacity coefficient charts for use in making manual VLE process calculations at high pressures. These charts gave f/x or f/Py as functions of temperature, pressure, and average normal boiling point of the components in the phase, liquid or vapor. The equation of state calculations required for preparing these charts were made on the best desktop calculators available in those days, and the results were recorded and then plotted by hand. Application of these charts to process engineering problems was very tedious, so K-charts with average boiling-point corrections for each phase were prepared. This was an improvement, but the resulting charts were destined to obsolescence, caused by the rapid development of the digital computer that made direct applications of the BWR Equation to VLE process calculations feasible. For more on the BWR Equation, the reader is referred to Chapters 4 and 5.

Another significant development in applied thermodynamics was the appearance in 1949 of the Redlich and Kwong (25) two-constant generalized equation of state, for which the constants can be evaluated from the critical constants of the mixture components, a great convenience. The BWR constants were determined by curve fitting to experimental P-V-T data for each component. Two other equations of the Redlich-Kwong type appeared later, namely the Soave (28) in 1972 and the Peng-Robinson (20) in 1976. The Redlich-Kwong, Soave, and Peng-Robinson equations have been widely used in VLE applications. Next we will look at the different ways these equations have been applied.

Corresponding States VLE Method

The Lee-Kesler (12) generalized correlations (see Chapter 6 in Volume 1) of thermodynamic properties were in equation form as well as in tabulations. The equations, which were based on the BWR, provide an analytical way of calculating the fugacities of components of the vapor and liquid phases. Recognizing this possibility, Ploecker, Knapp, and Prausnitz (22) applied the Lee-Kesler equations to vapor-liquid phase equilibria. Component combination rules, interaction coefficients, and pseudo-critical conditions were all a part of the program for this application. The method is rather large and not suitable for microcomputer applications. The method has not been widely accepted.

Method Structure

From a theoretical viewpoint, a single equation should be used for both vapor and liquid when making VLE prediction calculations, because each component has the same molecular structure, regardless of its phase or concentration in the vapor and liquid. This is equivalent to saying that the components are "R-K" fluids, "Soave" fluids, "P-R" fluids, or "BWR" fluids. Our ability to do this depends upon the equation of state used because it must apply to the liquid as well as to the vapor phase. It is easier to find an equation that fits vapor P-V-T data than it is to fit similar data for the liquid phase.

The BWR equation of state was powerful enough to fit vapor and liquid phases and could have been used to develop computer algorithms for VLE predictions for the past 25 years, except for complexity of the BWR, difficulty in evaluating the coefficients of the BWR, and inadequacy of digital computers until recently. In 1972, Starling and Han (29) generalized the BWR, making the coefficients easier to evaluate and suitable for mainframe applications. Computer hardware advances in recent years have increased capacity and speed so that using a BWR algorithm is now feasible on microcomputers, as well as on mainframes.

During the 1960s and 1970s the hydrocarbon process industries used hybrid methods in which vapor-phase calculations were made by an equation of state and the liquid-phase calculations were made by two other equations, both being semi-empirical. This three-equation technique was the basis of the Chao-Seader (6) method

that appeared in 1961. Cavett (5) modified the C-S method in 1962 and demonstrated the application of the method for distillation calculations. Grayson and Streed (10) also modified the C-S method in 1963 with emphasis on heavier hydrocarbons, such as petroleum fraction cuts. A similar three-equation method was proposed for cryogenic systems by Lee, Erbar, and Edmister (12) in 1973.

These four "three-equation" methods were all based upon the following expression for the vapor-liquid equilibrium distribution ratio:

$$K_i = y_i/x_i = (f/P)_i \, (\bar{f}/fx)_i/(\bar{f}/Py)_i \tag{16.1}$$

where $(f/P)_i$ = fugacity coefficient of i in pure liquid state

$(\bar{f}/fx)_i$ = activity coefficient of i in liquid mixture

$(\bar{f}/Py)_i$ = fugacity coefficient of i in vapor mixture

Separate analytical expressions are used for the three coefficients in Equation 16.1. The expression for the vapor-phase fugacity coefficient was derived from an equation of state, the Redlich-Kwong EOS being used in the Chao-Seader, the Cavett, and the Grayson-Streed methods, but a new EOS, of the Redlich-Kwong type, was derived for the Lee-Erbar-Edmister method. The mathematical expressions derived from these equations of state for calculating the fugacity coefficient are given in Chapter 5.

Expressions for the two liquid-phase terms in Equation 16.1, i.e., the fugacity and activity coefficients, were derived by a two-step procedure: (1) selecting mathematical models from theoretical considerations, and (2) determining the coefficients of the equations for these models via multiple regressions to obtain the best possible agreements with experimental values of $K_i = y_i/x_i$.

The resulting expressions were based upon using the acentric factor generalization of hydrocarbon pressure-volume-temperature data, as proposed by Pitzer et. al. (21), and the solubility parameter for regular liquid solutions (11,27), as proposed by Prausnitz (24). The expression for the liquid activity coefficient based upon the regular solution theory resembles the Van Laar (30) equation. Departures from regular solutions were not recognized. With the mathematical models using these two parameters, the authors of the four previously mentioned three-equation methods (5, 6, 10, 12) determined the required empirical constants.

For the pure liquid fugacity coefficient equation in the method of Lee-Erbar-Edmister (12), 17 empirical constants are required for each of eight cases, which are:

1. Hydrocarbons (containing nitrogen) at $T_r \leq 1.0$
2. Hydrocarbons (excluding methane) at $T_r > 1.0$
3. Methane   at $T_r > 0.93$
4. Nitrogen   at $1.0 < T_r \leq 2.2$
5. Nitrogen   at $T_r > 2.2$
6. Carbon dioxide   all temperatures
7. Hydrogen sulfide   all temperatures
8. Hydrogen   all temperatures

Although not elegant, and limited to the systems and to the conditions of the data used in the curve-fitting calculations, such empirical prediction methods are workable on digital computers and did serve the hydrocarbon industries well for several years.

Further details about the "three-equation" methods will not be presented here. An interested reader can refer to the original work described in the literature.

The present preference is to use the single-equation procedure, which was made practical with the appearance of two improved R-K type EOSs—the Soave (28) and the Peng-Robinson (20) equations. In addition to mathematical consistency, the single EOS method permits using interaction coefficients. Binary interaction coefficients are available for the Soave and the Peng-Robinson EOS and will be included in the vapor-liquid equilibrium program EQUIL.

EQUIL Program Options

New and improved computer programs are available for making these thermodynamic calculations on mainframe and micro, or personal, computers. Except for the input and output, i.e., the front-end and rear-end programs, the source programs are the same for both types of computers. Following is a general discussion of the layout and design of these new FORTRAN programs.

Vapor-liquid equilibrium separation calculation procedures, in the form of computer programs, are needed for each of the eleven equilibrium flash processes shown in Table 16.1. These programs and their subprograms, or subroutines, vary with the equations and techniques used. All of the flash calculations previously discussed are summarized in Table 16.1. In all cases, the composition of the feed mixture, i.e., the z_i values, are assumed to be known.

Each of the four bubble- and dew-point cases—Cases 2 through 5 in Table 16.1—requires a solution of Equations 11.1 and 11.2, whereas the isothermal flash and the two constant V/F flashes—Cases 1, 6, and 7—require solutions of Equations 11.1, 11.2, and 11.3. The last four flashes—Cases 8, 9, 10, and 11—require the determination of only one property, i.e., P or T, if the final condition is in the single-phase region. If, on the other hand, the final condition for any one of these last four flashes is in the two-phase region, then the solution must include one of the flashes, Cases 1, 6, or 7.

Solution Methods

The vapor-liquid equilibrium (VLE) calculations formulae, Equations 11.1 through 11.3 must be solved iteratively, because the equations are non-linear, particularly when equations of state are used to calculate the fugacities required by Equation 11.1. Many different iterative methods have been proposed for solving VLE problems, but no one method has a clear advantage over the others for all applications. The performance of these calculation methods is affected notably by:

Table 16.1
Flash Types and Variables of EQUIL

Case	Flash	Specified Variables	Unknown Variables
1	Isothermal flash	P, T	V/F, y_i, x_i
2	Bubble temperature	P, $x_i = z_i$ (note 1)	T, y_i
3	Bubble pressure	T, $x_i = z_i$ (note 1)	P, y_i
4	Dew temperature	P, $y_i = z_i$ (note 1)	T, x_i
5	Dew pressure	T, $y_i = z_i$ (note 1)	P, x_i
6	Constant V/F flash at P	P, V/F	T, y_i, x_i
7	Constant V/F flash at T	T, V/F	P, y_i, x_i
8	Isenthalpic flash at P	P, H (note 2)	T (note 3)
9	Isenthalpic flash at T	T, H (note 2)	P (note 3)
10	Isentropic flash at P	P, S (note 2)	T (note 3)
11	Isentropic flash at T	T, S (note 2)	P (note 3)

Note 1: x_i, y_i, and z_i are liquid, vapor, and feed mole fractions.
Note 2: If not known, H or S must be calculated at initial conditions.
Note 3: If the final condition happens to be in the two-phase region, V/F, y_i, and x_i are determined from Flashes 1, 6 or 7.

1. The methods of calculating fugacities—EOS, activity coefficient models, etc.
2. The type of mixtures in the problem—ideal, non-ideal, close-boiling, wide-boiling, etc.
3. The conditions of the problem—critical region, proximity to the phase boundary, etc.
4. The number of components in the mixture of the problem.
5. The type of equilibrium calculation—bubble point temperature or pressure, isothermal flash, etc.

The difficulty of the solutions of these VLE problems is clearly reflected by the enormous number of technical papers and by the variety of the titles (see References 1, 3, 15–19).

Despite the diversity of the titles, all of the iterative solution methods discussed in the papers may be classified into two groups—simultaneous improvement methods and direct substitution methods.

Most solution methods belong to the first group, which uses mathematical formulations for solving a set of nonlinear equations. These methods include the multivariate Newton Method and variations upon it.

When applied to EQUIL equations, the Newton Method converges faster than its variations, as well as the direct substitution methods, if sufficiently accurate initial estimates of the unknowns are given and the inverse of the Jacobian Matrix exists. Otherwise, the solution tends to diverge. To alleviate this tendency Powell (23), in his "Hybrid Method," implemented the Newton Method with the divergence-avoiding characteristics of steepest-descent method. The Powell implementation often unnecessarily slows down the computations in many applications, where the Newton method, or the direct substitution method, is otherwise satisfactory. For this reason, Nghiem et al. (18) suggested the use of Powell's method, only if the direct substitution method converges very slowly or diverges.

Another weakness of Newton's Method is that it requires at each iteration evaluating the inverse of the Jacobian Matrix, of which the size varies depending upon the number of components and also upon the flash types. For example, in bubble-point temperature calculations, fugacity derivatives with respect to each component's vapor-molar composition and also with respect to temperature are required to set up the matrix. Because of the necessity of handling matrices, the Newton Method requires considerably more storage in the computer than does the direct substitution method, for mixtures of many components, typically 20 or more. In addition, the evaluations of the fugacity coefficient derivatives are time-consuming even when a relatively simple equation of state is used.

Such evaluations become impractical for a complex EOS, such as the BWR (Benedict-Webb-Rubin) type or the perturbated hard-chain model equations. For such cases, Rubin (26) suggested the secant method for approximating the derivatives of the fugacity coefficients, thus avoiding the direct evaluations, but at the expense of slower convergence.

Another way of avoiding the evaluations of the fugacity derivatives is to use the more popular "Quasi-Newton Method" as proposed by Broyden (4). The Broyden Method successively updates the approximations to the Jacobian Matrix without direct calculations of the fugacity derivatives. The convergence rate of this method is generally slower than the rigorous Newton Method and also it is often more sensitive to initial estimates of the unknowns used to approximate the initial Jacobian Matrix. The Newton method is given in Appendix B of Volume 1.

The Newton type methods are also frequently used in the Gibbs Free Energy minimization approach for predicting phase equilibria, as well as for phase/chemical equilibria. A notable example in this approach is the one used by Gautan and Seider (9).

Direct Substitution Method

From a mathematical point of view, the direct substitution method is less sophisticated than the Newton type methods, but it has certain advantages over the latter, particularly for hydrocarbon systems:

1. First the direct substitution method is a simple extension of manual calculation methods, that have endured the tests of time.
2. Direct substitution requires less computer storage than do the Newton type methods, which is an advantage for microcomputer applications.
3. Physically unreasonable solutions are easily identified and corrected during the iterative process.
4. For most applications to hydrocarbon systems, direct substitution is as effective as the Newton type calculations, when used together with an equation of state.

From these considerations, the direct substitution calculation method was used in developing FORTRAN (VS) programs for the EQUIL processes given in Table 16.1.

EQUIL Files

A FORTRAN 77 program has been prepared for the eleven processes given in Table 16.1. A total of 61 callable subroutines, functions, and data bank are arranged in six separate files of the EQUIL (Vapor Liquid and Chemical Equilibrium) Program, as shown in Table 16.2. These files are essentially in the order that their subroutines are called, with one major exception, RE-SULT, which is called after the calculations have converged. The last five subprograms in PROPVT.FOR are for use in calculating enthalpy and entropy by the Lee-Kesler (13) method, which is the only application made of the L-K method in EQUIL. The options for vapor-liquid equilibrium calculations are the S-R-K and the P-R methods, which are also options for enthalpy and entropy calculations.

Table 16.2
FORTRAN Files of Program EQUIL

File Names	Subroutines, Functions, Data, etc.
EQUIL.FOR	MAIN, DFPREP
PETCUT.FOR	PFRACK, ASSAY, FGEN, MOLAR, SIST, CURFIT, ARRAY, SIMPSO
INOUT.FOR	AHT2, DATAIN, RESULT, INTEG, VXYZ, KHFIT, INVER, GASFLO, REACK
DATBK.FOR	BLOCK DATA
PROPVT.FOR	PHPROP, CUTPRO, EQCONS, EQN, CUBEQ3, XSRKF, XPRKF, MIXING, SRKEQN, PREQN, FUGACI, DELHS, IDEAHS, ENTHS1, ENTHS, LIQVOL, LKCONS, LKMIXG, LKEQN1, INITIA, LKEQN2
EQXYZ.FOR	IDEALK, APPROX, TSAT, TAVG, BUBT,BUBP, DEWT, DEWP, FLSHTP, FLSHTV, FLSHPV, FLSHPH, EXCOM1, EXCOM2, YCONV, XCONV, XYCONV, VFCONV, NEWTON, SINGH

As presented here, EQUIL is for a microcomputer, such as the IBM-AT or XT and IBM compatible machines having adequate memory and math processing capacities. For use on the mainframe computer, the first three files would differ slightly because of different input and output requirements.

Listings of these FORTRAN 77 programs are not included in this book. The programs have been compiled and linked via MicroSoft-FORTRAN and tested on a wide variety of problems, including mixtures containing petroleum. The machine language version of this program is called EQUIL.EXE and is available on a diskette. Chapter 17 describes in detail the subprograms listed in Table 16.2. At this point, the grouping and functioning of the program is discussed.

Function Groupings

For convenience and efficiency, the EQUIL programs and the operations may be divided into three major groups:

1. Pre-flash calculations
2. Equilibrium flash calculations
3. Post-flash calculations

The pre- and post-flash calculation steps are merely consequential needs for the equilibrium flash calculations. These needs are best explained by an illustration, which will be done after detailing the pre-flash and post-flash calculation steps.

Pre-Flash Calculations

The following preliminary calculations are performed to prepare for the flash calculations:

Input Data Entry

This is done by first preparing a data file for the problem to be solved, via the interactive programs in EQUIL.FOR, i.e., the computer prompts the user to provide the necessary data in the correct form and order. Control codes required are K-value method, H and S value method, output units, multiple runs, type of flash calculation, inlet conditions, outlet conditions, tolerances, and the characters and quantities of feed components.

Feed components are identified by component ID numbers, which are greater than 100 for pure discrete components, such as hydrocarbons; less than 50 if they are pseudo components made from petroleum fractions by the program; and between 50 and 100 if they are hypothetical components for which the properties are provided by the user.

If the feed contains a petroleum fraction and 11 or 21 ID numbers below 50, pseudo components have been identified. The required assays are then provided and the program calls the subprograms in PETCUT.FOR to divide the oil into pseudo components and calculate the properties.

Physical Properties Retrieval and Calculations

For the discrete components (i.e., ID > 100) characterizing properties are obtained from BLOCK DATA by DATAIN and PHPROP. BLOCK DATA contains all of the components and properties in Tables 7.1 and 7.5 of Volume 1. For the hypothetical and pseudo components (i.e., ID < 100) characterizing properties are obtained by DATAIN from CUTPRO, where the properties of the hypothetical and pseudo components are calculated by previously presented empirical mathematical equations, Equations 7.16, 7.17, 7.25, and 7.26.

Equation of State Constants

In order to make the calculations more efficient, the EOS constants that are not dependent upon composition, temperature, or pressure are calculated immediately after the required physical properties have been retrieved or calculated. These calculations need not be repeated during later iterations.

Grouping of Variables

The retrieved or calculated properties, the EOS constants, and the composition variables are grouped together into appropriate COMMON statements. For instance, noting that the AHT Program uses corresponding states methods for all thermodynamic properties calculations, the T_c, P_c, and the acentric factors plus the component ID numbers are stored in a labeled COMMON/CRIT to be shared by all of the subroutine involved in the corresponding states method computations.

Similarly, the ideal gas state enthalpy and entropy coefficients are stored in COMMON/HSTR and the liquid density coefficients are in COMMON/DLIQ, to be shared by the subroutines using them. The properties used only in MAIN, DATAIN, and RESULT are stored in COMMON/PRINT, which contains other variables that are transmitted to other subroutines by their arguments.

Flash Conditions (T and P)

Some flash conditions are specified by the user, others are found by the calculations. Conditions of interest are temperature, pressure, vapor fraction, enthalpy and entropy. When T or P is not specified in the problem, an estimated value can be provided to speed up the solution.

After the pre-flash calculation step is completed, the flash calculation step follows. A step-by-step analysis of the bubble-point calculation subroutine BUBT is given next to illustrate a computational concept adapted for EQUIL. All other flash calculation subroutines follow the same basic approach.

Now, we look at the "main event," i.e., equilibrium flash calculations, using an illustration (bubble-point temperature prediction) to aid in the descriptions.

Equilibrium Flash Calculations Example

A bubble-point temperature calculation procedure is illustrated in the following. As indicated in Table 16.1, the bubble-point temperature calculation for a mixture is made to determine the equilibrium temperature (T) and the vapor composition (y_i's) from a specified pressure (P) and the composition of the feed mixture (z_i's). This requires solving Equations 11.1 and 11.2. Recognizing that $z_i = x_i$ ($i = 1,2,...N$) at the bubble point, and rewriting Equations 11.1 and 11.2 in terms of K_i (K-value as defined by Equation 9.6 of Volume 1) gives

$$R_i = f^L_i (T,P,y_i) = f^V_i (T,P,y_i), \; i = 1,2,...N \qquad (16.2)$$

$$\Sigma K_i x_i = 1.0 \qquad (16.3)$$

The following computation steps are required to solve these ($N + 1$) equations for the ($N+1$) variables, the temperature (T), plus K_i ($i = 1...N$).

Step 1. Fix T (one of the ($N+1$) variables).
Step 2. Solve the first N equations (Equation 16.2) for the remaining N variables, K_i ($_i = 12,...N$). This solution is carried out by the following successive substitution approach:

   (a) approximate K_i's values;
   (b) calculate y_i's and the fugacity ratio, R_i, the left side of Equation 16.2;
   (c) find new K-values,
$$(K_i)_{new} = (K_i * R_i)_{old} \qquad (16.4)$$

   (d) go back to Step (b) to calculate new y_i and R_i's using the new K_i's until all $R_i = 1.0$, i.e., $(K_i)_{new} = (K_i)_{old}$.

Step 3. Check to see if Equation 16.3 is satisfied. If not, improve T and go back to Step 2. The improvement of T is generally made by the single variable Newton type methods or any other methods presented in Appendix B of Volume 1.

Note that in Step 2 the new K_i's are directly and successively substituted into Equation 16.2. This is why this type of approach is called the direct substitution or "successive substitution" method. Prior to the actual solution of the equations, however, five decisions must be made. Each of these requirements is stated below and followed by the decision that was made in creating the EQUIL programs that were developed.

Decision 1—How to make the best initial estimate of T in Step 1.

Users provide the best possible initial estimate of T. However, in case the user-supplied estimate of T fails to lead to a successful solution, the program will provide an estimate based on Raoult's Law K-value, i.e., Equation 10.3. This is done in routines APPROX and IDEALK. The vapor pressures required in this calculation are found by Equation 7.46.

Decision 2—How to approximate K_i's in Step 2a.

The initial approximations of K_i's are also obtained from the Raoult's Law K-values.

Decision 3—How to obtain the fugacities in Step 2b.

Two options for calculating the fugacities are available, namely: S-R-K EOS (Equation 5.97) and Peng-Robinson EOS (Equation 5.102). Users specify their preference, and solutions of the EOS give the fugacity coefficients rather than the fugacities. Fugacity coefficients are more convenient for updating the K_i's.

As defined by Equation 3.83, the fugacity coefficients are

$$\theta^L_i = f^L_i / x_i P$$
$$\theta^V_i = f^V_i / y_i P$$
$$\theta^L_i / \theta^V_i = (y_i / x_i)(\bar{f}^L_i / \bar{f}^V_i) = K_i R_i \quad (16.5)$$

Comparing Equations 16.4 and 16.5 shows that the fugacity coefficient ratio can be substituted directly for the new K_i's.

Decision 4—What tolerance to use for the convergence of K_i's in Step 2d, recognizing that all R_i's can never become exactly unity.

The default value of the composition tolerance in the EQUIL program is 0.00001, which the user can override by supplying another value.

Internal consistency of tolerances is an important factor in any iterative calculation methods, when an iterative calculation is included nested within another iterative calculation.

Decision 5—What tolerance to use for checking whether Equation 16.3 is satisfied in Step 3.

The default value of the tolerance for the composition summation is set at 0.00002. Obviously, this tolerance is closely related to that of the individual compositions.

Decisions 1 through 3 require the critical properties and the acentric factor for each component in the mixture. In addition, item 3 requires information of the K-value option. In some of the flash calculations the enthalpy/entropy option must also be chosen.

Decisions 4 and 5 require the tolerances for the compositions as well as their summation. All this information is supplied to the programs in the pre-flash calcula-

tion step, in which some other important calculations are also made for more efficient calculations.

Arguments. The K-value option code (IKV), the number of components (NC), the temperature (T,°R), the pressure (P, psia), the tolerance in T (TTOL), and the error code (IER) are all transmitted to Subroutine BUBT in the argument and the bubble point is calculated by the following four steps:

1. Set IER equal to zero. This value remains unchanged for a successful solution. Should the calculations fail to converge to an acceptable solution, however, IER is reset to -1 to retry with the program generated starting temperature, or reset to the value of the flash code (ICOD) to flag the failed flash option in the output report.

2. Save the input temperature for use in a second trial with the same initial temperature in case the first trial failed.

3. Set the high and low limits of the temperature range at ($\pm 300°$F of input T), to which temperature search is limited.

4. Set $TX = 40$. This is the maximum allowable temperature step.

   Set $TXX = -2$. This is provided only for the temperature search in 2°F steps near the critical region without recourse to the reguli-falsi method (Newton). If $TXX = -2$ fails, then it is set to $+2$ to search for T in the opposite direction.

   Set $F1 = 0$. $F1$ is temporarily set equal to zero at the beginning. This is redetermined in NEWTON.

   Set $T1 = 1E7$. This large number is used in NEWTON as a signal to initialize the reguli-falsi procedure.

   Set $TD = 0.7*TX$. TD is the temperature difference between two successive iterations. At the beginning it is arbitrarily set to this value.

Iterations

Now the reguli-falsi iterations start. Call EQN to calculate and save the liquid fugacity coefficients at T and XF (x_i, liquid composition). For the first iteration, $TD > 10$. Call IDEALK to get Raoult's Law K-values. IDEALK is also called when $0.8 < K_i < 1.2$ for all K-values calculated in YCONV ($K_1 = 1$ or $K_1 = 2$) to perturb the K-values, thus avoiding the premature convergence to unity K-values. Then calculate $y_i = K_i x_i$ and $SM = \Sigma y_i$ and call YCONV to get converged values of y_is by calculating K_is from Equation 11.7.

If $K_1 = 0$, i.e., the condition is such that at least one of the K-values is greater than 1.2 or less than -0.8, then a new T is calculated in NEWTON to get $SM = 1$. If convergence is not achieved or the temperature limit is reached, then restart with the program generated initial temperature. If $K_1 > 0$, i.e., if $0.8 < K_i < 1.2$ for all K-values and $TX > 7$, then TX is reduced to $TX/2$ or $TX = 7$, whichever is greater, to start the iteration from the beginning.

Otherwise, the reguli-falsi search continues unless $K_1 = 2$, i.e., $0.97 < K_i < 1.03$ for all K_i's, in which case the temperature search is continued in 2°F steps without calling NEWTON. If it does not lead to a converged solution, the search proceeds in the opposite direction with the same 2°F steps. If the program-generated initial value of T does not converge, then the assignment of IER = 2 (the ICOD value for BUBT) is made and the program returns.

Having completed the flash calculations, we are ready to see the results on the screen or on paper.

Post-Flash Calculations

After the calculations are finished, the action shifts to RESULTS, where an output file is created. This includes the input data, as well as the results. If the solution did not converge, a message to this effect is printed in place of the normal output.

When the user has asked for it in the control codes of the input, the output will include polynomials from curve fits made to the resulting data, giving K and H versus T relationships for the system and conditions of the problem. Such simple equations can be used in distillation calculations.

In Chapter 17, the subprograms of each VLE file will be described in more detail.

References

1. Asselineau, L., B. Bogdamic, and J. Vidal, *Fluid Phase Equilibria* 3, 273 (1979).
2. Benedict, M., G. B. Webb and L. C. Rubin, *J. Chem. Phys.*, 8, 334 (1940); 19, 747 (1942).
3. Boston, J. F. and H. I. Britt, *Comp. Chem. Eng.* 2, 109 (1978).
4. Broyden, C. G., *Math. Comp.* 19, 577 (1965); 21, 368 (1967); 25, 285 (1971).
5. Cavett, R. H., *Proc. of Am. Petr. Inst.*, Div. of Refg., 42-III, 351 (1962).
6. Chao, K. C. and J. D. Seader, *AIChE J.* 7, 598 (1961).
7. Edmister, W. C. and C. L. Ruby, *Chem. Engr. Prog.* 51, 95F (1955).
8. Fussell, D. D. and J. L. Yanosik, *Soc. Pet. Eng. J.* 18, 173 (1978).
9. Gautan, R. and W. D. Seider, *AIChE J.* 25, 991 (1979).
10. Grayson, H. G. and C. W. Streed, *Proc. 6th World Petr. Cong.*, Frankfurt, Sect. III, June 1963, p. 223.
11. Hildebrand and Scott, *Solubility of Non-Electrolytes*, Reinhold (1950).
12. Lee, B. I., J. H. Erbar and W. C. Edmister. *AIChE J.* 19, 349 (1973).
13. Lee, B. I. and M. C. Kesler, *AIChE J.*, 21, 510 (1975).
14. Mah, R. S. H. and T. D. Lin, *Comp. Chem. Eng.* 4, 75, (1980).
15. Mauri, C., *I & E C Process Des. Dev.* 19, 482 (1980).
16. Mehra, R. K., R. A. Heideman, and K. Aziz, *Soc. Pet. Eng. J.* 22, 61 (1982).
17. Michelson, M. L., *Fluid Phase Equilibria* 9, 1, and 21, (1982).
18. Nghiem, L. X., K. Aziz and J. L. Yanosik, *Soc. Pet. Eng. J.* 23, 521 (1983).
19. Nghiem, L. X. and R. A. Heideman, 2nd European Symp. on Enhanced Oil Recovery, Paris (1982).
20. Peng, D. Y. and D. B. Robinson, *IEC Fundam.* 15, 59 (1976).
21. Pitzer, K. S. et. al., *J.A.C.S.* 77, 3427 (1955); 79, 2369 (2369); *Ind. Eng. Chem.* 50, 265 (1958).
22. Ploecker, U., H. Knapp and J. M. Prausnitz, *I&EC Fundam.* 17, 324 (1978).
23. Powell, M. J. D., *Numerical Methods for Non-Linear Algebraic Equations*, P. Rabinowitz, editor, Gordon Breach, London (1970).
24. Prausnitz, J. M., *AIChE J.* 6, 78 (1960).
25. Redlich, O. and J. N. S. Kwong, *Chem. Rev.* 44, 233 (1949).
26. Rubin, D. I., CEP Symp. Ser. No. 37, 54 (1962).
27. Scatchard, *Trans.* Faraday Society 33, 160 (1937).
28. Soave, G., *Chem. Eng. Sci.*, 27, 1197 (1972).
29. Starling, K. E. and M. S. Han, *Hydrocarbon Processing*, 51 (5), 129 (1972).
30. Van Laar, Z. *Physik. Chem.* 72, 723 (1910); 83, 599 (1913).

17

Subprograms of Computer Algorithm for Applied Hydrocarbon Thermodynamics

The thermodynamic concepts and iterative procedures required for the different options were discussed in Chapter 16 and a new computer program, EQUIL (Vapor Liquid Phase and Chemical Reaction Equilibrium), was shown by listing the names of the 6 FORTRAN files and the 61 subprograms (subroutines, functions, data, etc.) contained therein. A method for dividing petroleum fractions into pseudo components so that the same thermodynamic properties and phase behavior calculations can be made for both discrete hydrocarbon components and petroleum fractions pseudo components has been presented in Chapter 15.

Now the functions and some of the details of these subprograms will be discussed. Listings of the FORTRAN coding are not included. Sample input and output files are included. The objective of this chapter is to inform the reader about contents and capabilities of EQUIL. The source code is not listed.

Program Description

From an operational viewpoint, this program may be considered as composed of the following functions:

1. Input data file preparation
2. Operations control
3. Data file reading
4. Properties calculations
5. Storage of data
6. VLE calculations
7. Writing of results

Note that these seven functions are somewhat overlapping. In these descriptions, the names of the subprograms, i.e., subroutines, functions, etc., are written in capital letters and grouped under the names of the FORTRAN file, in which they appear.

Information is stored for use in one or more subprograms via COMMON statements, all of which are labeled. The COMMON's used in this program are:

Common Label	Shared By
PRINT	AHT2, DATAIN, RESULT, KHFIT, GASFLOW, REACK, TAVG
COMP	AHT2, DATAIN, RESULT, KHFIT, GASFLO, ENTHS, LIQVOL, IDEALK APPROX, BUBT, BUBP, DEWT, DEWP, FLSHTP, FLSHPV, FLSHTV TSAT, TAVG, FLSHPH, XYCONV, VFCONV, XCONV, YCONV
EQNS	EQCONS, MIXING, FUGACI, DELHS, LKMIXG
DSDATA	BLOCK DATA, DFPREP, PHPROP

MIX	RESULT, FUGACI, MIXING, SRKEQN, PREQN, LKEQN1
CRIT	DATAIN, RESULT, EQCONS, LIQVOL, LKCONS, LKMIXG, IDEALK, REACK
HSTR	DATAIN, IDEAHS, ENTHS1, REACK
HFOR	DATAIN, REACK
SCRA	ENTHS1, BUBT, BUBP, DEWT, DEWP, FLSHTP, FLSHPV, FLSHTV, TSAT, FLSHPH, XYCONV, XCONV
DDD	DFPREP, PFRAK, ASSAY, MOLAR, CURFIT, SIMPSO
DLIQ	DATAIN, LIQVOL, RESULT, REACK
LEEKS	LKCONS, LKMIXG
XKHC	RESULT, KHFIT
TEL	RESULT, KHFIT
GFLO	DATAIN, RESULT, GASFLO

Both dimensioned and undimensioned items are stored and transmitted by these COMMON statements. Both control symbols (undimensioned integers) and data, which are real numbers and usually dimensioned, are included in the same COMMON statements in this program.

In many cases information is transferred from one subprogram to another by the calling arguments, which can differ at both ends of the call. For example: CALL CUBEQ3(AZ,BZ,1.0,B,Z) is a calling statement, and SUBROUTINE CUBEQ3(A,B,C,D,Z) is the corresponding subprogram name. The terms in the argument have a one-to-one correspondence, but are not identically the same symbols. Thus, CUBEQ3 can be used conveniently for solving any cubic equation.

When transferring data by arguments, dimension statements that give the indicies of variables are often necessary. Also, dimensions are given in COMMON statement variables, where required.

EQUIL.FOR

The two subprograms in this file are for preparing the input data file and directing the user's instructions for operations.

Main

The application of EQUIL is started here by the preparation of input data files. This is done by the user with promptings on the screen. User provided selections are stored in memory for use in running the program. Subroutine DFPREP is called for the preparation of the data

file. For execution of the problem, Subroutine AHT2 is called from MAIN, after which control is returned to MAIN for viewing and/or printing results and then terminating the run.

Subroutine DFPREP

Data files are created here and petroleum fractions are divided into pseudo components in Subroutine PFRAK and its satellite subroutines, which are in another file. The user is prompted to provide the following:

- Problem title
- VLE K-value method
- Enthalpy and entropy method
- Zero enthalpy datum choice
- Choice of mole percent, molar, or mass flows output units
- Tolerances for compositions, temperature, and pressure
- Number of pure components in the feed
- Number of petroleum components in the feed
- ID numbers and quantities of the pure components
- ID numbers and names of the petroleum components
- Quantity of the petroleum fraction
- Assays for the petroleum feed
- Problem conditions
- Multiple runs control

The quantities of discrete components, i.e., hydrocarbons, etc., in the feed are entered in moles. The quantity of petroleum fraction, on the other hand, might be entered in mass or liquid volume units. Conversions are made in this program to molar units. In this way, the rates of all of the feeds, i.e., both pure components and petroleum, are expressed in the same units. For these calculations, PFRAK and a few other subprograms are called.

The multiple runs control makes it convenient to make up to 20 runs on the same feed mixture, changing the K or H methods or other conditions, as desired. This feature will be illustrated later in this chapter.

With the run file EQUIL.EXE, made by linking the compiled versions of the FORTRAN files shown in Table 16.2, the command "EQUIL" starts the user-interactive input data file procedure.

PETCUT.FOR

The eight subprograms in this file are for petroleum fraction assay and breakdown calculations, as required in dividing petroleum fractions into pseudo components and estimating the assays of liquid and vapor products from flashing petroleum.

Subroutine PFRAK

When the feed contains a petroleum fraction, PFRACK is called for converting the distillation and gravity assays of the oil into pseudo components. The petroleum assays, which have been entered in DFPREP, are read and used in PFRAK. A file is created to receive the assay calculation results, for later viewing and/or printing.

The purpose of these assay calculations is to divide the oil fraction into 11 or 21 pseudo components and then find the normal boiling points and specific gravities of these components. In this form, the petroleum fractions components are ready for properties calculations along with the other, i.e., gas and hydrocarbons, components for VLE calculations. Six smaller subroutines are called for parts of these calculations, as will be described in the following section. Reference should be made to Chapter 15 for more detailed information.

Subroutine ASSAY

In this subroutine, which is called in DFPREP, the ASTM distillation assays are converted to TBP assays, and vice versa, using an empirical equation from the API-TDB. A correction for thermal cracking of the oil at vapor (in laboratory flask) temperatures in excess of 475°F is included.

Function FGEN

This function is used in PFRAK in computing temperature and gravity values at point on the TBP assays.

Subroutine MOLAR

In this subroutine volumetric or weight TBP curves are converted into molar TBP assays.

Subroutine CURFIT

In this subroutine polynomial equations are fitted to points on the molar TBP and specific gravity assays. The size of the derived equation depends upon the number of points in the starting assay. Seven point assays give third-degree polynomials; eleven point assays give fourth-degree polynomials; and twenty-one point assays give eighth-degree polynomials. Subroutines ARRAY and SIST are called in CURFIT.

Subroutine ARRAY

This subroutine is called by CURFIT to set up the matrix arrays for the curve-fit operation.

Subroutine SIST

This subroutine is called by CURFIT to solve for the coefficients of the polynomials.

Subroutine SIMPSO

In this subroutine, which is called in PFRAK, the petroleum fraction feed stock is divided into 11 or 21 pseudo components and the mole fractions are found by applying Simpson's rule to the molar TBP curve of the oil. As shown in Chapter 15, pseudo components selected this way make it possible, and convenient, to apply the same kind of multicomponent equilibrium flash calculations that are used for mixtures of discrete components to predict the vaporization characteristics of the petroleum.

These pseudo component data are entered along with the pure component information. The resulting input data are filed under an assigned name for current or later use. In addition, a record of the petroleum assay and pseudo component data is made available in an identified file.

INOUT.FOR

The eight subroutines here are for reading the input data file, controlling the operations, and printing the results. In addition to doing these things for the basic eleven types of equilibrium flash operations, as shown in Table 16.1, three additional calculations are made and reported by programs in this file.

Subroutine AHT2

The sequence of calculations is controlled in this subroutine, to which control may be directed in the user-interactive front end. The problem title is read and then Subroutine DATAIN is called for entering the control codes and the data for the problem. After this, the subroutines required for the thermodynamic properties and equilibrium flashes are called and the calculations made. When the calculations have been completed, a results file is created. This file can be viewed and/or printed at the user's instructions.

Table 17.1, which is an expansion of Table 16.1 to provide additional information, summarizes the required specifications for the different flash calculations, which are identified by their ICODs, between 1 and 11.

An ICOD specification of 12 signals a curve fit to the K & H versus T values from series of three flash runs on the same feed. These three flashes must be in the ICOD = 1 to 7 range and might all be at ICOD = 6, i.e., a specified constant pressure and V/F values of 0.1, 0.5,

Table 17.1
Calculational Options of the EQUIL Program

ICOD	Calculation Type	Req'd Spec	Subroutine Name
1	Flash at constant T and P	TI, PI	FLSHTP
2	Bubble T at constant P	$TI*, PI$	BUBT
3	Bubble P at constant T	$TI, PI*$	BUBP
4	DEW T at constant P	$TI*, PI$	DEWT
5	DEW P at constant T	$TI, PI*$	DEWP
6	Flash at constant P and V/F	$TI*, PI, EF = V/F$	FLSHPV
7	Flash at constant T and V/F	$TI, PI*, EF = V/F$	FLSHTV
8	Flash at constant P and H	$TI, PI, EF = Q**$ $TO*, PO$	FLSHPH***
9	Flash at constant P and S	TI, PI $TO*, PO$	FLSHPH***
10	Expansion or compression	$TI, PI, EF = (0<\epsilon<1)$ $TO*, PO$	EXCOM1
11	Expansion or compression (single phase only)	$TI, PI, EF = (0<\epsilon<1)$ $TO*, PO$	EXCOM2
12	Three flashes and curve fits to K & H vs T results	$TI*, PI, E/F = V/F$	FLSHPV
13	Two flashes and gas flow calcs for nozzle, pipeline & transfer flows with the resulting T, P & exponents	TI, PI TO, PO Constant H Constant S	FLSHTP FLSHPH GASFLO
14	Chemical heat of reaction Equilibrium product mixture	Reaction formula Temp. & press.	REACK

where: ϵ is expander or compressor efficiency between 0 and 1.
    * initial estimate by user
    ** $+ Q$: heat to be added to the feed enthalpy
    ** $- Q$: heat to be removed from the feed enthalpy
    *** same subroutine is used for ICOD = 8 with enthalpy, and for ICOD = 9 with entropy.

and 0.9, for example. The control is KHF, which equals 1, 2, and 3 for the three successive runs.

An ICOD specification of 13 calls for flow calculations for one of three types of compressible fluid flow calculations. Preliminary VLE runs are necessary to provide thermodynamic data for the flow calculations. The KHF, KYF and KTL coding for the preliminary runs before ICOD = 12 or 13 is necessary so as to have the results tagged and available for recall and use.

Subroutine DATAIN

All of the data required to run the program are entered here. To maintain simplicity and clarity, this program does not provide any validity checks of input data or offer a wide variety of options, being unlike many of the commercial simulation programs in this respect. Input data are entered by formatted lines, or cards, of 80 columns. In this description, the term "card" may be used to define each line of input data, even though the data may read from a file that is stored on a disk. In this way, the reader is less likely to confuse a line of data with a line of text.

The following describes the input cards, where the 2nd and last cards have the same formats and differ in some respects. The last card will contain the first five items with MULT = 1 if another run is desired on the same feed mixture. If no more runs are wanted, a STOP command is issued by making MULT = 2 on this last card. In this case, the remainder of the card can be blank. Input cards consist of:

1st card:
Problem Title
2nd card:
Control Codes
  K-Value Option:
    IKV = 1, Soave-Redlich-Kwong (SRK)
       = 2, Peng-Robinson (PR)
  Enthalpy/Entropy Option:
    IHS = 1, Soave-Redlich-Kwong (SRK)
       = 2, Peng-Robinson (PR)
  Zero Enthalpy Datum Option:
    IHD = 1, ideal gas state at 0°R
       = 2, liquid at − 200°F
       = 3, elements at 0°R

For the entropy there are no options; Standard entropy units are referred to ideal gas state at 0°R.

Output Print Option for Results:
IOUT = 0, mole % plus molar and mass flow rates
= 1, mole percent only
= 2, molar flow rates only
= 3, mass flow rates only
Properties, quantities, and conditions of feed and products are included for all four options. Total flows are given in both molar and mass units.

Multiple Run Option:
MULT = 0, Reads all (n + 2) cards for first run, i.e., Case 1; Another card (n + 3) will be required later to STOP or Rerun. See instructions later.

Tolerances:
XTOL, Default = 0.00005 in molar composition
TTOL, Default = 0.1 in °F
PTOL, Default = 0.001 in logarithm of P in psi

3rd + n cards:
Feed input for "n" components (pure plus petroleum);
ID numbers and flow rates of pure components, having ID numbers greater than 100; and/or
Name and quantity of petroleum fraction to be divided by computer program into 11 or 21 pseudo components that will have ID numbers of 1 to 11 or 1 to 21; and/or
ID numbers, flow rates, boiling points, specific gravities, and names of petroleum components already prepared and having ID numbers 50 < ID < 100;
Plus one blank card.

Next-to-last card:
Equilibrium flash option plus conditions.

ICOD = Type of flash to be run, from 1 to 11
ICOD = 12 must be preceded by three preparatory runs in the ICOD = 1 to 7 range; suggest ICOD = 6, with V/F = 0.1, 0.5 and 0.9
ICOD = 13 must be preceded by two preparatory runs
when KYT = 1 use ICOD = 1 and ICOD = 9
KYT = 2 use ICOD = 1 and ICOD = 1
KYT = 3 use ICOD = 1 and ICOD = 8
KHF = K & H vs. T curve-fit control
= 1 for 1st run at lowest T of K&H curve-fit range
= 2 for 2nd run at intermediate T in K&H range
= 3 for 3rd run at highest T of K&H curve-fit range
= 0 for no duplicate runs for curve fits
KYT = Type of gas flow calculations
= 1 for nozzle flow
= 2 for pipeline flow

= 3 for transfer line flow
= 0 no gas flow calculations
KTL = Preliminary VLE run control for flowing gas
= 1 for VLE calcs on gas at flow starting point
= 2 for VLE calcs on gas estimated final point of flow
= 0 no gas flow calculations
TI = Inlet temperature, °F
PI = Inlet pressure, psia
EF = Fraction vaporized, or Compression/Expansion efficiency, or Heat added or removed
TO = Outlet temperature, °F
PO = Outlet pressure, psia

Last Card for Stop or Rerun Controls:
This card is necessary whether or not another run is desired. It must have same format as 2nd card. If a duplicate run on same feed is desired, the (n + 3) card should contain four items: IKV, IHS, IHD, IOUT, and MULT = 1, in same format.

Then EQUIL reads only the first and last cards, skipping the component cards for subsequent runs on the same component feed rates.

If no more runs are desired terminate by making MULT = 2, in proper format.

DATAIN calls PHPROP and CUTPRO to obtain the physical and thermal property constants and calls needed in the calculations and then calls EQCONS or LKCONS to set up the equation of state calculations.

Subroutine RESULT

All of the results from ICODs 1 through 11 are collected in this subroutine, formatted into tabulations for reports. Two kinds of reports are prepared for viewing or printing, namely:

1. Tabulations of the assays of the petroleum and the properties of the pseudo component of the oil
2. Equilibrium flash and thermodynamic properties results

Also, some of these results are sent on to other calculations under ICOD's 12, 13 and 14. Thus Subroutine RESULT performs dual services in EQUIL.

Subroutine KHFIT

This subroutine is called in AHT2 after a series of three flashes have been run on the same feed mixture and when ICOD = 12. The fourth run on the same feed is polynomial equation fitting to the lnK_i and H values obtained in the three flashes. Subroutine INVER is called for matrix inversions. The output includes the coefficients of the resulting third order polynomials.

Subroutine INVER

Inverts matrices for KHFIT.

Subroutine INTEG

This subroutine, which is called after RESULT in AHT2, generates TBP assays for petroleum products of flash calculations. This assay generation starts with the pseudo component compositions, which are in molar units, and finds the molar and the liquid-volume-percent vaporized at each temperature for the corresponding TBP assay. Subroutine VXYZ is called to assist in these calculations.

Subroutine VXYZ

Assists in calculations of INTEG.

Subroutine GASFLO

The subroutine, which is called in AHT2 after a series of two preparatory flash calculations have been run on the gas feed mixture and when ICOD = 13, makes gas flow calculations according to the type and conditions selected by the user. The types of compressible fluid flow are adiabatic or polytropic nozzle flow, isothermal pipeline flow, irreversible transfer-line flow, and critical flow controls for nozzle and transfer-line flows.

There are two nozzle cases: (1) given diameter, find flow rate; and (2) given flow rate, find diameter. There are three pipeline cases: (1) given pressures, length, and flow rate, find diameter; (2) given pressures, flow rate and diameter, find length; and (3) given pressures, length and diameter, find flow rate. There are three transfer-line cases: (1) given flow rate, find diameter and length; (2) given diameter, find flow rate and length; and (3) given length, find flow rate and diameter.

Temperatures and pressures are specified, not calculated. A feed quantity is specified in the preliminary runs and is used with a feed rate multiplier, *FDX*, that is specified in the ICOD = 13 conditions, to find the given feed rate, where needed. All cases are run so the user must provide all conditions, specified and approximated values.

Subroutine REACK

This subroutine, which is called in AHT2, calculates the heat of reaction and the equilibrium constant for the distribution of components in the reactor effluent. The reaction equilibrium constant found in these calculations is K_a, the activity cofficient K value that is applicable for perfect mixtures of ideal gases. Fugacity coefficient corrections are necessary to convert the K_m values to K_P values that can be applied with partial pressures.

DATBK.FOR

This file contains BLOCK DATA only. Data are stored in mass units for convenience and efficient use of space. For the pure components included in Tables 7.1 and 7.5 of Volume 1, *ID*(I) and *FD*(I) only are needed to specify components and their quantities. All of the property data in Tables 7.1 and 7.5 have been assembled into BLOCK DATA for convenient use in this program. Mole percents, mole fraction, or moles per hour are acceptable values for the flow rate. The component names and the required physical properties are then retrieved from the BLOCK DATA via Subroutine PHPROP.

BLOCK DATA

This subprogram contains DIMENSION, EQUIVALENCE, and COMMON statements and a matrix, named and dimensioned DS(33,216), which contains 33 items of data for 215 substances. The substances and most of the data are those of Tables 7.1 and 7.5 in Volume 1. Eight of the 33 data items are the 3-digit ID numbers of the components and their names given as seven 4-letter words. A total of 28 alphanumeric characters are permitted in the name of a feed component. The remaining 25 data items are thermal and physical constants.

Two hundred and four of the substances are hydrocarbons (70 paraffins, 39 monoolefins, 18 diolefins, 38 cycloparaffins, and 39 aromatics), nine are non-hydrocarbon gases, and two are carbon and sulfur. The data missing in Tables 7.1 and 7.5 have also been entered to complete BLOCK DATA.

PROPVT.FOR

The file contains twenty-one subprograms, all of which are concerned with basic data and derived properties of components and their mixtures, including the pseudo component of petroleum. Also, solutions and uses of equations of state are included.

Subroutine PHPROP

This subroutine, which is called by DATAIN, shares COMMON/DSDATA/DS(33,216) with BLOCK DATA. This makes it possible for the DS matrix data to be read directly into PHPROP, using the ID number to identify the desired component. These readings are made, names and symbols assigned for each component in the problem of interest. The new names and symbols, acquired here, serve to transmit the names and data to DATAIN via the calling argument of PHPROP. The data retrieved are:

CNAME = component name
WM = molecular weight
BP = normal boiling point, °F
SG = specific gravity, 60°F/60°F
TC = critical temperature, °R
PC = critical pressure
ZC = critical compressibility factor
WC = acentric factor
RKW = acentric factor of Hankinson-Thomson
VST = characteristic volume of Hankinson-Thomson
HOV = heat of vaporization at normal boiling point
HOF = heat of formation at 77°F (equiv. std. ref. temp. of 25°C)
FOF = free energy of formation at 77°F
HCOEF = nine coefficients of H (and S) in ideal gas state

Subroutine CUTPRO

For cuts, or pseudo components, of petroleum, three additional items of data are required and must be entered on these cards. These additional properties are the normal atmospheric boiling point, *BP*, the specific gravity, SG, and the alphanumeric name of the petroleum component, CNAME, which is also the term used for the pure component names. Up to a total of 28 alphanumeric characters are permitted in the names of the components, both pure and pseudo.

The values of *ID*(I) for the petroleum components must be less than 100, being assigned by the user. All pure components in BLOCK DATA have *ID*(I) numbers greater than 100. From the normal boiling points and the specific gravities, the other properties required are calculated in CUTPRO from empirical equations.

This subroutine is called by DATAIN to obtain the physical and thermal constants that will be needed for the petroleum components, i.e., those components having ID numbers less than 100. The calling argument of CUTPRO brings in the boiling point and specific gravity of the petroleum components and takes the following properties back to DATAIN:

TC = critical temperature by Equation 7.16
PC = critical pressure by Equation 7.17
WM = molecular weight by Equation 7.2
WC = acentric factor by Equations 7.25 and 7.26
RKW = WC

VST = critical volume ml/gmole
CF = nine coefficients of H in ideal gas state (equiv to HCOEF)

Subroutine EQCONS

This subroutine is called by DATAIN when the Soave-RK (SRK) or the Peng-Robinson (PR) methods are being used in the calculations. The job done here is to combine the critical constants, the acentric factor, and the binary interaction coefficients to get the generalized parameters of the equations of state. Also the binary interaction coefficients, AIC (notation used for k_{ij} and also for the composition-independent part of Equations 5.68 and 5.69) values, are retrieved from XSRKF for the Soave-Redlich-Kwong or from XPRKF for Peng-Robinson, depending upon the method selected. Then equation of state constants are calculated.

It should be noted that it is not acceptable to use the PR method for the entropy/enthalpy calculations while using the SRK method for the *K*-value, calculations, or vice versa. However, the Lee-Kesler method is acceptable with either SRK or PR being used for the *K*-values. Therefore, *IHS* may equal 3 when *IKV* is equal to either 1 or 2 in the problem specifications.

For the SRK method Equations 5.28 and 5.29 give a_i and a_j values for binary pairs, while Equations 5.39 and 5.40 do the same for the PR. With these a values, the following are calculated:

$$AIC(\text{I,J}) = (a_i a_j)^{0.5} * (1 - k_{ij})$$

AIC is the composition independent part of Equation 5.68 or 5.69. In this expression, a_i is not exactly the same as Equation 5.10 for SRK or Equation 5.34 for PR, disregarding the difference in numerical constants (0.42748 for SRK and 0.457235 for PR) and the gas constant, *R*, which will eventually be cancelled out in Equation 5.13 when calculating *A*. Later, the numerical constants are put back into SRKEQN and in PREQN. The a_i constant is temporarily defined without the gas constant and the numerical constant in the program as *BI*(I) in the calculation of AIC.

However, *BI*(I) is later redefined as b_i again without the numerical and gas constants for evaluating Equation 5.9 for the SRK or Equation 5.32 for the PR. That is

$$BI(\text{I}) = T_{ci}/P_{ci}$$

It should be noted again that *ALF*, *AIC*, and *BI* are independent of compositions, temperature, pressure, or phase identity and can be calculated only once as long as the component order remains unchanged in the mixture of interest.

Subroutine LKCONS

This subroutine calculates the generalized parameter of critical constants that are used in the Lee-Kesler method.

Subroutine LKMIXG

Combination of constants for mixtures for the Lee-Kesler method is the purpose of this subroutine.

Function XSRKF

This function gives the values of the binary interaction coefficients for the Soave-RK equation of state.

Function XPRKF

This function gives the values of the binary interaction coefficients for the Peng-Robinson equation of state.

Subroutine EQN

Called by the flash or the property program, this subroutine directs the calculations to Soave-RK, Peng-Robinson, or Lee-Kesler, as needed for fugacities, enthalpy or entropy, depending upon the values of *IC* where:

IC = 1, for fugacity coefficient calculations
IC = 2, for enthalpy calculations
IC = 3, for entropy calculations

When the control *LIQ* = 0, a vapor phase is specified; *LIQ* = 1 for a liquid. In EQN the appropriate equation of state is called, depending upon the value of *IKV*. When *IKV* = 1, SRKEQN is called. When *IKV* = 2, PREQN is called. Fugacity coefficient calculations are made in FUGACI by SRK or PR methods, as controlled by *IC*.

The enthalpy calculation control is *IHS*, which code may be 1, 2, or 3, designating the SRK, PR, or LK method for enthalpy/entropy calculations. With *IHS* = 1 or 2 and with *IC* = 2, Subroutine EQN calls DELHS for enthalpy difference calculations; and with *IC* = 3, DELHS is called for entropy difference calculations. When *IHS* = 3, the call is for LKMIXG and LKEQN1 for either enthalpy or entropy, depending upon the value of *IC*.

Subroutine SRKEQN

This subroutine, which is called by EQN, calls MIXING to get the EOS coefficients for solving Equation 5.12. Constant *A* is found by Equation 5.13 and constant *B* is found by Equation 5.14.

Next, Equation B.20 is solved for the compressibility factor, *Z*, in CUBEQ3. Also calculated in this subroutine are the following two terms for later use in calculating the fugacity coefficient in FUGACI plus enthalpy and entropy departures in DELHS:

$$A1 = \ln(Z - B)$$

$$A2 = A/B \ln(1 + B/Z)$$

The functions of the subroutine are identical to those of PRKEQN, except they are for the Soave-Redlich-Kwong equation of state.

Subroutine PRKEQN

Also called by EQN, this subroutine calls MIXING to get the EOS coefficients for solving Equation 5.35. Constant *A* is found by Equations 5.13 and constant *B* is found by Equation 5.14.

Next, Equation B.20 is solved for the compressibility factor, *Z*, in CUBEQ3. The following two terms are also calculated here for later use in the calculations of fugacity coefficients in FUGACI and enthalpy/entropy departures in DELHS:

$$A1 = \ln(Z - B)$$

$$A2 = A/(2^{1.5}B) \ln((Z + (2^{0.5} + 1)B)/(Z - (2^{0.5} - 1)B))$$

The functions of the subroutine are identical to those of SRKEQN, except they are for the Peng-Robinson equation of state.

Subroutine MIXING

This subroutine, which is called by SRKEQN or by PRKEQN, combines the generalized constants calculated in EQCONS and finds constant values for mixtures.

This subroutine performs the calculations of Equations 5.68 and 5.9 for the SRK method or those by Equations 5.69 and 5.32 for the PR method to obtain constants *a* and *b* without the numerical and gas constants.

Subroutine CUBEQ3

This subroutine is called by SRKEQN or by PRKEQN. It uses the FORTRAN program previously given in Volume 1 as Table B.6 and solves the cubic equation of state. The parameters are transferred by the calling argument.

In this subroutine Equation 5.12 for SRK or Equation 5.35 for PR is solved for the compressibility factor, using the values of A and B via the modified Richmond equation, i.e., Equation B.20 in Appendix B of Volume 1.

Subroutine FUGACI

This subroutine, which is called by EQN, computes the fugacities of components in both vapor and liquid phases for both Soave-RK and Peng-Robinson EOS's. This is done when the control $IC = 1$.

In this routine, the fugacity coefficient is calculated via Equation 5.97A for SRK or via Equation 5.102A for PR, using the values of $A1$ and $A2$ previously evaluated in SRKEQN or in PREQN.

Subroutine DELHS

Called by EQN, this subroutine calculates the isothermal effects of pressure on the enthalpy, or the entropy, for both vapor and liquid phases by either Soave-RK or Peng-Robinson EOS. When the control $IC = 2$, the calculation is for the enthalpy. When $IC = 3$, the calculation is for the entropy. The liquid or vapor phase has already been specified previously.

These isothermal departure of enthalpy or entropy calculations are made via Equation 5.83A or 5.93A for SRK or via Equation 5.85A or 5.94A for PR, using the values of $A1$ and $A2$ previously evaluated.

Subroutine IDEAHS

This subroutine is called by ENTHS and RESULT, and finds the ideal gas state values of enthalpy or entropy, depending upon the value of control IC ($= 2$ for H^*; $= 3$ for S^*). Equations 7.29 and 7.31 are used. The enthalpy is computed with three alternate zero H datum states, as described under DATAIN.

The ideal gas state enthalpy and entropy values calculated here are stored as HSI values in COMMON/HSTR/ for use later in ENTHS1. Thus, IDEAHS must be called before ENTHS1.

Subroutine ENTHS1

Also called by ENTHS and RESULT, this subroutine calculates the enthalpy of a vapor or a liquid, depending upon the value of control LIQ, using the ideal gas H values from IDEAHS and calling EQN to get the delta H values for pressure effect. EQN is called and then DELHS when $IHS = 1$ or 2, or the calls are for LKMIXG and LKEQN1 when $IHS = 3$.

Subroutine ENTHS

This subroutine, which is called by AHT2, SINGH, EXCOM1, EXCOM2, FLSHPH, and RESULT, calculates the enthalpy or entropy of a given system (two-phase or single). It bypasses the insignificant phase to save computer time and directs the enthalpy and entropy calculations that are required when making isentropic or isenthalpic flashes.

In this routine, the enthalpy is calculated when $IC = 2$ or the entropy is calculated when $IC = 3$. These calculations are made for a given mixture in either or both liquid and vapor phases as indicated by the value of VF, the vapor fraction of the mixture. If $VF > 0.9999$, the liquid enthalpy calculation is bypassed and if $VF < 0.0001$ the vapor enthalpy calculations are bypassed, thus avoiding unnecessary calculations.

When VF is between these limits, enthalpy calculations are made for both vapor and liquid phases and the enthalpy of the mixed phase system is found by proportioning the two single-phase values. ENTHS first calls IDEAHS where the pure component ideal gas state enthalpies or entropies are calculated and stored; then ENTHS1 is called to complete the H and S calculations.

Subroutine LIQVOL

This subroutine is called by RESULT to calculate the liquid volumes by the Hankson-Thomson method, via Equations 6.49 through 6.54 as follows:

$TCM = T_c$ from Equation 6.50
$VCM = V^*$ from Equation 6.51
$WCM = w_{SRK}$ from Equation 6.52
$VR0 = V^{(o)}$ in Equation 6.49
$VR1 = V^{(\delta)}$ in Equation 6.49
$VOL = V_m$ by Equation 6.49
PR = reduced pressure
VPR = exponential of Equation 6.54
$CAPC = C$ in Equation 6.53
$ECON = e$ in expression for B in Equation 6.53
$BPC = B/P_c$ in expression for B for Equation 6.53
$VOL = V$ by Equation 6.53

Subroutine LKEQN1

This subroutine is called by EQN for making generalized property calculations via the Lee-Kesler method. It calls INITIA and LKEQN2.

Subroutine INITIA

Called by LKEQN1, this subroutine calculates the reduced volume for liquid and vapor phases.

Subroutine LKEQN2

This subroutine is called by LKEQN1 and completes the calculations of the isothermal effects of pressure on the entropy and the enthalpy.

EQXYZ.FOR

All eleven of the calculation options shown in Table 16.1, and the first eleven options in Table 17.1, require the computation of thermodynamic properties, such as fugacity coefficient, enthalpy, entropy, and density, for the vapor and liquid phases. The fugacity coefficients are required in all eleven options; enthalpies are needed in options 8, 10, and 11; and entropies are needed in options 9, 10, and 11. Densities and compressibility factors are calculated only for the output reports. The subroutines involved in making these thermodynamic properties calculations are as follows:

At the beginning of each flash routine, i.e., those subprograms for which ICOD = 1 through 11, the error code IER is set equal to zero at the start. If the equilibrium flash calculations fail to converge, IER is reset to the ICOD value of the failed flash option to prepare for a printout of the appropriate message for the user. This error message is printed in RESULT. All flash routines share COMMON/COMP/ and COMMON/SCRA/ for convenient storing and updating compositions and K-values. All other variables are transmitted by arguments of the subroutine name. These argument variables are:

IKV = K-value option code
IHS = enthalpy/entropy option
NC = number of components
T = current temperature, °R
P = current pressure, psia
EF = efficiency fraction
VF = vapor/feed ratio
HF = feed enthalpy, Btu/lb-mole (plus Q for ICOD = 8)
SF = feed entropy, Btu/lb-mole °R
TTOL = temperature convergence tolerance, °F
PTOL = logarithm of pressure convergence tolerance, log (psia)
XTOL = composition tolerance, mol fraction
IER = error code (= 0 when converged, = ICOD for failure)

There are ten primary plus ten secondary subroutines in the EQXYZ.FOR file. These are programmed to perform the eleven types of equilibrium flash calculations. The following descriptions will be in the order of ICOD controls, i.e., from 1 through 11, with the secondary subroutines inserted where needed. It will be noted that three subroutines are called for two ICOD's.

Subroutine FLSHTP (ICOD = 1)

This is the first equilibrium flash subroutine called in AHT2. It is called first regardless of the ICOD control number. For other than an ICOD = 1 call, the calculations are made at TI and TO and the results are available as aids in subsequent calculations. For the ICOD = 1 calculations at a given T and P, the system may be all vapor, mixed vapor/liquid, or all liquid. Thus, ICOD = 1 computes three phase conditions and reports the results.

The calculation starts with $VF = 0.5$ ($V/F = 0.5$), calls IDEALK to get the Raoult's Law K-values at P and T, after which it stores $\ln K_i$'s in YY for use in XYCONV to moderate the changes in the fugacity K-values. Note that the YY's are always reset to $\ln K_i$'s obtained in the previous iteration in XYCONV. Using Raoult's Law K-values, X and Y are calculated in XYCONV in the first iteration.

The iterative calculations start by calling VFCONV to get a "new" VF, which is used to calculate "new" X and Y, from which new K-values are calculated via SRK or PR, depending upon the IKV value. This iteration is repeated until VF, X, and Y are fully converged. If all X's are converged, all Y's are automatically converged also. To avoid premature abortion, this routine is forced to iterate a minimum of four times. It will abort if $0.97 < K_i < 1.03$ for all K_i's after eight iterations. If $VF \geq 1$ or $VF \leq 0$, all vapor or all liquid is assumed.

Subroutine XYCONV

This subroutine, which is called by FLSHTP, calculates Y and X from Equations 9.35A and 9.36A, then checks to determine if the changes in the X's from the previous values (saved in XX) are smaller than the tolerance $XTOL$. If not, it normalizes Y and X and calculates the logarithms of the fugacity coefficients of the vapor FY and of the liquid FX, from which the logarithm of the K-value, $\ln(YOX) = FX - FY$, is calculated. However, to moderate the changes in K-values from previous values, $\ln(YOX)$ is set equal to the arithmetic average of the current and the previous $\ln(YOX)$ values. The new YOX values are used in the next iteration.

Subroutine VFCONV

In this subroutine, VF is calculated from the XF's, the feed composition, and the YOX's, the K-values, by solving Equation 9.39A. The Newton-Raphson Method, Equation B.16 (Appendix B of Volume 1), is used to make this solution, using Equation 9.40 for the derivative. Note that $f(x) = SM1$ (Equation 9.39A) and that $f(x) = SM2$ (Equation 9.40).

Subroutine IDEALK

In this subroutine, the Raoult's Law K-values are calculated by Equation 10.3, using vapor pressure values from Equation 7.46. For the components in the supercritical region, the vapor pressures are calculated from $\ln(P^m{}_r) = a(1 - T^{-1}{}_r)$, where a is determined by Equation 7.46 at $T_r = 1.0$.

Subroutine APPROX

This subroutine is called by BUBT, DEWT, BUBP, and DEWP, if and when the user-supplied initial estimate of the unknown condition does not lead to a converged solution. The initial temperature or pressure is approximated by using Raoult's Law K-values from IDEALK, guided by the control code INT where:

INT = 1 for bubble-point temperature, in which a search is made for a temperature that satisfies $0.7 < \Sigma XF_i K_i < 1.3$

INT = 2 for dew-point temperature, in which a search is made for a temperature that satisfies $0.7 < \Sigma XF_i/K_i < 1.3$

INT = 3 for bubble-point pressure, where $P = PI*\Sigma XF_i K_i$, but P is not allowed to be greater than ten times TI, the user-supplied estimate

INT = 4 for dew-point pressure, where $P = PI/(\Sigma XF_i K_i)$, but P is not allowed to be less than one-tenth of PI, the user-supplied estimate

Subroutine BUBT (ICOD = 2)

This subroutine, which is called in AHT2, calculates the bubble-point temperature T of a feed at a given pressure. In this case, the composition of the bubble-point liquid is the same as the feed mixture so the composition of the equilibrium liquid is the same as the feed. An initial estimate of T that has been provided by the user is saved for a possible use in a second trial, should the first trial fail to converge. Also, minimum and maximum limits of temperature, over which the bubble-point temperature search is made are set. These temperature limits prevent the search from going outside the initial estimate $\pm 300°F$, or going below $100°R$ or over $1,500°R$.

The iterative calculations start with the calculations of the liquid fugacity coefficients in EQN and the Raoult's Law K-values in IDEALK, all at given P and estimated

T. In this case, liquid fugacity coefficients, being dependent upon T, P, and X, will vary only with T because P and X values are fixed, as $X = XF$ at the bubble point. Vapor compositions, i.e., values of Y, are next calculated by Equation 9.6 and the values of Y are summed to check if Equation 9.42 is satisfied. The Y values are converged in YCONV and returned with the value of the convergence code K1.

If K1 = 0, the common case, checks are made to see if SD (sum departure) and TD (temperature departure) are within their respective tolerances or if SD changes sign, while TD is sufficiently small, i.e., $0.01°F$. If these do not check, NEWTON is called to get an improved value of T and reset the values of SD, TD, and/or TX (temperature increment).

Iterations are repeated until either one of the above K1 conditions is met. If the calculations hit one of the temperature limits, the search direction is reversed, starting from the original estimate of T. If this second round also fails, then the calculation returns to its own initial estimate and starts from the beginning. However, in this last trial, the step size TX is reduced to half of its previous value, but is not made less than $7°F$, in each iteration where $0.8 < K_i < 1.2$ until all K-values are satisfied. This is to minimize the chance of skipping over the correct temperature.

Subroutine YCONV

This subroutine, called by BUBT and BUBP, finds the vapor compositions, i.e., Y values. This is done by normalizing the calculated Y values, calculating the logarithms of the vapor fugacity coefficients, FX and FY, and the logarithms of the K-values, $\ln(YOX) = FX - FY$, using the values of FX that had been calculated previously. Then, Y values are recalculated by averaging the new ($Y = FX*YOX$) and previous values of Y. It should be noted that the new values are before normalization. This calculation is repeated until all Y's have converged to within 1E-5. After the convergence, or thirty iterations, the convergence ranges are checked and the values of K1 are set depending upon the range.

K1 = 0. (Not all of the K-values are in the range $0.8 < K_i < 1.2$.)

K1 = 1. (All of the K-values are in the range $0.8 < K_i < 1.2$. This is an indication that the condition is close to the critical region.)

K1 = 2. (All of the K-values are in the range $0.97 < K_i < 1.03$. This indicates that the condition is in the critical region or the single-phase region.)

Subroutine BUBP (ICOD = 3)

In this subroutine, which is called in AHT2, the bubble-point pressure P of the feed is calculated at a given temperature T. In this case the composition of the equilibrium liquid is the same as the feed, as it was in BUBT. A user-supplied initial estimate of pressure is saved for possible use in the second round, should convergence fail in the first round. Minimum and maximum pressures are set to establish the pressure range over which the search is made. These pressure limits prevent the search from going outside of 5 times or one-fifth of the user-supplied initial pressure.

The bubble-point pressure prediction iterations start with calculating the liquid fugacity coefficient in EQN and the Raoult's Law K-values in IDEALK, both at the given T and the estimated P. As T and X are fixed by the specifications of the problem, the only changeable variable affecting the fugacity coefficients is the pressure. Then the vapor compositions are calculated by Equation 9.6 and these Y's are summed to check later if Equation 9.42 is satisfied. When all Y's have converged in YCONV, so that the $SD = 0$, the Y's returned with the appropriate convergence code, K1.

If K1 = 0, the common case, a check is made to see if SD (sum departure) and PD (pressure departure) are within their respective tolerances, or if SD changes sign while PD is sufficiently small, e.g., 0.0001. If these do not check, NEWTON is called to get improved values of pressure and reset the values of SD, PD, and/or PX.

Iterations continue until either one of the previous conditions is met. If one of the limits is hit in the iterations, the search direction is reversed, starting from the original pressure estimate. If this second round also fails, the calculations are repeated from the beginning with the computer estimate of initial pressure. However, in this last trial, the step size PX is reduced by half, but not less than 0.2 of its previous value in each iteration, in which $0.8 < K_i < 1.2$ occurs. This is to minimize the chance of skipping the feasible pressure region.

Subroutine DEWT (ICOD = 4)

In this subroutine, which is called in AHT2, the dew-point temperature T of a feed is calculated at a given pressure P. The equilibrium vapor composition is equal to the feed composition. The user-supplied initial estimate of the dew temperature T is saved for a possible use in the second round, should the convergence fail in the first round, and the minimum and maximum limits of the temperature search are set to prevent the search from going outside $\pm 300°F$ of the supplied initial temperature, or going below $100°R$ or above $1{,}500°R$.

The iterative calculations start in EQN with the calculation of the vapor fugacity coefficient FY and the calculation of the Raoult's Law K-values YOX in IDEALK, both calculations being at T. Of the conditions affecting the vapor fugacity coefficient, i.e., P, T, and Y, T is the only that can vary, the others are fixed by the definitions of the problem. Next, liquid compositions X's are calculated by Equation 9.6, the Y's are summed to check to see if Equation 9.43 is satisfied, i.e., sum departure $SD = 0$. When all X's are converged in XCONV, their sum is returned from XCONV with K-value range code K1 set.

If K1 = 0, the common case, a check is made to see if SD and TD, i.e., sum and temperature departures, are within their respective tolerances or if SD changes sign, while TD is sufficiently small, i.e., 0.01. If this test is not passed, NEWTON is called to get improved values of T and to reset the values of SD, TD, and/or TX.

The iteration is repeated until either one of these conditions is met. If the iterative calculation hits one of the temperature limits, it reverses the search direction, starting from the original estimate of the temperature. If this second round also fails, it starts over from the beginning with the program estimated temperature. In this last trial, however, the step size TX is reduced to half, but no less than $7°F$, of its previous value in each iteration, in which all K_i's are in the range $0.8 < K_i < 1.2$. This is to minimize the chance of skipping the feasible region of temperature.

Subroutine XCONV

This subroutine, called by DEWT and DEWP, calculates liquid compositions, i.e., X values. This is done by first normalizing the calculated X values, calculating the logarithm of the liquid fugacity coefficients. Then the logarithms of the K-values, $\ln(YOX) = FX - FY$, are calculated, using the FY values found for the vapor phase in the calling routine. Then X's are recalculated by averaging the new, i.e., $X = XF/YOX$, and the previous value of X, the new values being before normalization. This calculation is repeated until all X's are converged within 1E-5. After convergence, or thirty iterations, the ranges of the K-values are checked and the convergence code K1 is set as follows:

K1 = 0. (Not all of the K-values are in the range $0.8 < K_i < 1.2$.)

K1 = 1. (All of the K-values are in the range $0.8 < K_i < 1.2$, indicating the proximity of the critical region.)

K1 = 2. (All of the K-values are in the range $0.97 < K_i < 1.03$, indicating that the condition is in the critical or single-phase regions.)

Subroutine DEWP (ICOD = 5)

In this subroutine, which is called by AHT2, the dew-point pressure is found for a feed at a constant temperature T. For this calculation the composition of the dew-point vapor is that of the feed. The first step is to save the user-supplied initial estimate of the pressure P for a possible use in the second round, should convergence fail in the first round. Minimum and maximum pressures, over which the bubble pressure will be searched are set. These pressure limits prevent the search from going outside of five times to one-fifth of the user-supplied pressure.

The iterative calculations start in EQN with the calculation of FY's, the logarithms of the fugacity coefficients of the components in the vapor phase, followed by the calculations, in IDEALK, of the Raoult's Law K-values YOX at P. In a dew-point pressure calculation, the FY values will change only with pressure from one iteration to another because the vapor composition and the temperature are fixed. Liquid compositions, the X's, are calculated by Equation 9.6 and then summed up for checking to see if Equation 9.43 is satisfied, i.e., if $SD = 0$, when all X's have converged in XCONV, which returns the sum of X's and the K-value range code.

When $K1 = 0$, the usual case, SD and PD, the sum and temperature differences, are checked to see if they are in their respective tolerances or if SD changes sign, while PD is sufficiently small, i.e., 0.0001. If there is no check, NEWTON is called to get improved values of pressure and reset the values of SD, PD, and/or PX.

The iterative calculations are repeated until either one of the previous conditions is satisfied. If one of the pressure limits is hit, the direction of the search is reversed and started with the original pressure estimate. If this second round also fails, a program estimate of the initial pressure is made and the calculations are started over from the beginning. In this last trial the step size PX is reduced to half, but no less than 7°F, of its previous value in each iteration in which the K-value is in the range $0.8 < K_i < 1.2$. This is to minimize the chance of skipping the correct pressure.

Subroutine NEWTON

This subroutine is used in all iteration convergence calculations, except VFCONV, for solving a single-variable equation by the numerical Newton-Raphson Method, i.e., Equation B.21 in Volume 1 of AHT. In NEWTON new values of T or P and of SD, TD or PD are found and returned to the phase equilibrium subroutine from which the NEWTON call had come.

Subroutine FLSHPV (ICOD = 6)

In this subroutine, which is called in AHT2, the calculation objectives are: the equilibrium flash temperature and the compositions of the coexisting equilibrium vapor and liquid, given the pressure and the vapor/feed ratio, i.e., VF. If the tolerance is negative in the calculation of the temperature, $TSAT$ is called and this calculates the dew and bubble temperatures. If the dew and bubble temperatures fail to converge, this routine makes no attempt to find the equilibrium flash temperature. Otherwise, it calculates the initial estimate in three different ways, depending upon the $TSAT$ results:

$T = TBUB$ (bubble temperature) $+ 2$ if the bubble temperature converged but the dew temperature did not

$T = TDEW$ (dew temperature) $- 2$ if dew temperature converged but the bubble temperature did not

$T = TBUB + VF (TDEW - TBUB)$ if both dew and bubble temperatures converged

Once the initial estimate is set, the calculation is the same as in the case where the tolerance is positive, i.e., the user-supplied initial estimate of the temperature is satisfactory.

The maximum step size TX is set to half of the temperature tolerance and TD is set at 0.7 TX. $XX(1)$ is set at an unreasonable value of 2 to avoid the highly unlikely coincidence of the convergence of the liquid composition calculation. Note that XX is the value of X in the previous iteration, and only one XX value needs to be unreasonable in the first iteration. $K1$ is set to 1 to make sure that IDEALK is called in the first iteration.

The iterative calculations start with calculating the Raoult's Law K-values in IDEALK. At this point, FLSHTP could be called to calculate X, Y, and VF; but it is not necessary nor efficient to find completely converged X, Y, and VF values in the first iteration. Accordingly, a "loose" isothermal flash is made. The logarithm of the K-values, $\ln(YOX)$, is saved in YY for use in XYCONV. The YOX values could be from either IDEALK (if $K1 = 1$ or $TD = 10$) or from XYCONV (if $K1 = 0$ and $TD = 10$). XYCONV is called a maximum of five times to calculate new X and Y values, then VFCONV is called to find VF. This iteration repeats until the calculated VF is close to the specified VF within 0.0001 and also TD is smaller than the tolerance, or FN changes sign, while TD is less than 0.01°F.

Subroutine FLSHTV (ICOD = 7)

In this subroutine, which is called in AHT2, the calculation objectives are the equilibrium flash pressure and the compositions of the coexisting equilibrium vapor and liquid, given the temperature and the vapor/liquid ratio, i.e., *VF*. If the tolerance is negative, the initial estimate of the pressure is calculated by calling PSAT, which calculates the dew and bubble pressures. If both the bubble and dew pressures fail to converge, this routine does not attempt to calculate the equilibrium pressure. Once the initial estimate is set, then it is the same as the case where the tolerance is positive, i.e., the user-supplied initial estimate is satisfactory.

The maximum step size *PX* is set at half of the pressure tolerance and *PD* is set at 0.7 *PX*. *XX*(1) is set to an unreasonable value of 2 to avoid the highly unlikely coincidence of the convergence of the liquid composition calculations for *X*. Note that *XX* is the value of *X* in the previous iteration, and that only one value of *XX* needs to be unreasonable in the first iteration. *K*1 is set to 1 to make sure that IDEALK is called in the first iteration.

The iterative calculations start in IDEALK with the calculations of the Raoult's Law *K*-values. At this point, FLSHTP could be called for calculations of *X*, *Y*, and *VF*. However, it is not necessary nor efficient to calculate fully converged values of *X*, *Y*, and *VF* in the first iteration. Accordingly, a "loose" isothermal flash is made. The logarithm of the *K*-values, ln(*YOX*), are saved in *YY* for use in XYCONV. The *YOX* values could be either from IDEALK, if *K*1 = 1 and *PD* = 0.02, or from XYCONV, if *K*1 = 0 and *TD* = 0.02. XYCONV is called to get *x* and *y* values. This iteration repeats until the calculated *VF* is close to the specified *VF* within 0.0001 and also *PD* is smaller than the tolerance, or *FN* changes sign, while *TD* is less than 0.0001.

Subroutine TSAT

Called by FLSHPV and FLSHPH, this subroutine calculates dew and bubble temperatures to bracket the temperature range in searching for the temperature, at given *P* and *V/F* values, assigning the following codes:

IHP = 1, if only the dew-point temperature converged
    = 2, if only the bubble-point temperature converged
    = 3, if both dew- and bubble-point temperatures converged
    = 4, if neither dew- or bubble-point temperatures converged

Subroutine TAVG

In this subroutine, which is called by FLSHPV, a new trial temperature is found for further iterations.

Subroutine FLSHPH (ICOD = 8 and 9)

In this subroutine, which is called in AHT2, the objective is to find the temperature at a given pressure and enthalpy or entropy. The symbol *HF* is used to designate the feed enthalpy or the entropy, depending upon the value of *IC* (*IC* = 2 for enthalpy; *IC* = 3 for entropy).

Dew and bubble temperatures, i.e., *TDEW* and *TBUB*, are calculated in TSAT. If only *TDEW* converges, ENTHS is called to calculate the enthalpy *HV* at *TDEW*. If *HV* is less than *HF* then SINGPH is called to search for the isenthalpic, or isentropic, temperature in the single-phase region. Otherwise, the initial value *T* is set to *TDEW* − 5 and the isenthalpic, or isentropic, temperature is sought in the two-phase region. If only bubble temperature, *TBUB*, is converged, ENTHS is called to calculate the enthalpy *HL* at *TBUB*. If *HL* is greater than *HF*, then SINGPH is called to search for the isenthalpic, or isentropic, temperature in the single-phase region. Otherwise, the initial temperature is set to *TBUB* + 5 and the isenthalpic, or isentropic, temperature is sought in the two-phase region.

If both dew and bubble temperature calculations have converged, both dew- and bubble-point enthalpies, or entropies, are calculated to give *HV* and *HL*. If *HF* is outside of *HV* and *HL*, SINGPH is called to search for the isenthalpic, or isentropic, temperature in the single-phase region. Otherwise, the initial estimate of the temperature is set by the following: $T = TBUB + (TDEW - TBUB)(HF - HL)/(HV - HL)$, then the temperature is sought in the two-phase region. If (*TDEW* − *TBUB*) is less than 30°F, i.e., a narrow boiling mixture, then *VF* becomes the iterate instead of the temperature to find the temperature that will give the same enthalpy, or entropy, as *HF* in the two-phase region.

Before the start of the iterative calculations, the following variables are set to get the Newton-Raphson calculations started;

TX: Maximum step size for temperature change in an iteration, e.g., if a larger temperature change than *TX* is found in NEWTON, the absolute value of the change is set at *TX*, to moderate excessive changes, thus minimizing the chances of divergence or oscillation.

*F*1: Previous value of *FN*, returned from NEWTON. Before starting iterative calculations, *F*1 is set to zero to avoid problems associated with an undefined variable.

*T*1: Previous value of *T*, returned from NEWTON. At the beginning of the iterative calculations, *T*1 is set equal to 1E7 as a signal of the first iteration in NEWTON.

TD: The absolute value of the difference between *T*1 and *T*. This provides a criterion of convergence; $FN = 0$ is another criterion. At the beginning, *TD* is arbitrarily set at 0.7 *TX*.

The temperature search in the two-phase region starts with calculating *X*, *Y*, and *VF* at *T* and *P* using FLSHTP, and is followed by enthalpy, or entropy, *H*1, calculations to obtain $FN = H1 - HF$, from which a new *T* is calculated in NEWTON. This iteration repeats until the calculated enthalpy, or entropy, is within 0.1 Btu/lb-mole or Btu/lb-mole °F, *T* converges within 0.1 or the sign of *FN* changes in two successive iterations, and the temperature change is less than 0.01 °F. However, for the case of narrow boiling mixtures, $(TDEW - TBUB) < 30°F$, the convergence criteria are $FN < 0.1$ Btu/lb-mole, or Btu/lb-mole °F and $VF - VF1$, the change in V/F in two successive iterations, < 0.00005, or the sign of *FN* changes in two successive iterations, and the $(VF - VF1)$ value is very small, 0.00001.

Subroutine SINGPH

Called by FLSHPH and EXCOM2, this subroutine finds the temperature in the single-phase region at a given pressure and enthalpy, or entropy, *HF*, which will be enthalpy when $IC = 2$ or entropy when $IC = 3$. Before the start of the iterative calculations, the following variables are set to get the Newton-Raphson method started:

TX: Maximum step size of temperature change in an iteration, e.g., if a change larger than *TX* is found in NEWTON, the absolute value of the change is set equal to *TX*, to moderate the excessive change, thus minimizing the chance of divergence or oscillation.

*F*1: Previous value of *FN*, returned from NEWTON. Before starting iterative calculations *F*1 is set to zero to avoid problems associated with an undefined variable.

*T*1: Previous value of *T*, returned from NEWTON. At the beginning of the iterative calculations, *T*1 is set equal to 1E7 as a signal of the first iteration to NEWTON.

TD: The absolute value of the difference between *T*1 and *T*, which provides one criterion of convergence; $FN = 0$ is the other. At the beginning, *TD* is arbitrarily set to 0.7 *TX*.

The temperature search in the single-phase region starts with the calculations of enthalpy, or entropy to get *H*1 and find $FN = H1 - HF$, from which a new *T* is calculated in NEWTON. The iteration repeats until the calculated enthalpy is within 0.1 Btu/lb-mole and *T* con-

verges within 0.1 or the sign of *FN* changes in two successive iterations, and the temperature change is less than 0.01 °F.

Subroutine EXCOM1 (ICOD = 10 and 11)

This subroutine, called by AHT2, is for expansion or compression calculations where equilibrium vapor and liquid are expected to coexist at either or both ends of the process. The calculations find the final temperature and the entropy and/or enthalpy for isentropic, polytropic, or isenthalpic processes, with a specified efficiency, which is defined for present purposes by the following: The isentropic process, which is reversible and adiabatic, has an efficiency of 100%, while the isenthalpic process is irreversible and adiabatic and has an efficiency of 0%. In this context, a polytropic process has an efficiency $100\% < \epsilon < 0\%$. For polytropic compression/expansion processes, the value of ϵ is usually between 70 and 90%, the reduction in ϵ being due to both irreversible and nonadiabatic conditions. There are both isentropic or polytropic compression and expansion processes but only expansion isenthalpic process.

The calculations for the isentropic and polytropic processes are made by using FLSHPH with $IC = 3$ to calculate the isentropic final temperature, at 100% efficiency, for the change from the given initial conditions of temperature and pressure and final pressure. Then the enthalpy at the final condition is calculated from enthalpies at the initial conditions and the final pressure and the isentropic final temperature as $HH = HF + (HH - HF) EF$, where the *HH* on right side is the value at the isentropic final and the *HH* on left is the value at the polytropic final.

It should be noted here that the same expression using the efficiency *EF*, as read in MAIN, can be applied to both polytropic compression and polytropic expansion by making *EF* in the previous expression for compression the inverse of the usual efficiency entered, i.e., *EF* is greater than 1.0 for polytropic compression but is less than 1.0 for polytropic expansion.

From the final value of *HH*, the polytropic final temperature, and other properties can be found. For an isenthalpic expansion, the final temperature can be found in FLSHPH with $IC = 2$ with $HH = HF$.

Subroutine EXCOM2 (ICOD 10 and 11)

This subroutine, called in AHT2, calculates compression and expansion of single-phase systems. Practically all compression processes are this kind. EXCOM2 is identical to EXCOM1 except that it calls SINGPH instead of FLSHPH.

Executable Program

An executable program is created by compiling the 6 FORTRAN files of 61 subprograms, obtaining object code versions. Then these .OBJ files are linked with the necessary library files, i.e., FORTRAN.LIB, MATH.LIB, or 8087.LIB. The resulting EQUIL.EXE will run on data from the designated data file and store the results in the designated result and assay files, all in the same directory of the computer disk.

Examples

Applications of EQUIL are given in Tables 17.2 through 17.13 for two feed mixtures that are typical of systems for which the program was developed. The feed mixture used for nine example calculations was a typical LNG plant feed gas from Chappelear et al (1). The feed mixture used for two example calculations was a Light Sumatran Crude Oil from a crude oil evaluation report of Chevron (2). The calculation results in Tables 17.2 through 17.13 illustrate many features of EQUIL for the user.

Input data files for both gas and oil feeds are prepared in the same format in Subroutine DFPREP for reading in DATAIN, the difference between gas and oil being that the oil must first be divided into 11 or 21 pseudo components, for which the properties are found differently. For the discrete components of the gas, the properties are from BLOCK DATA via Subroutine PHPROP at the beginning of the actual run.

For the oil the properties are calculated in Subroutine CUTPRO, as functions of the *BP* (boiling point) and *SG* (specific gravity) values of the pseudo components. The *BP* and *SG* values are from the starting assays, i.e., boil-

ing point and gravity. Pseudo component names are assigned when the oil is divided in DFPREP. Tables 15.6, 15.7, and 15.8 illustrated these assay-to-pseudo component calculations. Except for these differences, the two types of feeds are handled in the same manner.

The run-report from EQUIL includes a summary of the controls and feed mixture information in the input data file. Table 17.2 illustrates this for the gas problems. The feed mixture portion of an oil feed summary is shown in Table 15.8. The controls portion would look the same for both oil and gas. The BP and SG values shown in Table 7.12 are not required in EQUIL calculations for discrete components, but they are included by the same program that reports BP and SG values for pseudo components.

Nine runs were made on the gas mixture to illustrate the nine basic vapor/liquid equilibrium calculations that might be encountered. These are one flash at given *T* and *P*; four bubble or dew point *T* or *P* predictions; and four flashes with constant *V/F*, *H*, or *S*, as described in Table 17.1. These are identified as ICOD = 1 through 9. The results of nine runs for these ICOD's, Cases 1 through 9, are given in Tables 17.3 through 17.11. ICOD's 1, 6, and 7 all work with the variables *P*, *T*, and *V/F*, fixing two and solving for the third. There are many two-variable combinations of these three variables that do not fall in the two-phase, vapor/liquid region.

With ICOD = 1, this means all liquid or all vapor possibilities are properly handled by the program, which identifies the fluid as liquid or vapor and gives the properties. For ICOD's of 6 or 7, an initial specification of *V/F* and *P* or *T* that falls outside the two-phase boundary will usually result in a failure message. Running both ICOD = 6 and ICOD = 7 as a multiple run and/or mak-

(text continued on page 96)

Table 17.2
Starting Data for Series of Computer Runs on Gas Plant Feed Using
Soave Redlich-Kwong Method for *K*, *H*, and *S* Predictions

CONTROLS							
IKV	IHS	IHD	IOUT	MULT	XTOL	TTOL	PTOL
1	1	1	1	0	.00001	.10000	.00010

Feed Mixture Comp. ID	Feed Mols/hr	Boil Pt °F	Specific Gravity	Component Name
603	30.750	− 320.40	.0000	Nitrogen
606	39.958	− 109.30	.0000	Carbon dioxide
101	6929.500	− 258.70	.3000	Methane
102	360.458	− 127.50	.3564	Ethane
103	142.167	− 43.70	.5077	Propane
105	60.208	10.90	.5631	2-Methylpropane (*I*-Butane)
104	39.208	31.10	.5844	*N*-Butane
107	20.750	82.10	.6247	2-Methylbutane (*I*-Pentane)
106	13.833	96.90	.6310	*N*-Pentane
114	48.417	209.20	.6882	*N*-Heptane

Table 17.3

(Case 1) Flash *V/F* Ratio at Given Temperature and Pressure for LPG Plant Feed Mixture, Using *K* and *H* Values from Soave-Redlich-Kwong Method

		Problem Conditions for Calculations that Follow						
ICOD	KHF	KYF	KTL	TI	PI	VF	TO	PO
1	0	0	0	− 100.00	600.00	.00	.00	.00

Material Balance Basis	Mole Percent			
Component Name	Feed	Vapor	Liquid	K-Value
Nitrogen	.400	.496	.103	4.797E + 00
Carbon Dioxide	.520	.405	.878	4.608E − 01
Methane	90.166	96.587	70.226	1.375E + 00
Ethane	4.690	2.189	12.458	1.757E − 01
Propane	1.850	.261	6.784	3.851E − 02
2-Methylpropane (*I*-Butane)	.783	.040	3.091	1.307E − 02
N-Butane	.510	.017	2.041	8.494E − 03
2-Methylbutane (*I*-Pentane)	.270	.003	1.099	2.710E − 03
N-Pentane	.180	.001	.735	1.863E − 03
N-Heptane	.630	.000	2.586	1.004E − 04
Totals	100.000	100.000	100.000	
Temperature, °F	− 100.00	− 100.00	− 100.00	
Pressure, psia	600.00	600.00	600.00	
Molecular Weight	18.74	16.62	25.32	
Compressibility Factor		.6326	.1436	
Density, lb-mols/ft³		.2457	1.0826	
Molar volume, ft³/lb-mole		4.0699	.9237	
Total flows, lb-moles/hour	7.685E + 03	5.813E + 03	1.872E + 03	
Total flows, pounds/hour	1.440E + 05	9.664E + 04	4.739E + 04	
Enthalpy, Btu/lb-mole*	1221.79	2021.10	− 1260.64	
Enthalpy, Btu/hour*	.939E + 07	.117E + 08	− .236E + 07	
Mole percents		75.64	24.36	

* Zero enthalpy datum is ideal gas to 0°R

Table 17.4

(Case 2) Bubble-Point Temperature at Given Pressure for LPG Plant Feed Mixture, Using *K* and *H* Values from Soave-Redlich-Kwong Method

		Problem Conditions for Calculations that Follow						
ICOD	KHF	KYF	KTL	TI	PI	VF	TO	PO
2	0	0	0	− 110.00	600.00	.00	.00	.00

Material Balance Basis	Mole Percent			
Component Name	Feed	Vapor	Liquid	K-Value
Nitrogen	0.400	0.000	0.400	2.665E + 00
Carbon Dioxide	0.520	0.000	0.520	4.545E − 01
Methane	90.166	0.000	90.166	1.082E + 00
Ethane	4.690	0.000	4.690	2.125E − 01
Propane	1.850	0.000	1.850	6.133E − 02
2-Methylpropane (*I*-Butane)	0.783	0.000	0.783	2.448E − 02
N-Butane	0.510	0.000	0.510	1.775E − 02
2-Methylbutane (*I*-Pentane)	0.270	0.000	0.270	6.782E − 03
N-Pentane	0.180	0.000	0.180	5.044E − 03
N-Heptane	0.630	0.000	0.630	4.370E − 04
Totals	100.000	100.000	100.000	
Temperature, °F	− 110.00	− 115.44	− 115.44	
Pressure, psia	600.00	600.00	600.00	
Molecular weight	18.74	.00	18.74	
Compressibility factor		.0000	.1518	
Density, lb-mols/ft³		.0000	1.0696	
Molar volume, ft³/lb-mole		4.0699	.9350	
Total Flows, lb-moles/hr	7.685E + 03	.000E + 00	7.685E + 03	
Total Flows, lb/hr	1.440E + 05	.000E + 00	1.440E + 05	
Enthalpy, Btu/lb-mole*	624.69	.00	− 206.39	
Enthalpy, Btu/hr*	.480E + 07	.000E + 00	− .159E + 07	
Mole percents		.00	100.00	

* Zero enthalpy datum is ideal gas at 0°R

Table 17.5
(Case 3) Bubble Point Pressure at Given Temperature for LPG Plant Feed Mixture, Using *K* and *H* Values from Soave-Redlich-Kwong Method

Problem Conditions for Calculations that Follow								
ICOD	KHF	KYF	KTL	TI	PI	VF	TO	PO
3	0	0	0	− 100.00	600.00	.00	.00	.00

Material Balance Basis		Mole Percent		
Component Name	Feed	Vapor	Liquid	K-Value
Nitrogen	.400	.000	.400	1.984E + 00
Carbon Dioxide	.520	.000	.520	6.034E − 01
Methane	90.166	.000	90.166	1.073E + 00
Ethane	4.690	.000	4.690	3.642E − 01
Propane	1.850	.000	1.850	1.610E − 01
2-Methylpropane (*I*-Butane)	.783	.000	.783	8.860E − 02
N-Butane	.510	.000	.510	7.132E − 02
2-Methylbutane (*I*-Pentane)	.270	.000	.270	3.810E − 02
N-Pentane	.180	.000	.180	3.134E − 02
N-Heptane	.630	.000	.630	6.360E − 03
Totals	100.000	100.000	100.000	
Temperature, °F	− 100.00	− 100.00	− 100.00	
Pressure, psia	600.00	750.49	750.49	
Molecular weight	18.74	.00	18.74	
Compressibility factor		.0000	.1975	
Density, lb-mols/ft³		.0000	.9842	
Molar volume, ft³/lb-mole		4.0699	1.0160	
Total flows, lb-moles/hr	7.685E + 03	.000E + 00	7.685E + 03	
Total flows, lb/hr	1.440E + 05	.000E + 00	1.440E + 05	
Enthalpy, Btu/lb-mole*	1221.79	.00	119.64	
Enthalpy, Btu/hr*	.939E + 07	.000E + 00	.919E + 06	
Mole percents		.00	100.00	

* Zero enthalpy datum is ideal gas at 0°R

Table 17.6
(Case 4) Dew Point Temperature at Given Pressure for LPG Plant Feed Mixture, Using *K* and *H* Values from Soave-Redlich-Kwong Method

Problem Conditions for Calculations that Follow								
ICOD	KHF	KYF	KTL	TI	PI	VF	TO	PO
4	0	0	0	− 80.00	600.00	.00	.00	.00

Material Balance Basis		Mole Percent		
Component Name	Feed	Vapor	Liquid	K-Value
Nitrogen	.400	.400	.000	1.153E + 00
Carbon Dioxide	.520	.520	.000	9.335E − 01
Methane	90.166	90.166	.000	1.029E + 00
Ethane	4.690	4.690	.000	8.725E − 01
Propane	1.850	1.850	.000	7.806E − 01
2-Methylpropane (*I*-Butane)	.783	.783	.000	7.275E − 01
N-Butane	.510	.510	.000	6.991E − 01
2-Methylbutane (*I*-Pentane)	.270	.270	.000	6.473E − 01
N-Pentane	.180	.180	.000	6.276E − 01
N-Heptane	.630	.630	.000	5.132E − 01
Totals	100.000	100.000	100.000	
Temperature, °F	− 80.00	− 104.72	− 104.72	
Pressure, psia	600.00	600.00	600.00	
Molecular weight	18.74	18.74	.00	
Compressibility factor		.1714	.0000	
Density, lb-mols/ft³		.9187	.0000	
Molar volume, ft³/lb-mole		1.0885	1.0160	
Total flows, lb-moles/hr	7.685E + 03	7.685E + 03	.000E + 00	
Total flows, lb/hr	1.440E + 05	1.440E + 05	.000E + 00	
Enthalpy, Btu/lb-mole*	1780.93	64.22	.00	
Enthalpy, Btu/hr*	.137E + 08	.494E + 06	.000E + 00	
Mole percents		100.00	.00	

* Zero enthalpy datum is ideal gas at 0°R

Table 17.7
(Case 5) Dew Point Pressure at Given Temperature for LPG Plant Feed Mixture, Using *K* and *H* Values from Soave-Redlich-Kwong Method

Problem Conditions for Calculations that Follow								
ICOD	KHF	KYF	KTL	TI	PI	VF	TO	PO
5	0	0	0	− 90.00	600.00	.00	.00	.00

Material Balance Basis	Mole Percent			
Component Name	Feed	Vapor	Liquid	*K*-Value
Nitrogen	.400	.400	.000	$1.210E + 00$
Carbon Dioxide	.520	.520	.000	$9.186E − 01$
Methane	90.166	90.166	.000	$1.041E + 00$
Ethane	4.690	4.690	.000	$8.371E − 01$
Propane	1.850	1.850	.000	$7.214E − 01$
2-Methylpropane (*I*-Butane)	.783	.783	.000	$6.556E − 01$
N-Butane	.510	.510	.000	$6.225E − 01$
2-Methylbutane (*I*-Pentane)	.270	.270	.000	$5.611E − 01$
N-Pentane	.180	.180	.000	$5.389E − 01$
N-Heptane	.630	.630	.000	$4.111E − 01$
Totals	100.000	100.000	100.000	
Temperature, °F	− 90.00	− 90.00	− 90.00	
Pressure, psia	600.00	856.96	856.96	
Molecular weight	18.74	18.74	.00	
Compressibility factor		.2472	.0000	
Density, lb-mols/ft³		.8739	.0000	
Molar volume, ft³/lb-mole		1.1443	1.0160	
Total flows, lb-moles/hr	$7.685E + 03$	$7.685E + 03$	$.000E + 00$	
Total flows, lb/hr	$1.440E + 05$	$1.440E + 05$	$.000E + 00$	
Enthalpy, Btu/lb-mole*	1540.70	334.34	.00	
Enthalpy, Btu/hr*	$.118E + 08$	$.257E + 07$	$.000E + 00$	
Mole percents		100.00	.00	

* Zero enthalpy datum is ideal gas at 0°R

Table 17.8
(Case 6) Flash Temperature at Given Pressure and *V*/*F* Ratio for LPG Plant Feed Mixture, Using *K* and *H* Values from Soave-Redlich-Kwong Method

Problem Conditions for Calculations that Follow								
ICOD	KHF	KYF	KTL	TI	PI	VF	TO	PO
6	0	0	0	− 110.00	600.00	.60	.00	.00

Material Balance Basis	Mole Percent			
Component Name	Feed	Vapor	Liquid	*K*-Value
Nitrogen	.400	.609	.089	$5.386E + 00$
Carbon Dioxide	.520	.053	1.216	$2.310E − 01$
Methane	90.166	98.851	77.220	$1.250E + 00$
Ethane	4.690	.469	10.983	$1.047E − 01$
Propane	1.850	.017	4.583	$2.064E − 02$
2-Methylpropane (*I*-Butane)	.783	.001	1.949	$6.776E − 03$
N-Butane	.510	.000	1.270	$4.433E − 03$
2-Methylbutane (*I*-Pentane)	.270	.000	.672	$1.407E − 03$
N-Pentane	.180	.000	.448	$9.731E − 04$
N-Heptane	.630	.000	1.569	$5.365E − 05$
Totals	100.000	100.000	100.000	
Temperature, °F	− 110.00	− 108.62	− 108.62	
Pressure, psia	600.00	600.00	600.00	
Molecular weight	18.74	16.20	22.52	
Compressibility factor		.6147	.1430	
Density, lb-mols/ft³		.2591	1.1137	
Molar volume, ft³/lb-mole		3.8602	.8979	
Total flows, lb-moles/hr	$7.685E + 03$	$4.611E + 03$	$3.074E + 03$	
Total flows, lb/hr	$1.440E + 05$	$7.471E + 04$	$6.931E + 04$	
Enthalpy, Btu/lb-mole*	624.69	1927.72	− 917.26	
Enthalpy, Btu/hr*	$.480E + 07$	$.889E + 07$	$− .282E + 07$	
Mole percents		60.00	40.00	

* Zero enthalpy datum is ideal gas at 0°R

Table 17.9

(Case 7) Flash Pressure at Given Temperature and *V/F* Ratio for LPG Plant Feed Mixture, Using *K* and *H* Values from Soave-Redlich-Kwong Method

Problem Conditions for Calculations that Follow								
ICOD	KHF	KYF	KTL	TI	PI	VF	TO	PO
7	0	0	0	− 110.00	600.00	.60	.00	.00

Material Balance Basis	Mole Percent			
Component Name	Feed	Vapor	Liquid	*K*-Value
Nitrogen	.400	.570	.146	3.904E + 00
Carbon Dioxide	.520	.337	.795	4.234E − 01
Methane	90.166	97.313	79.447	1.225E + 00
Ethane	4.690	1.576	9.361	1.684E − 01
Propane	1.850	.167	4.374	3.818E − 02
2-Methylpropane (*I*-Butane)	.783	.025	1.921	1.307E − 02
N-Butane	.510	.011	1.259	8.699E − 03
2-Methylbutane (*I*-Pentane)	.270	.002	.672	2.811E − 03
N-Pentane	.180	.001	.449	1.957E − 03
N-Heptane	.630	.000	1.575	1.095E − 04
Totals	100.000	100.000	100.000	
Temperature, °F	− 110.00	− 110.00	− 110.00	
Pressure, psia	600.00	579.27	579.27	
Molecular weight	18.74	16.49	22.12	
Compressibility factor		.6052	.1390	
Density, lb-mols/ft^3		.2550	1.1108	
Molar volume, ft^3/lb-mole		3.9209	.9002	
Total flows, lb-moles/hr	7.685E + 03	4.611E + 03	3.074E + 03	
Total flows, lb/hr	1.440E + 05	7.604E + 04	6.798E + 04	
Enthalpy, Btu/lb-mole*	624.69	1909.22	− 853.43	
Enthalpy, Btu/hr*	.480E + 07	.880E + 07	− .262E + 07	
Mole percents		60.00	40.00	

* Zero enthalpy datum is ideal gas at 0°R

Table 17.10

(Case 8) Isenthalpic Flash Temperature at Given Pressure for LPG Plant Feed Mixture, Using *K* and *H* Values from Soave-Redlich-Kwong Method

Problem Conditions for Calculations that Follow								
ICOD	KHF	KYF	KTL	TI	PI	QF	TO	PO
8	0	0	0	− 80.00	800.00	.00	− 80.00	600.00

Material Balance Basis	Mole Percent			
Component Name	Feed	Vapor	Liquid	*K*-Value
Nitrogen	.400	.488	.097	5.014E + 00
Carbon Dioxide	.520	.415	.884	4.691E − 01
Methane	90.166	96.453	68.379	1.411E + 00
Ethane	4.690	2.298	12.982	1.770E − 01
Propane	1.850	.280	7.290	3.840E − 02
2-Methylpropane (*I*-Butane)	.783	.043	3.348	1.297E − 02
N-Butane	.510	.019	2.214	8.384E − 03
2-Methylbutane (*I*-Pentane)	.270	.003	1.195	2.660E − 03
N-Pentane	.180	.001	.799	1.823E − 03
N-Heptane	.630	.000	2.812	9.678E − 05
Totals	100.000	100.000	100.000	
Temperature, °F	− 80.00	− 98.22	− 98.22	
Pressure, psia	800.00	600.00	600.00	
Molecular weight	18.74	16.65	25.99	
Compressibility factor		.6403	.1439	
Density, lb-mols/ft^3		.2416	1.0751	
Molar volume, ft^3/lb-mole		4.1397	.9301	
Total flows, lb-moles/hr	7.685E + 03	5.964E + 03	1.721E + 03	
Total flows, lb/hr	1.440E + 05	9.929E + 04	4.474E + 04	
Enthalpy, Btu/lb-mole*	1288.73	2048.76	− 1345.20	
Enthalpy, Btu/hr*	.990E + 07	.122E + 08	− .232E + 07	
Mole percents		77.61	22.39	

* Zero enthalpy datum is ideal gas at 0°R

Table 17.11
(Case 9) Isentropic Flash Temperature at Given Pressure for LPG Plant Feed Mixture, Using *K* and *H* Values from Soave-Redlich-Kwong Method

				Problem Conditions for Calculations that Follow				
ICOD	KHF	KYF	KTL	TI	PI	EF	TO	PO
9	0	0	0	− 80.00	800.00	.00	− 80.00	600.00

Material Balance Basis	Mole Percent			
Component Name	Feed	Vapor	Liquid	*K*-Value
Nitrogen	.400	.610	.176	3.460E + 00
Carbon Dioxide	.520	.320	.733	4.363E − 01
Methane	90.166	97.374	82.477	1.181E + 00
Ethane	4.690	1.489	8.105	1.837E − 01
Propane	1.850	.166	3.647	4.545E − 02
2-Methylpropane (*I*-Butane)	.783	.026	1.591	1.649E − 02
N-Butane	.510	.012	1.042	1.129E − 02
2-Methylbutane (*I*-Pentane)	.270	.002	.556	3.885E − 03
N-Pentane	.180	.001	.371	2.768E − 03
N-Heptane	.630	.000	1.302	1.822E − 04
Totals	100.000	100.000	100.000	
Temperature, °F	− 80.00	− 109.83	− 109.83	
Pressure, psia	800.00	600.00	600.00	
Molecular weight	18.74	16.48	21.15	
Compressibility factor		.5835	.1450	
Density, lb-mols/ft^3		.2739	1.1023	
Molar volume, ft^3/lb-mole		3.6514	.9072	
Total flows, lb-moles/hr	7.685E + 03	3.967E + 03	3.718E + 03	
Total flows, lb/hr	1.440E + 05	6.537E + 04	7.865E + 04	
Enthalpy, Btu/lb-mole*	1288.73	1854.98	− 653.57	
Enthalpy, Btu/hr*	.990E + 07	.736E + 07	− .243E + 07	
Entropy, Btu/lb-mole-F	30.98	32.19	29.69	
ΔH, Btu/lb-mole				− 647.42
Mole percents		51.62	48.38	

* Zero enthalpy datum is ideal gas at 0°R

Table 17.12
Flash at 500°F and 15 psia of Sumatran Light Crude Oil (Top 75%) Previously Divided into 21 Pseudo Components Using *K*, *H* and *S* Values from Soave-Redlich-Kwong Method

Material Balance Basis	Mole Percent			
Component Name	Feed	Vapor	Liquid	*K*-Value
Sumatran Light Oil Comp 01	1.667	3.229	0.051	6.324E + 01
Sumatran Light Oil Comp 02	6.667	12.737	0.391	3.260E + 01
Sumatran Light Oil Comp 03	3.333	6.227	0.342	1.820E + 01
Sumatran Light Oil Comp 04	6.667	12.234	0.912	1.342E + 01
Sumatran Light Oil Comp 05	3.333	5.967	0.611	9.767E + 00
Sumatran Light Oil Comp 06	6.667	11.569	1.598	7.239E + 00
Sumatran Light Oil Comp 07	3.333	5.559	1.033	5.381E + 00
Sumatran Light Oil Comp 08	6.667	10.498	2.706	3.880E + 00
Sumatran Light Oil Comp 09	3.333	4.836	1.780	2.718E + 00
Sumatran Light Oil Comp 10	6.667	8.700	4.565	1.906E + 00
Sumatran Light Oil Comp 11	3.333	3.765	2.887	1.304E + 00
Sumatran Light Oil Comp 12	6.667	6.142	7.209	8.520E − 01
Sumatran Light Oil Comp 13	3.333	2.362	4.338	5.445E − 01
Sumatran Light Oil Comp 14	6.667	3.349	10.097	3.317E − 01
Sumatran Light Oil Comp 15	3.333	1.021	5.724	1.784E − 01
Sumatran Light Oil Comp 16	6.667	1.151	12.369	9.303E − 02
Sumatran Light Oil Comp 17	3.333	0.317	6.451	4.918E − 02
Sumatran Light Oil Comp 18	6.667	0.263	13.287	1.981E − 02
Sumatran Light Oil Comp 19	3.333	0.048	6.730	7.068E − 03
Sumatran Light Oil Comp 20	6.667	0.025	13.533	1.840E − 03
Sumatran Light Oil Comp 21	1.667	0.001	3.389	3.140E − 04
Totals	100.000	100.000	100.000	

(table continued on next page)

Table 17.2
Continued

Temperature, °F	500.00	500.00	500.00
Pressure, psia	15.00	15.00	15.00
Molecular weight	232.84	150.15	318.31
Compressibility factor		0.9730	0.0089
Density, lb-mols/ft³		0.0015	0.1637
Molar volume, ft³/lb-mole		668.0684	6.1105
Total flows, lb-moles/hr	1.214E + 03	6.171E + 02	5.970E + 02
Total flows, lb/hr	2.827E + 05	9.267E + 04	1.900E + 05
Enthalpy, Btu/lb-mole*	57621.34	46157.69	69472.41
Enthalpy, Btu/hr*	.700E + 08	.285E + 08	.415E + 08
Mole percents		50.83	49.17

* Zero enthalpy datum is ideal gas at 0°R

Table 17.13
True Boiling Point Assays of Petroleum Products from Flashing Sumatran Light Crude Oil

Boiling Point °F	Overhead		Bottoms	
	Mole %	Volume %	Mole %	Volume %
0.00	0.00	0.00	0.00	0.00
105.01	8.11	5.36	0.22	0.08
188.41	17.75	12.26	0.60	0.23
228.89	27.13	19.81	1.24	0.50
268.39	36.38	27.53	2.01	0.85
303.67	45.30	35.67	3.13	1.39
337.02	54.00	43.91	4.47	2.07
371.95	62.16	52.14	6.37	3.10
407.83	69.95	60.31	8.66	4.38
441.86	76.83	68.11	11.88	6.33
476.41	83.17	75.55	15.68	8.72
512.75	88.20	81.98	20.81	12.24
548.69	92.52	87.78	26.69	16.47
586.38	95.43	92.10	34.03	22.34
630.75	97.65	95.58	42.08	29.14
674.50	98.75	97.52	51.28	37.83
715.00	99.50	98.90	60.86	47.35
770.06	99.79	99.51	70.90	58.77
829.25	99.95	99.87	81.08	71.29
902.25	99.99	99.96	91.39	86.35
993.00	100.00	100.00	100.00	100.00

ing more than one trial for the one of most interest increases the chances of getting a converged solution.

The TBP assays of the equilibrium overhead and bottoms products from flashing the crude oil are given in Table 17.13. These are prepared from the pseudo-component compositions of the equilibrium vapor and liquid products. These assays are given in both molar and volumetric units. It should be noted that the percent-off points are given for the original assay temperatures. This results in uneven percent-off intervals for the products. Curve fitting and interpolations are necessary to find the temperature that corresponds to even percent-off intervals. This is not done in the program.

References

1. Chappelear, P. S., R. J. J. Chen, and D. G. Elliot, *Proceedings of the Fifty-Sixth Annual Convention*, Gas Processors Association, (1977).
2. Chevron Oil Trading Company, "35.3°API Sumatran Light Crude Oil," November (1967).

18

Alternate Datum States for Pure Component Enthalpies

The development and selection of the equations and coefficients for calculating enthalpy and entropy of pure components and their mixtures, including pseudo components of petroleum, was covered in Chapter 7 of Volume 1. This included the effects of temperature, pressure, phase condition, and composition. Computer subroutines for calculating entropies and enthalpies were described in Chapter 17. Now, alternate datum states at which the enthalpy are zero will be defined.

As explained in Chapter 17, enthalpies and entropies of both liquid and vapor mixtures are found in EQUIL by combining the ideal gas state values from Subroutine IDEAHS with the isothermal effects of pressure and phase change from Subroutine ENTHS. These subroutines are preceded by PHPROP for discrete components and by CUTPRO for pseudo or hypothetical components of petroleum. Computer applications of the equations and coefficients are the ultimate objectives.

Enthalpies are always values above an arbitrary datum state where the value of the enthalpy is zero. For the steam tables and Mollier Chart this zero enthalpy datum point is saturated liquid at 32°F. Such a zero H datum may be best for steam-power calculations, which are concerned with only one molecular species, but it would be awkward for hydrocarbon process engineering work, which is concerned with an almost infinite variety of components and their mixtures.

Enthalpy and Entropy Reference States

Alternate zero datum states and units are of interest for enthalpy, but not for entropy, because entropy units are standardized and the same in metric and U.S. engineering units. Entropy units (eu) are the same numerical value in g-cal/g-mol °K and in Btu/lb-mol °R. The Third Law of Thermodynamics, which states "the entropy of a perfect crystalline substance is zero at absolute zero temperature," defines the internationally accepted zero entropy reference state. Entropy values at the standard reference state of 25°C have been obtained by API Research Project 44 and are available. Determining them is beyond the scope of this book.

For the enthalpy, on the other hand, there are three alternate zero enthalpy datum states that will be considered: (1) the ideal gas state of the compound at absolute zero T; (2) the saturated liquid of the compound at −200°F; and (3) the ideal gas state of the elements at absolute zero T.

Number 1 zero enthalpy datum state is the one used in the compilations of Research Project 44 of the American Petroleum Institute. Enthalpies referred to this datum state are obtained for pure compounds in the ideal gas by using Equation 7.29 and the values of coefficients, A, B, C, D, E, and F from Table 7.5, in Volume 1. The ideal gas state entropy is found for all components by Equation 7.31 and coefficients B thru G from Table 7.5.

Number 2 zero enthalpy datum state is the one used in the *Technical Data Book* of the American Petroleum Institute. This −200°F liquid zero enthalpy datum state is arbitrary and is believed to have been chosen to avoid negative enthalpy values in the API-TDB charts, tabulations, and equations. I do not know of any technical merit in a −200°F liquid datum state. Ideal gas state enthalpies referred to this datum are also found by Equation 7.29 with constants *B* through *F* from Table 7.5 and constant *A* being modified by the addition of EE(I), which will be described in more detail later. This correction will be evaluated for the components in Table 7.5 and in BLOCK DATA of the computer program.

Number 3 zero enthalpy datum state is the ideal gas state at 0°K or 0°R for the elements, i.e., hydrogen, oxygen, nitrogen, carbon (graphite), and sulfur (rhombic). This is the zero enthalpy datum used by Starling in preparing the pressure-enthalpy charts in Chapter 8, Volume 1. Ideal gas state enthalpies relative to this datum are computed by Equation 7.29, using the same *B* through *F* constants as in Numbers 1 and 2, and an *A* constant that is modified by the addition of the heat of formation of the compound at 0°R, which will be identified by the symbol HOFZ(I) below. With enthalpies referred to this datum base, heat balances on reactors can be readily made because each enthalpy value is referred to the elements at 0°R, a universal datum state.

It is of interest to point out that Number 3 zero enthalpy datum uses the same datum conditions, i.e., the elements at absolute zero temperature, that define the zero entropy datum state. For this reason it is logical to define both enthalpy and entropy above this zero point as "absolute."

Absolute Enthalpy

By combining the enthalpy of the elements, the heat of formation, and the enthalpy of the compound, it is possible to obtain the absolute enthalpy of chemical substances at any condition, relative to some datum point. Although the datum point is arbitrary and can be any condition, i.e., temperature, pressure, and state, the most natural zero enthalpy datum, and the one used here, is the ideal gas state at absolute zero temperature.

At this point the absolute enthalpies of the elements are assumed to be zero. It is convenient to use "graphite" as the datum state for carbon and "rhomboid" as the state for sulfur. For hydrogen the zero enthalpy is for the molecular form, H_2; for nitrogen it is N_2, and for oxygen it is O_2.

Absolute enthalpies of the reactants and products of chemical reactions can be combined directly in making heat balances on reactors, without a separate evaluation of the heats of reaction and the effects of temperature and pressure at the conditions in the reactor. This is because the absolute enthalpies include the heats of formation and the effects of temperature and pressure. Examples of such absolute enthalpy charts for reacting system are shown in Chapter 21.

Theoretically, absolute enthalpy charts can be constructed for any substance, even a multicomponent mixture such as a petroleum fraction, which might be di-

vided into pseudo components, as illustrated in Chapter 15. Then the ideal gas state enthalpies of these pseudo components would include their heats of formation. The pressure and composition effects would next be calculated and combined with the ideal gas *H* values. A method of evaluating the heats of formation for the pseudo is required for this procedure.

In the past "heat content charts" would be prepared for petroleum fractions for use in some process design job. A typical chart of this kind is shown here for illustration only. Figure 18.1 is a temperature-enthalpy diagram for a cracked gasoline having the following assay:

ASTM D-86 Distillation Assay Temperatures, °F.

API Gr.	IBP	10% Pt	30% Pt	50% Pt	70% Pt	90% Pt	FBP
55.5	117	182	237	276	306	356	400

This *T-H* diagram was prepared in two steps. First, a phase diagram for the mixture was made by the method described in Chapter 12. Then enthalpies were found for several points on the phase boundary curve and also in the vapor phase. The enthalpies were relative to liquid at

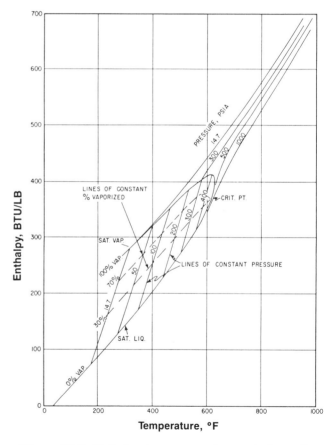

Figure 18.1. Enthalpy-temperature for cracked gasoline.

32°F, the zero enthalpy datum point that has been used for water and steam for many years. Then the enthalpy and vapor-liquid phase points were plotted, as shown.

The enthalpy scale of Figure 18.1 could be converted to an absolute enthalpy by adding the heat of formation and the enthalpy between 0°R and 0°F, but this is not recommended. A better method for computer applications is to follow the same procedure used for discrete components and their mixtures, i.e., make the same phase and enthalpy calculations for pseudo components that are made for discrete components.

Any other zero enthalpy datum state, including the "absolute enthalpy" datum, can be used for petroleum fractions in this way.

Conversion of API-44 to API-TDB Enthalpies

The additional constants needed to convert Number 1 datum enthalpies to Number 2 datum enthalpies can be calculated from the data already contained in Tables 7.1 and 7.5, which data are also in BLOCK DATA of the EQUIL program. These conversion constants could be calculated each time they are needed, if necessary. It is more convenient to make these constant calculations beforehand and add an extra data value to the original 31 entries and thereby save some calculation time in running the program. This was done.

The enthalpy-temperature diagrams in Figure 18.2 show the relationship between the API-44 and the API-TDB zero enthalpy datum states, where $H^{o'}_o = 0$ and $H^o_o = 0$ are the two reference states. The two H-T diagrams, labeled "heavier components" and "lighter components" show that there is a small difference in H values of ideal gas and saturated vapor at $-200°F$ for the lighter component, as designated by Y on the lower H-T plot in Figure 18.2. The value of Y is about 14 Btu/lb for methane, which has a normal boiling point of $-258.7°F$. Other hydrocarbons of interest have boiling points above $-200°F$ and the value of Y is insignificant.

The difference between API-44 and API-TDB enthalpies is shown as EE in Figure 18.2, from which it can be seen that

$$EE = HOVT + Y - X \qquad (18.1)$$

where HOVT = heat of vaporization at $-200°F$
X and Y are defined by Figure 18.2
and Y is negligible except for methane.

In the API-TDB the value of HOVT was found as follows:

$$HOVT = HOV \left[(T_c - 259.7)/(T_c - T_b)\right]^{0.38} \qquad (18.2)$$

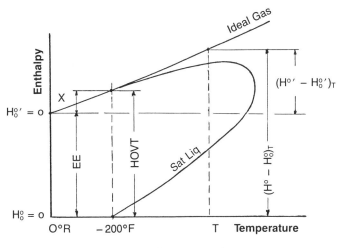

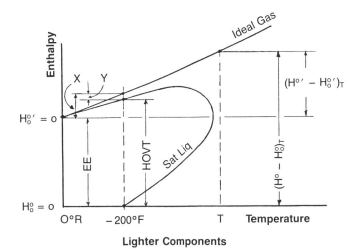

Figure 18.2. Enthalpy-temperature diagrams illustrating API-44 and API-TDB zero enthalpy datum states.

where HOV = heat of vaporization at normal atmospheric boiling point
T_c = critical temperature, °R
T_b = normal boiling point, °R
$259.7 = -200 + 459.7$

The value of the remaining term X is the ideal gas state enthalpy difference at $-200°F$, or

$$X = (H^{o'} - H^{o'}_o)_{-200} \qquad (18.3)$$

where the H values are calculated by Equation 7.29

All of the data required in evaluating EE for all of the components in BLOCK DATA are there already so values of EE could be calculated as needed. This is not convenient as having EE values available in BLOCK DATA. The alternative of combining the heat of vaporization and the ideal gas enthalpy difference, as in Equation 18.1, calculating EE and entering into BLOCK DATA has been followed for pure components.

Calculating X by Equation 18.3 presents a problem in that the ideal gas state enthalpy by Equation 7.29 is not zero at $T = 0°R$. Equation 7.29 was not intended for applications between absolute zero and $-200°F$ and would probably not give the same values of X that would be obtained by a more correct extrapolation. This problem is not considered very serious, however, as small errors would cancel out in normal heat balance applications.

A FORTRAN program is discussed later for calculating HOVT and EE values by the previous equations for discrete components and also the value of HOFZ, the heat of formation at $0°R$.

Conversion of API-44 to Absolute Enthalpies

Number 3 datum state enthalpies are "absolute enthalpies," i.e., enthalpies referred to the elements at absolute zero. API-44 enthalpies plus the heat of formation at $0°R$ is this absolute enthalpy.

Standard heats of formation at $25°C$ ($77°F$ or $536.7°R$) are given in Table 7.5 and in BLOCK DATA of the computer program. These standard heats of formation are the bases for calculating the heats of formation at absolute zero temperature. This calculation is made by Equation 7.32 (Chapter 7 of Volume 1). The enthalpy equation coefficients and the standard heats of formation required in that calculation are from Table 7.5 or BLOCK DATA.

The notation used for these items is:

HOF(I) = the standard heat of formation (at $536.7°R$) in Btu/lb

HOFZ(I) = the heat of formation at $0°R$ in Btu/lb

HCF(J,I) = coefficients for ideal gas state enthalpy equation, in which the units are Btu/lb-mol

Having the first two items on a weight basis and the third on a molar basis is for data storage convenience only. There is no technical reason and it is not an error.

As was true for the EE conversion computations, all of the data required in calculating HOFZ are present in BLOCK DATA, so that this conversion term could be evaluated each time it is needed. Here also, it seems more efficient to calculate all of the HOFZ values in advance and have them available in BLOCK DATA for use when needed.

vance and have them available in BLOCK DATA for use when needed.

Computer Program and Results for Discrete Components

A FORTRAN program named AZHDS (Table 18.1), for Alternate Zero H Datum States, was compiled and linked to create AZHDS.EXE. The program was made user-interactive for convenient use on an IBM compatible microcomputer. In the linking step, some files from EQUIL and the library of the MicroSoft FORTRAN Compiler were included, creating an executable program.

In the heat of formation calculations it is necessary to make enthalpy calculations for the elements, as well as the compounds. For this, the amounts of the elements must be found. This is done with a datum item CHONS, which appears as DS(24,I) for each compound in BLOCK DATA. Program AZHDS reads and interprets the CHONS (for Carbon, Hydrogen, Oxygen, Nitrogen, and Sulfur) and then proceeds with the calculations.

These calculations were run for over 200 components in BLOCK DATA. The results are given Tables in 18.2, 18.3, 18.4, and 18.5, each tabulation containing about 52 components. All of the HOVT(I) and HOFZ(I) values are in Btu/lb units, to be compatible with the other constants in BLOCK DATA. The use of specific, rather than molar units, minimizes data storage space.

It will be noted that some of the data required in these calculations are missing from Table 7.5. Except for one component, i.e., 531, the HCOEF blanks have been filled. All of the HOF blanks in Table 7.5 have been filled in BLOCK DATA. Component 531 is flagged in Table 18.5.

Note that components 602 (hydrogen), 603 (nitrogen), and 604 (oxygen) are not included in these tables. Alternate zero enthalpy datum states are of no interest for these gases, which are in reference state form to begin with. Note also that only the $0°R$ heat of formation is included for carbon monoxide because the elements datum can be used for CO gas but the liquid datum cannot be applied.

Zero Enthalpy Datum States for Pseudo Components of Petroleum

The previous calculations of the heat of vaporization at $-200°F$ and the heat of formation at $0°R$ are for discrete components only. A different technique is required for the pseudo components of petroleum. This subject is covered in Chapter 25.

Table 18.1
Computer Program for Calculating HOFZ, HOVT, and EE Values for Alternate Zero Enthalpy Datum States of Discrete Components

```fortran
      REAL JE(30,5)
      COMMON /PRINT/ TITLE(20),CNAME(7,30),FD(30),WM(30),BP(30),
     *   SG(30),ICOD,KHF,KYF,KTL,IKV,IHS,IHD,IOUT,NC,
     *   TI,PI,EF,TO,PO,TTOL,PTOL,XTOL,SUMF,HF,SF
      COMMON /CRIT/ ID(30),TC(30),PC(30),WC(30)
      COMMON /HSTR/ HCF(9,30),HSI(30),HOF(30)
      COMMON /HFOR/ FOF(30),CHONS(30)
      COMMON /DLIQ/ RKW(30),VST(30),HOV(30)
      COMMON /DSDATA/DS(33,216)
      DIMENSION KE(30,5),HONS(30),ONS(30),NS(30),S(30),TT(30),
     * HCR(30),HCT(30),HER(30,5),HET(30,5),DLH(30,5),SDLH(30),
     * DELH(30),HOFZ(30),HOVT(30),HV200(30),EE(30)
      OPEN(5,FILE = 'AZHDS.DAT',STATUS = 'NEW')
      OPEN(6,FILE = '       ',STATUS = 'NEW')
      WRITE(*,9)
9     FORMAT(//,' Enter number of components.')
      READ(*,*) NC
      WRITE(*,10) NC
10    FORMAT(//,' Enter 3-digit ID numbers for ', I3, 'components.'/)
      READ(*,*) (ID(I),I = 1,NC)
      WRITE(5,5) (ID(I),I = 1,NC)
      REWIND 5
      OPEN(5,FILE = 'AZHDS.DAT',STATUS = 'OLD')
      READ(5,5) (ID(I),I = 1,NC)
5     FORMAT(20I4)
C     CALCULATE HOF AT ZERO R FOR ELEMENTS AT 0 R ENTHALPY DATUM
      DO 100 I = 1,NC
        CALL PHPROP(ID(I),CNAME(1,I),WM(I),BP(I),SG(I),TC(I),PC(I),
     *   WC(I),RKW(I),VST(I),HOV(I),HOF(I),FOF(I),CHONS(I),HCF(1,I))
C     FIND IDEAL GAS STATE H AT -200 F VALUES.
        HSI(I) = HCF(1,I) + 259.7*(HCF(2,I) + 259.7*(HCF(3,I) +
     *   259.77*(HCF(4,I) + 259.7*(HCF(5,I) + 259.7*HCF(6,I)))))
        HV200(I) = HSI(I)/WM(I)
C     FIND NUMBERS OF ELEMENTS (C,H,O,N,S) FOR EACH COMPOUND
        KE(I,1) = 610
        KE(I,2) = 602
        KE(I,3) = 603
        KE(I,4) = 604
        KE(I,5) = 611
C     AMOUNTS OF ELEMENTS IN REACTANTS
        JE(I,1) = INT(CHONS(I)/1E5)
        HONS(I) = CHONS(I) - JE(I,1)*1E5
        JE(I,2) = INT(HONS(I)/1E3)
        ONS(I) = HONS(I) - JE(I,2)*1E3
        JE(I,3) = INT(ONS(I)/1E2)
        NS(I) = ONS(I) - JE(I,3)*1E2
        JE(I,4) = INT(NS(I)/1E1)
        S(I) = NS(I) - JE(I,4)*1E1
        JE(I,5) = INT(S(I)/1E0)
C     CORRECT FROM MOLAR TO ATOMIC BASIS FOR ELEMENT GASES H2, O2, & N2
        JE(I,2) = JE(I,2)/2.
        JE(I,3) = JE(I,3)/2.
        JE(I,4) = JE(I,4)/2.
        DO 101 J = 1,5
          CALL PHPROP(KE(I,J),CNAME(1,I),WM(I),BP(I),SG(I),TC(I),PC(I),
     *     WC(I),RKW(I),VST(I),HOV(I),HOF(I),CHONS(I),HCF(1,J))
101     CONTINUE
C     CALCULATE IDEAL H VALUES FOR ELEMENTS: HER(I,J) = VALUE AT 536.7,
C     HET(I,J) = VALUE AT 0.0.
        SDLH(I) = 0.0
        CALL IDEAHS(1,1,5,536.7,2)
        DO 102 J = 1,5
102       HER(I,J) = JE(I,J)*HSI(J)
        CALL IDEAHS(1,1,5,0.0,2)
        DO 103 J = 1,5
103       HET(I,J) = JE(I,J)*HSI(J)
        DO 104 J = 1,5
104       DLH(I,J) = HET(I,J) - HER(I,J)
        DO 105 J = 1,5
105       SDLH(I) = SDLH(I) + DLH(I,J)
100   CONTINUE
C     CALCULATE H VALUES FOR COMPOUNDS AND THEN FIND HOF VALUES
      DO 200 I = 1,NC
        CALL PHPROP(ID(I),CNAME(1,I),WM(I),BP(I),SG(I),TC(I),PC(I),
     *   WC(I),RKW(I),VST(I),HOV(I),HOF(I),FOF(I),CHONS(I),HCF(1,I))
        CALL IDEAHS(1,1,NC,536.7,2)
200   CONTINUE
      DO 201 I = 1,NC
201     HCR(I) = HSI(I)
        DO 203 I = 1,NC
          DELH(I) = -HCR(I)
          HOFZ(I) = HOF(I) + (DELH(I) - SDLH(I))
          HOFZ(I) = HOFZ(I)/WM(I)
C     CALCULATE HOVT AT -200 F
          TT(I) = BP(I) + 459.7
          TCBR = (TC(I) - 259.7)/(TC(I) - TT(I))
          PTEH = (TCBR)**0.38
          HOVT(I) = HOV(I)*PTEH
          HOVT(I) = HOVT(I)/WM(I)
C     FIND EE VALUES FOR COMPONENTS
          EE(I) = HOVT(I) - HV200(I)
203   CONTINUE
C     WRITE RESULTS
      WRITE(6,6)
6     FORMAT(3X,'ID',4X,'COMPONENT NAME',16X,'HOFZ VALUES',2X,'HOVT VALU
     *ES',1X,'EE VALUES',/,42X,'BTU/LB',6X,'BTU/LB,6X,'BTU/LB')
      DO 300 I = 1,NC
        WRITE(6,7) ID(I),(CNAME(J,I),J = 1,7),HOFZ(I),HOVT(I),EE(I)
7       FORMAT(3X,I3,3X,7A4,3(2X,F10.4))
300   CONTINUE
      REWIND 5
      REWIND 6
      STOP
      END
```

Table 18.2
Constants for Alternate Zero Enthalpy Datum States for Components 101 Thru 152

ID	Component Name	HOFZ Values Btu/lb	HOVT Values Btu/lb	EE Values Btu/lb
101	Methane	− 1797.7850	179.0185	50.6326
102	Ethane	− 994.6902	234.6948	163.0555
103	Propane	− 801.8870	220.3669	166.6303
104	n-Butane	− 737.6689	209.1017	157.1940
105	2-Methylpropane (i-Butane)	− 789.7183	196.9826	150.6663
106	n-Pentane	− 684.3046	201.0775	146.4271
107	2-Methylbutane (i-Pentane)	− 724.1567	190.9345	141.4765
108	2,2-Dimethylpropane	− 786.3714	173.7296	133.9878
109	n-Hexane	− 656.0507	194.2253	140.7769
110	2-Methylpentane	− 679.8553	185.6257	138.4859
111	3-Methylpentane	− 665.2068	187.1917	140.4590
112	2,2-Dimethylbutane	− 728.3377	173.1993	126.8630
113	2,3-Dimethylbutane	− 689.4377	180.6534	134.3171
114	n-Heptane	− 631.9293	188.2397	134.3196
115	2-Methylhexane	− 652.6844	181.0254	133.1792
116	3-Methylhexane	− 639.3331	181.3087	133.7909
117	3-Ethylpentane	− 616.5609	181.7198	147.1703
118	2,2-Dimethylpentane	− 687.2609	170.0169	135.4674
119	2,3-Dimethylpentane	− 657.7609	177.7038	143.1543
120	2,4-Dimethylpentane	− 669.5610	172.7336	138.1841
121	3,3-Dimethylpentane	− 667.5609	172.7820	138.2325
122	2,2,3-Trimethylbutane	− 681.6609	167.2710	132.7214
123	n-Octane	− 614.2170	183.5057	129.5475
124	2-Methylheptane	− 631.9752	179.0653	130.1669
125	3-Methylheptane	− 621.1752	179.2137	130.3152
126	4-Methylheptane	− 621.5751	179.0090	130.1105
127	3-Ethylhexane	− 614.5751	177.0953	128.1968
128	2,2-Dimethylhexane	− 660.2668	168.9810	125.6596
129	2-Methyl-3-Ethylpentane	− 609.9669	173.5551	130.2337
130	2,2,4-Trimethylpentane	− 652.2919	161.0698	122.6253
131	n-Nonane	− 599.9646	179.0060	125.0531
132	2-Methyloctane	− 623.9645	176.6465	122.6936
133	3-Methyloctane	− 614.4645	176.7762	122.8233
134	4-Methyloctane	− 614.8645	175.8201	121.8672
135	3-Ethylheptane	− 604.8646	175.9263	121.9734
136	4-Ethylheptane	− 604.8646	175.5686	121.6157
137	2,2-Dimethylheptane	− 659.8645	165.7144	111.7615
138	3,3-Diethylpentane	− 609.8645	169.3179	115.3650
139	2,2,3-Trimethylhexane	− 640.8645	163.8515	109.8987
140	2,2-Dimethyl-3-Ethylpentane	− 631.1645	163.5646	109.6117
141	2-Methyl-3-Ethylhexane	− 612.5645	171.0092	117.0563
142	2,2,3,3-Tetramethylpentane	− 627.5645	164.5828	110.6300
143	n-Decane	− 588.8028	175.1884	121.5925
144	2-Methylnonane	− 610.4028	173.7849	120.1890
145	3-Methylnonane	− 601.8029	172.8521	119.2562
146	4-Methylnonane	− 601.8029	171.3294	117.7335
147	5-Methylnonane	− 601.8029	170.8455	117.2496
148	3-Ethyloctane	− 593.1028	170.4206	116.8247
149	4-Ethyloctane	− 593.1028	168.5537	114.9578
150	2,2 Dimethyloctane	− 642.7028	165.2234	111.6275
151	4-n propylheptane	− 593.0027	167.5426	113.9467
152	4-Isopropylheptane	− 593.0027	163.6482	110.0523

Table 18.3
Constants for Alternate Zero Enthalpy Datum States for Components 153 Thru 234

ID	Component Name	HOFZ Values Btu/lb	HOVT Values Btu/lb	EE Values Btu/lb
153	2-Methyl-3-Ethylheptane	− 600.0027	164.9349	111.3390
154	2,2,3-Trimethylheptane	− 625.6028	160.7469	107.1510
155	2,2-Dimethyl-3-Ethylhexane	− 616.9028	158.5509	104.9550
156	2,4-Dimethyl-3-Isopropylpent	− 613.9028	156.8783	103.2824
157	2,2,3,3-Tetramethylhexane	− 613.6028	155.3152	101.7193
158	2,2,3,3,4-Pentamethylpentane	− 581.3029	154.2601	100.6642
159	2-Methyl-3-Isopropylhexane	− 607.0028	159.2606	105.6647
160	3,3-Diethylhexane.	− 607.5028	162.6558	109.0599
161	2,2,3-Trimethyl-3-Ethylpentane	− 597.7028	155.2952	101.6993
162	2-Methyl-3,3-Diethylpentane	− 586.5028	161.0058	107.4099
163	*n*-Undecane	− 571.4376	171.4684	120.2135
164	*n*-Dodecane	− 563.2251	168.2339	117.8644
165	*n*-Tridecane	− 556.3381	165.2890	114.9741
166	*n*-Tetradecane	− 550.4796	162.8747	112.8884
167	*n*-Pentadecane	− 545.4706	160.4290	110.5854
168	*n*-Hexadecane	− 540.8993	158.1096	108.5880
169	*n*-Heptadecane	− 536.9249	155.9618	106.6150
170	*n*-Octadecane	− 532.8928	153.8228	104.6751
201	Ethene	927.3956	224.2415	148.0782
202	Propene	356.8032	224.0650	164.7198
203	Isobutene	26.0786	212.2423	164.4464
204	1-Butene	153.6992	210.2001	155.1086
205	*cis*-2-Butene	106.4030	225.4649	166.8194
206	*trans*-2-Butene	66.5098	219.9737	166.6914
207	1-Pentene	23.7238	200.5157	145.2926
208	*cis*-2-Pentene	− 20.1762	208.4750	153.2518
209	*trans*-2-Pentene	− 42.7762	207.9256	152.7024
210	2-Methyl-1-Butene	− 70.6762	203.2895	148.0663
211	3-Methyl-1-Butene	− 25.5762	190.4108	135.1876
212	2-Methyl-2-Butene	− 108.9762	212.2257	157.0025
213	1-Hexene	− 59.6514	193.3950	140.2147
214	*cis*-2-Hexene	− 114.1514	198.5903	145.4100
215	*trans*-2-Hexene	− 122.1514	197.4596	144.2793
216	*cis*-3-Hexene	− 90.0514	195.6198	142.4395
217	*trans*-3-Hexene	− 124.9514	197.4921	144.3119
218	2,3-Dimethyl-1-Butene	− 131.5514	184.6260	131.4457
219	3,3-Dimethyl-1-Butene	− 67.1514	171.0557	117.8754
220	4-Methyl-1-Pentene	− 72.0514	182.7298	129.5495
221	2-Methyl-2-Pentene	− 152.0514	197.3368	144.1565
222	3-Methyl-*cis*-2-Pentene	− 141.8514	196.0895	142.9092
223	3-Methyl-*trans*-2-Pentene	− 146.5514	199.5637	146.3834
224	2,3-Dimethyl-2-Butene	− 149.3514	202.0166	148.8363
225	1-Heptene	− 118.6885	188.0374	135.8335
226	4,4-Dimethyl-1-Pentene	− 193.3885	165.8367	113.6329
227	2,3-Dimethyl-2-Pentene	− 232.5885	184.9808	132.7770
228	1-Octene	− 163.0716	181.9664	130.3308
229	*cis*-2-Octene	− 207.4715	184.6785	133.0429
230	*trans*-2-Octene	− 207.5715	184.3803	132.7447
231	1-Nonene	− 197.5719	178.0835	126.8170
232	2,3-Dimethyl-2-Heptene	− 333.9720	157.5613	106.2948
233	1-Decene	− 225.0797	174.2147	123.8553
234	1-Undecene	− 247.7165	170.3568	120.2090

Table 18.4
Constants for Alternate Zero Enthalpy Datum States for Components 235 Thru 414

ID	Component Name	HOFZ Values Btu/lb	HOVT Values Btu/lb	EE Values Btu/lb
235	1-Dodecene	− 266.4543	166.8270	117.3588
236	1-Tridecene	− 282.2426	164.0573	114.5979
237	1-Tetradecene	− 295.5453	161.7176	112.3950
238	1-Pentadecene	− 307.4926	158.9820	110.0683
239	1-Hexadecene	− 317.4323	157.2103	108.3019
251	Propadiene	2138.1080	239.6084	182.2303
252	1,2-Butadiene	1392.5690	242.3037	192.1895
253	1,3-Butadiene	985.9484	225.0587	175.8080
254	1,2-Pentadiene	1033.5300	224.5010	171.0814
255	1,4-Pentadiene	777.5465	200.8804	159.8647
256	2-Methyl-1,3-Butadiene	596.6074	210.3115	160.3441
257	3-Methyl-1,2-Butadiene	937.5250	220.9392	181.2932
258	1-*trans*-3-Pentadiene	612.1728	222.1740	178.4274
259	1,*cis*-3-Pentadiene	621.8300	224.6501	176.8845
260	1,5-Hexadiene	549.1282	173.9927	112.0169
261	2,3-Dimethyl-1,3-Butadiene	334.4320	185.8295	133.2789
262	4-Methyl-1,3-Pentadiene	316.6526	192.0570	125.7785
281	Ethyne	3750.2470	316.2785	243.1631
282	Propyne	2062.6790	289.5106	233.0411
283	1-Butyne	1416.5000	245.5562	199.5848
284	2-Butyne	1266.5000	266.4191	220.4478
285	1-Pentyne	1022.3970	225.2457	177.7464
286	2-Pentyne	931.6061	238.7345	192.2461
301	Cyclopentane	− 279.5378	215.3808	171.4194
302	Methylcyclopentane	− 360.9326	196.0061	150.0202
303	Ethylcyclopentane	− 376.3952	190.9329	143.2539
304	*n*-Propylcyclopentane	− 389.5995	179.5206	132.2403
305	Isopropylcyclopentane	− 423.0995	178.4167	131.1364
306	1-Methyl-*cis*-2-Ethylcyclopentane	− 410.8995	182.1431	134.8629
307	1,1,2-Trimethylcyclopentane	− 464.5995	170.2422	122.9619
308	*n*-Butylcyclopentane	− 397.0284	173.0814	125.9765
309	Isobutylcyclopentane	− 429.0284	163.3680	116.2631
310	*sec*-Butylcyclopentane	− 429.0284	166.6731	119.5682
311	*tert*-Butylcyclopentane	899.6716	160.2892	113.1843
312	*n*-Pentylcyclopentane	− 405.3681	172.2607	125.2240
313	*n*-Hexylcyclopentane	− 411.0603	169.6697	122.6742
314	*n*-Heptylcyclopentane	− 416.2581	166.7604	119.8760
315	*n*-Octylcyclopentane	− 420.4971	164.5219	117.7991
316	*n*-Nonylcyclopentane	− 424.2265	161.4176	114.1955
317	*n*-Decylcyclopentane	− 427.5414	159.0895	112.3564
318	*n*-Undecylcyclopentane	− 430.2810	156.8911	110.2180
401	Cyclohexane	− 433.3987	201.1064	163.3721
402	Methylcyclohexane	− 487.4969	182.7747	145.4641
403	Ethylcyclohexane	− 469.6726	179.7335	147.3398
404	1,1-Dimethylcyclohexane	− 501.6925	170.0741	139.5113
405	1,*cis*-2-Dimethylcyclohexane	− 469.9366	176.1181	141.9995
406	1,*trans*-2-Dimethylcyclohexane	− 501.4352	171.9337	135.3483
407	1,*cis*-3-Dimethylcyclohexane	− 519.0809	171.3461	136.9857
408	1,*trans*-3-Dimethylcyclohexane	− 488.1418	176.9015	143.9016
409	1,*cis*-4-Dimethylcyclohexane	− 487.9417	176.4552	143.4553
410	1,*trans*-4-Dimethylcyclohexane	− 518.2026	170.2037	138.5212
411	*n*-Propylcyclohexane	− 470.6907	170.4274	135.1741

Table 18.5
Constants for Alternate Zero Enthalpy Datum States for Components 415 Thru 539 and 601 thru 609

ID	Component Name	HOFZ Values Btu/lb	HOVT Values Btu/lb	EE Values Btu/lb
412	Isopropylcyclohexane	− 506.0907	165.7278	130.4744
413	*n*-Butylcyclohexane	− 468.8585	166.1546	130.9263
414	Isobutylcyclohexane	− 503.1585	163.2011	127.9728
415	*n*-Pentylcyclohexane	− 468.6948	167.1027	129.6449
416	*n*-Hexylcyclohexane	− 469.1703	163.6536	126.7655
417	*n*-Heptylcyclohexane	− 469.3580	161.0071	122.1917
418	*n*-Octylcyclohexane	− 469.5726	157.9138	119.3593
419	*n*-Nonylcyclohexane	− 469.6738	155.6566	116.9260
420	*n*-Decylcyclohexane	− 469.9485	153.1971	112.7723
501	Benzene	548.4061	220.4696	188.8632
502	Toluene	337.0445	207.2360	174.9191
503	Ethylbenzene	231.0809	196.9631	163.2414
504	*o*-Xylene	183.2639	204.2285	172.4566
505	*m*-Xylene	180.4526	202.4347	169.4674
506	*p*-Xylene	182.6618	200.2686	168.0437
507	*n*-Propylbenzene	143.8624	190.7523	154.8553
508	Isopropylbenzene	133.6102	186.3553	155.1457
509	1-Methyl-2-Ethylbenzene	107.6671	193.3422	159.1064
510	1-Methyl-3-Ethylbenzene	108.7576	193.1834	160.2036
511	1-Methyl-4-Ethylbenzene	103.4903	192.1593	158.6834
512	1,2,3-Trimethylbenzene	77.9877	200.2501	169.3461
513	1,2,4-Trimethylbenzene	62.1401	197.1619	166.7393
514	1,3,5-Trimethylbenzene	58.8085	196.8911	162.1894
515	*n*-Butylbenzene	47.5465	179.0078	141.4774
516	Isobutylbenzene	49.6842	171.0399	133.5095
517	*sec*-Butylbenzene	62.8842	169.2184	131.6880
518	*tert*-Butylbenzene	50.6842	167.2697	129.7393
519	1-Methyl-2-Isopropylbenzene	36.9842	171.5519	134.0215
520	1-Methyl-3-Isopropylbenzene	24.8842	170.1373	132.6069
521	1-Methyl-4-Isopropylbenzene	26.2542	173.1707	135.6403
522	1,4-Diethylbenzene	47.4842	180.1075	142.5771
523	1,2,4,5-Tetramethylbenzene	− 26.2158	209.1938	171.6634
524	*n*-Pentylbenzene	22.7109	173.3063	135.2892
525	*n*-Hexylbenzene	− 20.0299	168.3678	129.3477
526	*n*-Heptylbenzene	− 55.9731	165.4706	123.6537
527	*n*-Octylbenzene	− 86.7163	162.2015	121.8724
528	*n*-Nonylbenzene	− 113.2990	160.4714	120.1743
529	*n*-Decylbenzene	− 136.3744	158.1964	117.2565
530	Naphthalene	573.7560	200.3747	179.7041
531	1,2,3,4-Tetrahydronaphthalene*	205.6426	180.3659	157.9193
532	2-Methylnaphthalene	428.3891	196.8769	173.7241
533	Ethenylbenzene	692.1084	203.5844	172.6197
534	Isopropenylbenzene	503.0317	190.7964	160.2436
535	*cis*-1-Propenylbenzene	533.5314	189.1777	158.6246
536	*trans*-1-Propenylbenzene	519.6667	192.5712	164.3684
537	1-Methyl-2-Ethenylbenzene	522.9294	192.3503	161.7970
538	1-Methyl-3-Ethenylbenzene	515.6141	188.2010	161.0323
539	1-Methyl-4-Ethenylbenzene	509.2309	189.7429	159.1899
601	Water	− 5706.3900	1222.8080	1108.9860
605	Carbon Monoxide	− 1749.6520	0.0000	0.0000
606	Carbon Dioxide	− 3834.4380	193.6334	152.7658
607	Hydrogen Sulfide	− 257.5788	269.6522	209.2976
608	Sulfur Dioxide	− 1996.8640	205.0710	172.6926
609	Ammonia	− 979.1828	700.8979	580.1900

* Error because of no HCF values in data bank

19

Gas Compression and Expansion Calculations

This chapter covers applications of thermodynamics to solving compression and expansion problems in gas processing. These applications involve pressures, temperatures, flow rates, heat and work duties, and liquefaction; not hardware design, construction, efficiency, or operation.

Gas compression and expansion machines may be of reciprocating or centrifugal design; the smaller units are often the piston type while the larger units are usually the turbo type. The thermodynamics of compression and expansion are essentially the same. Computer programs are available for making these thermodynamic calculations. These are subroutines of EQUIL, which was described in Chapters 16 and 17. Some of the background technology of the compression and expansion processes will be reviewed now.

H-S or *P-H* Charts as Design Tools

A well established procedure for steam power and ammonia refrigeration calculations is to use *H-S* (Mollier) or *P-H* charts for predicting the isentropic, i.e., reversible and adiabatic, heat and work quantities and using these as the basis for design. Such charts have been available for steam, ammonia, carbon dioxide, sulfur dioxide, and some of the hydrocarbons and Freons. Similar charts can be prepared for gas mixtures but this procedure is not a normal practice because gas mixtures vary so widely in composition.

These chart-aided calculations are illustrated by Figure 19.1, where two isotherms, T_1 and T_2, are shown on

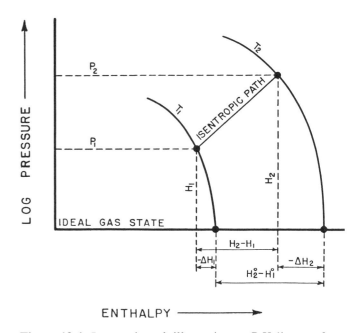

Figure 19.1. Isentropic path illustration on *P-H* diagram from Mollier chart.

log P versus H coordinates. An isentropic path connecting points 1 and 2 is shown, as are the enthalpy differences that are computed in the process of evaluating $H_2 - H_1$ for the process. The isentropic path shown on Figure 19.1 could be for a compression process or for an expansion process.

An adiabatic and irreversible, i.e., polytropic, compression path starting at point 1 and going to the same

106

final pressure of P_2 would be flatter so that the value of $H_2 - H_1$ would be larger than the isentropic ΔH. Likewise, a polytropic expansion path starting at point 2 and going to the final pressure of P_1 would be steeper so that the value of $H_1 - H_2$ would be smaller than the isentropic ΔH.

Definitions

The following terms, which will be used in this chapter, are defined at the beginning to help clarify the discussion. Some of these terms have already been defined on page 4 of Volume 1.

Isenthalpic Path. In an isenthalpic path, temperature and pressure change at constant enthalpy; other properties such as molar volume and entropy will change also. The process is adiabatic and irreversible and is not relevant to a discussion of compression and expansion calculations.

Isentropic Path. In an isentropic path, temperature and pressure change at constant entropy; other properties such as molar volume and enthalpy will change also. The process is reversible and adiabatic.

Isentropic Exponents. An exponent n', when available, approximates temperature change with pressure at constant entropy, and an exponent n, when available, approximates volume change with pressure at constant entropy.

Reversible Process. The classical definition of a reversible process is: an ideal process in which the work done as it proceeds forward equals the work done as it proceeds backward along an identical path.

Polytropic Process. A polytropic path differs from an isentropic path due to heat addition or removal and/or internal molecular friction, i.e., irreversibility. Final temperatures may be greater or less than the final isentropic value in compression and expansion. There is no thermodynamic way to find which or how much.

Adiabatic Processes. As the name indicates, an adiabatic process is one with no heat addition or removal from the system. An adiabatic process may be reversible or irreversible. If reversible, it follows a constant entropy path. The departure of an adiabatic process from an isentropic path is expressed in terms of an efficiency.

Adiabatic Efficiencies. Adiabatic efficiencies will be defined here as the less than unity fractions that are to be applied to the reversible and adiabatic (isentropic) enthalpy differences to calculate the actual enthalpy differences for polytropic compression and expansion processes.

Adiabatic Compression Efficiency. Adiabatic efficiency is defined for compression as the ratio of the enthalpy difference for the equivalent isentropic process to the enthalpy difference for the actual polytropic compression process. Mathematically this is

$$\epsilon_c = (\Delta H_s / \Delta H_p)_a$$

where ϵ_c = adiabatic compression efficiency
   ΔH_s = enthalpy difference for the isentropic process
   ΔH_p = enthalpy difference for the polytropic process

If the compression is reversible, as well as adiabatic, efficiency $\epsilon_c = 1.0$.

Adiabatic Expansion Efficiency. Adiabatic efficiency is defined for expansion as the ratio of the enthalpy change for the actual polytropic process to the enthalpy change for the equivalent isentropic process. Mathematically, this is

$$\epsilon_e = (\Delta H_p / \Delta H_s)_a$$

where ϵ_e = adiabatic expansion efficiency
   ΔH_p = enthalpy difference for the polytropic process
   ΔH_s = enthalpy difference for the equivalent isentropic process

If the expansion is reversible, as well as adiabatic, efficiency $\epsilon_e = 1.0$.

Volumetric Efficiency. Volumetric efficiency of a reciprocating gas compressor is defined as the ratio of the volume of the gas intake to the piston displacement. Thus

$$\epsilon_v = (V_d + V_c - V_e)/V_d$$

where the numerator is the net volume of the intake gas for one stroke of the piston, and V_d, in the denominator, is the displacement volume. V_c is the clearance volume, and V_e is the volume to which the clearance gases expand on the intake stroke.

Lost Work. Defined as LW, this is the difference between the isentropic and the corresponding polytropic paths, if both are adiabatic. When Q is not zero, the difference between isentropic and polytropic is QW

where $QW = Q + LW$

For compression, QW is the extra work required for the process due to irreversibility. For expansion, QW is work lost by the process due to irreversibility.

Isentropic and Polytropic Path Calculations

Although the compression and expansion processes are not reversible, thermodynamic reversibility is an important step in making the process calculations. There are two ways of making these isentropic predictions for isentropic paths, namely the "exponent" and the "entropy/enthalpy" methods. These methods will be described after deriving the isentropic exponent relationships.

Isentropic Exponent

An exponent that predicts changes of T and V with P for an ideal gas is first derived by starting with the following equation from page 6 of Volume 1:

$$dH = TdS + VdP \tag{2.9}$$

For a reversible and adiabatic process, i.e., constant entropy path:

$$dH = VdP \tag{19.1}$$

For an ideal gas

$$dH = C_P dT$$

Combining with Equation 19.1 gives

$$VdP = C_P dT \tag{19.2}$$

For an ideal gas

$$PV = RT$$

Combining with Equation 19.2 gives

$$C_P dT = (RT/P)dP, \text{ or } dT/T = R/C_P \, dP/P \tag{19.3}$$

For an ideal gas

$$C_P - C_V = R, \text{ and } k = C_P/C_V \text{ by definition}$$

Combining these ideal gas expressions with Equation 19.3 gives

$$dT/T = \{(k-1)/k\} \, dP/P \tag{19.4}$$

Integrating Equation 19.4 between point 1 and point 2 gives

$$T_2/T_1 = (P_2/P_1)^{(k-1)/k}, \text{ or } TP^{(1-k)/k} = \text{constant} \tag{19.5}$$

For an ideal gas

$$T_2/T_1 = P_1 V_1/P_2 V_2$$

Combining with Equation 19.5 gives

$$P_1 V_1^k = P_2 V_2^k, \text{ or } PV^k = \text{constant} \tag{19.6}$$

Equations 19.5 and 19.6 are expressions for the isentropic changes of temperature and volume with temperature for an ideal gas. Similar expressions have been written by inspection for real gases in terms of exponents that are defined as n and n' by the following expressions:

$$TP^{(1-n')/n'} = \text{constant} \tag{19.7}$$

$$PV^n = \text{constant} \tag{19.8}$$

Using different symbols than k, the heat capacity ratio for ideal gas, for real gases in Equations 19.7 and 19.8 provides for the possibility that expressions might be useable with different exponents for real gases.

Equations 19.7 and 19.8 have been derived by analogy from similar ideal gas expressions. Their validity for real gases will be tested by using isentropic paths on Mollier, or P-H Diagrams. Figure 19.2 shows the paths chosen to demonstrate this testing, which included isenthalpic paths as well as isentropic paths so as to determine if an exponent of this type might be applied to polytropic paths as well as isentropic paths.

Figure 19.2 is a log-log plot of pressure versus volume for isentropic and isenthalpic paths for steam from Keenan and Keys (6) and for ethane from Barkeley et al. (2). Referring to Figure 19.2, note that straight lines have been drawn through the points for both isentropic and isenthalpic paths. Straight lines on these coordinates are represented by Equation 19.8, and values of n are tabulated in the box on Figure 19.2. Values of the temperature change exponent n' are also shown for comparison.

Note that the temperature change exponents are appreciably larger than the corresponding volume change exponents. Also, the isentropic exponents are appreciably larger than the corresponding isenthalpic exponents.

With isenthalpic and isentropic PV data forming straight lines on these scales for both steam and ethane, it appears reasonable to postulate straight lines for intermediate polytropic paths, and also for similar paths of

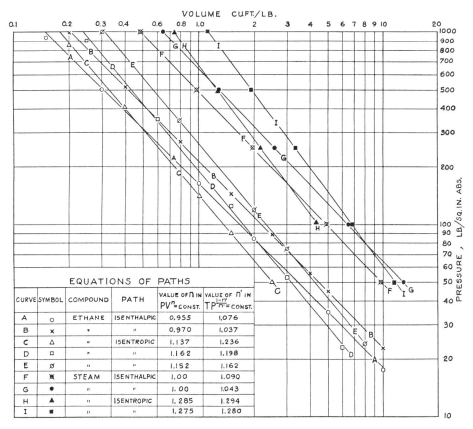

Figure 19.2. Plots of data for isentropic and isenthalpic paths for steam and ethane.

other substances. Equations 19.7 and 19.8 may be considered applicable over limited ranges but not generally applicable for the entire range of pressure covered by the *P-H* diagrams. It appears that the normal pressure range of compressors and expanders are well within the limits of validity of Equations 19.7 and 19.8, when using properly evaluated exponents.

At constant entropy, Equation 2.9 becomes $dH = VdP$, which indicates that changes in enthalpy may also be evaluated by integrating VdP, which gives for the isentropic process:

$$\Delta H = \int_1^2 VdP \tag{19.9}$$

With a plot or an equation giving V as a function of P for the gas being compressed or expanded, it is a simple matter to solve Equation 19.9 for the enthalpy difference between point 1 and point 2. This is the basis of the exponent method.

This exponent method has been applied to estimating isentropic changes for gases for which no Mollier Chart or thermodynamic properties tables were available. The calculation procedure is based on exponents that express the changes in temperature (or volume) with changes in

pressure. Although this exponent relationship has long been the basis for the manual method of compression calculations, it turns out that these exponents also have applications to compressible fluid flow calculations, as will be shown in Chapter 20.

Heat and Work Values

Gases may follow different paths in being compressed or expanded from one state to another. A look at some of the possibilities is of interest. Figure 19.3 is a section of a *P-H* diagram for ethane with five compression paths shown thereon. Figure 19.4 is a similar graph showing five expansion paths. These paths are labeled *A, B, C, D,* and *E* for identification in comparing information that has been read and calculated for the alternate paths and tabulated on the graphs. Note that the isentropic paths are the same on both graphs. With the isenthalpic path as the lower limit and the isentropic path as a guide, or reference line, three polytropic paths were drawn so that two were between the isenthalpic and isentropic paths and one was beyond the isentropic path.

For each path shown in Figures 19.3 and 19.4, values of ΔH, Q, and $\int VdP$ have been computed and tabulated on the diagrams. These calculations are made as follows:

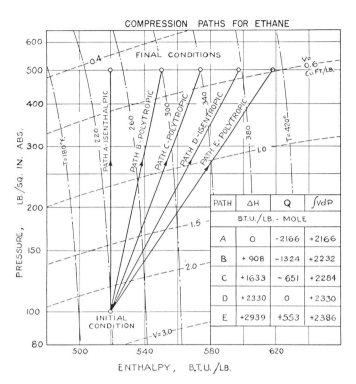

Figure 19.3. Section of ethane *P-H* diagram illustrating five compression paths.

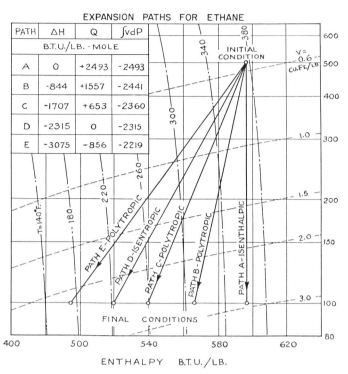

Figure 19.4. Section of ethane *P-H* diagram illustrating five expansion paths.

Step 1. ΔH is read from the *P-H* diagram.

Step 2. $\int VdP$ is calculated by a direct integration of VdP along a PV^n = constant path, using the exponent n from Equation 19.8. Mathematically, this integral is

$$\int VdP = \{n/(n-1)\}(\Delta PV)$$
$$= \{n/(n-1)\}P_1V_1[(P_2/P_1)^{(n-1)/n} - 1] \quad (19.10)$$

Step 3. Q is found by difference, i.e.,

$$Q = \Delta H - \int VdP$$

It is of interest to note the wide variations in the values of ΔH and Q for the five paths and the near constancy of the $\int VdP$ values. From this it can be seen that serious errors would result from taking ΔH as the work for a polytropic path that is far removed from the isentropic path.

From the information on Figures 19.3 and 19.4, it is evident that Mollier chart readings alone are not adequate for calculating the polytropic work of compression or expansion. Adiabatic efficiencies are also necessary. An important observation is that Paths *D* and *E* of Figure 19.3, i.e., isentropic and polytropic, are the paths of interest in compression calculations, while Paths *C* and *D* of Figure 19.4, i.e., isentropic and polytropic, are the ones of interest in expansion calculations.

Paths *C*, *D*, and *E* were next analyzed in a similar manner for nine substances, for which suitable compilations of thermodynamic properties were available in the technical literature. Terminal conditions were chosen and *P-H* chart readings are for two compression paths and two expansion paths for each substance. For each initial point, there were two final points, one terminating the isentropic path and one terminating the polytropic path. These conditions and the chart readings are given in Table 19.1 for compression and in Table 19.2 for expansion.

Using these terminal conditions, the temperature and volume change exponents were computed for all paths. These calculations are given in Table 19.3 for compression and in Table 19.4 for expansion. Note that in Table 19.3 the exponents for polytropic Paths *E* are greater than those for the corresponding isentropic Paths *D*. Likewise, in Table 19.4 the exponents for polytropic Paths *C* are less than those for the corresponding isentropic Paths *D*.

Next, the heat, work and adiabatic efficiencies were calculated for these sample compression and expansion paths for these nine substances. These results are given in Table 19.5 for the compression paths and in Table 19.6 for the expansion paths. ΔH values were calculated from the enthalpy chart readings in Tables 19.1 and 19.2. The integral of VdP was calculated by Equation 19.10, using

(*text continued on page 114*)

Table 19.1

Terminal Conditions for Compression Paths from *P-H* Diagrams

Gas & Ref.	Pres. psia		Temperature, °F	Final		Volume, CF/lb	Final		Enthalpy, Btu/lb	Final	
	Init.	Final	Init.	D	E	Init.	D	E	Init.	D	E
H$_2$O	500	2000	570	955	1000	1.1122	0.3758	0.3935	1279.5	1445.0	1474.5
(6)	150	1300	400	955	1100	3.2230	0.6023	0.6816	1219.4	1468.4	1553.9
NH$_3$	28	100	150	347	390	13.56	5.00	5.25	693.7	798.0	823.0
(1)	50	300	50	305	360	6.20	1.52	1.639	636.0	761.5	795.3
SO$_2$	200	1000	180	440	500	0.464	0.1190	0.1345	222.0	244.2	267.1
(8)	60	300	200	446	500	1.78	0.475	0.509	231.2	269.5	279.7
CH$_4$	80	400	−170	−20	20	2.252	0.664	0.744	276.1	340.0	363.9
(7)	80	250	−70	20	57	3.167	1.334	1.402	329.3	390.0	403.4
C$_2$H$_4$	294	882	20	154	180	0.503	0.200	0.219	242.5	276.0	291.2
(10)	147	735	100	290	340	1.382	0.355	0.389	286.5	355.5	382.4
C$_2$H$_6$	100	800	0	209	230	1.484	0.2324	0.2478	418.4	482.5	496.3
(2)	100	500	220	377	410	2.366	0.5637	0.5910	520.2	597.7	617.9
CH$_3$Cl	100	260	171	297	340	1.246	0.557	0.601	225.6	248.7	261.4
(9)	26	200	89	344	440	4.353	0.802	0.918	213.0	264.5	264.5
CCl$_2$F$_2$	36	200	100	230	260	1.332	0.272	0.290	92.2	108.6	114.2
(3)	50	220	231	350	380	1.202	0.309	0.325	112.8	130.4	136.2
CHClF$_2$	55	240	60	202	240	1.092	0.2936	0.3196	113.3	131.1	138.3
(5)	50	260	200	366	400	1.578	0.3666	0.3858	136.1	162.6	169.6

Table 19.2

Terminal Conditions for Expansion Paths from *P-H* Diagrams

Gas & Ref.	Pres. psia		Temperature, °F	Final		Volume, CF/lb	Final		Enthalpy, Btu/lb	Final	
	Init.	Final	Init.	C	D	Init.	C	D	Init.	C	D
H$_2$O	2000	500	955	660	570	0.3758	1.2478	1.1220	1445.0	1334.2	1279.5
(6)	1300	150	955	540	400	0.6023	3.8560	3.2230	1468.4	1295.0	1219.4
NH$_3$	100	28	347	220	150	0.1190	0.531	0.464	253.9	233.5	222.0
(1)	300	50	305	120	50	0.4750	1.980	1.780	269.5	243.5	231.2
SO$_2$	1000	200	440	240	180	0.1190	0.5310	0.4640	253.9	233.5	222.0
(8)	300	60	446	270	200	0.4750	1.980	1.780	269.5	243.5	231.2
CH$_4$	400	80	−20	−130	−170	0.664	2.663	2.252	340.0	297.6	276.1
(7)	250	80	57	−30	−70	1.334	3.521	3.167	390.0	350.1	329.3
C$_2$H$_4$	882	294	154	40	20	0.200	0.545	0.503	276.0	252.1	242.5
(10)	735	147	290	160	100	0.355	1.556	1.382	355.5	311.6	286.5
C$_2$H$_6$	800	100	200	50	0	0.2324	1.700	1.484	482.5	440.3	418.4
(2)	500	100	377	260	220	0.5637	2.517	2.366	597.7	540.9	520.2
CH$_3$Cl	260	100	297	200	171	0.557	1.318	1.246	248.7	232.5	224.6
(9)	200	26	344	160	89	0.802	4.969	4.353	264.5	227.9	213.0
CCl$_2$F$_2$	200	36	230	130	100	0.272	1.412	1.332	108.6	96.8	92.2
(3)	220	50	350	270	231	0.309	1.277	1.202	130.4	119.4	112.8
CHClF$_2$	240	55	202	100	60	0.2936	1.1902	1.0921	131.1	119.5	113.3
(5)	260	50	366	250	200	0.3666	1.7070	1.5780	162.6	144.8	136.1

Table 19.3
Calculations of Exponents for Compression Paths

Gas Ref.	Press Ratio	Temp. Ratio		Vol. Ratio		Temp. Exp. n′		Vol. Exp. n	
		D	E	D	E	D	E	D	E
H_2O	4.00	1.372	1.418	2.960	2.825	1.295	1.336	1.278	1.336
(6)	8.67	1.642	1.814	5.350	4.730	1.298	1.380	1.288	1.390
NH_3	3.57	1.323	1.394	2.712	2.583	1.282	1.354	1.278	1.340
(1)	6.00	1.500	1.607	4.079	3.783	1.293	1.360	1.274	1.346
SO_2	5.00	1.406	1.500	3.900	3.450	1.268	1.336	1.183	1.300
(8)	5.00	1.373	1.455	3.727	3.476	1.245	1.304	1.227	1.292
CH_4	5.00	1.518	1.655	3.390	3.025	1.350	1.455	1.320	1.455
(7)	3.12	1.326	1.385	2.375	2.260	1.330	1.402	1.315	1.395
C_2H_4	3.00	1.278	1.332	2.515	2.296	1.287	1.354	1.190	1.321
(10)	5.00	1.292	1.378	3.890	3.550	1.189	1.249	1.185	1.271
C_2H_6	8.00	1.455	1.500	6.385	5.990	1.221	1.243	1.122	1.162
(2)	5.00	1.231	1.279	4.198	4.002	1.148	1.181	1.122	1.160
CH_3Cl	2.60	1.200	1.268	2.238	2.073	1.236	1.330	1.186	1.311
(9)	7.70	1.465	1.640	5.300	4.634	1.227	1.320	1.224	1.330
CCl_2F_2	5.56	1.232	1.286	4.898	4.591	1.139	1.172	1.080	1.125
(3)	4.40	1.173	1.215	3.890	3.700	1.120	1.151	1.091	1.133
$CHClF_2$	4.37	1.273	1.346	3.720	3.419	1.196	1.252	1.122	1.199
(5)	5.20	1.251	1.303	4.402	4.090	1.156	1.191	1.112	1.171

Table 19.4
Calculations of Exponents for Expansion Paths

Gas Ref.	Press Ratio	Temp. Ratio		Vol. Ratio		Temp. Exp. n′		Vol. Exp. n	
		C	D	C	D	C	D	C	D
H_2O	0.25	0.791	0.728	0.301	0.338	1.240	1.297	1.155	1.278
(6)	0.12	0.707	0.608	0.156	0.187	1.191	1.299	1.166	1.289
NH_3	0.28	0.843	0.756	0.329	0.369	1.155	1.282	1.143	1.277
(1)	0.17	0.758	0.667	0.214	0.245	1.184	1.294	1.161	1.273
SO_2	0.20	0.778	0.711	0.224	0.257	1.185	1.269	1.075	1.183
(8)	0.20	0.806	0.729	0.240	0.267	1.186	1.245	1.128	1.220
CH_4	0.20	0.750	0.659	0.252	0.295	1.218	1.350	1.169	1.319
(7)	0.32	0.832	0.754	0.379	0.421	1.192	1.330	1.173	1.316
C_2H_4	0.33	0.814	0.782	0.367	0.398	1.230	1.288	1.008	1.191
(10)	0.20	0.826	0.773	0.228	0.257	1.135	1.191	1.090	1.185
C_2H_6	0.13	0.763	0.688	0.137	0.157	1.150	1.220	1.046	1.122
(2)	0.20	0.860	0.813	0.224	0.238	1.103	1.149	1.076	1.123
CH_3Cl	0.38	0.872	0.834	0.423	0.447	1.167	1.235	1.110	1.188
(9)	0.13	0.771	0.683	0.162	0.184	1.146	1.230	1.119	1.208
CCl_2F_2	0.18	0.855	0.812	0.193	0.204	1.101	1.139	1.041	1.080
(3)	0.23	0.902	0.853	0.242	0.257	1.075	1.160	1.045	1.091
$CHClF_2$	0.23	0.846	0.786	0.247	0.269	1.128	1.195	1.054	1.121
(5)	0.19	0.866	0.799	0.215	0.232	1.095	1.151	1.071	1.130

Table 19.5
Heat, Work, and Efficiency Calculations for Compression Paths

Gas & Ref.	Pres. psia		Temperature, °F			Isentropic Path D	Polytropic Path E			$\epsilon_c = \Delta H_D/\Delta H_E$
	Init.	Fin.	Init.	Final D	Final E	ΔH	ΔH	$\int VdP$	Q	
H_2O	500	2000	570	955	1000	+ 2980	+ 3510	+ 3038	+ 472	0.849
(6)	150	1300	400	955	1100	+ 4490	+ 6030	+ 4806	+ 1224	0.745
NH_3	28	100	150	347	390	+ 1775	+ 2200	+ 1824	+ 377	0.807
(1)	50	300	50	305	360	+ 2138	+ 2711	+ 2243	+ 468	0.752
SO_2	200	1000	180	440	500	+ 2044	+ 2890	+ 2154	+ 736	0.707
(8)	60	300	200	446	500	+ 2453	+ 3109	+ 2528	+ 581	0.789
CH_4	80	400	− 170	− 20	20	+ 1025	+ 1505	+ 1073	+ 332	0.730
(7)	80	250	− 70	20	57	+ 925	+ 1185	+ 1095	+ 90	0.821
C_2H_4	294	882	20	154	180	+ 940	+ 1366	+ 961	+ 405	0.688
(10)	147	735	100	290	340	+ 1936	+ 2690	+ 2010	+ 680	0.720
C_2H_6	100	800	0	209	230	+ 1926	+ 2342	+ 1952	+ 390	0.822
(2)	100	500	220	377	410	+ 2330	+ 2939	+ 2380	+ 559	0.793
CH_3Cl	100	260	171	297	340	+ 1167	+ 1805	+ 1195	+ 610	0.647
(9)	26	200	89	344	440	+ 2602	+ 3916	+ 2770	+ 1146	0.664
CCl_2F_2	36	200	100	230	260	+ 1979	+ 2641	+ 2018	+ 623	0.749
(3)	50	220	231	350	380	+ 2131	+ 2831	+ 2163	+ 668	0.717
$CHClF_2$	55	240	60	202	240	+ 1545	+ 2166	+ 1583	+ 583	0.713
(5)	50	260	200	366	400	+ 2206	+ 2900	+ 2343	+ 557	0.792

Table 19.6
Heat, Work, and Efficiency Calculations for Expansion Paths

Gas & Ref.	Pres. psia		Temperature, °F			Isentropic Path D	Polytropic Path C			$\epsilon_c = \Delta H_C/\Delta H_D$
	Init.	Fin.	Init.	Final D	Final C	ΔH	ΔH	$\int VdP$	Q	
H_2O	2000	500	955	660	570	− 2980	− 1996	− 3086	+ 1090	0.670
(6)	1300	150	955	540	400	− 4490	− 3123	− 4791	+ 1668	0.696
NH_3	100	28	347	220	150	− 1775	− 1141	− 1848	+ 707	0.642
(1)	300	50	305	120	50	− 2138	− 1474	− 2261	+ 787	0.689
SO_2	1000	200	440	240	180	− 2044	− 1318	− 2097	+ 779	0.645
(8)	300	60	446	270	200	− 2453	− 1666	− 2543	+ 877	0.679
CH_4	400	80	− 20	− 130	− 170	− 1025	− 678	− 1169	+ 491	0.661
(7)	250	80	57	− 30	− 70	− 973	− 638	− 1021	+ 393	0.656
C_2H_4	882	294	154	40	20	− 940	− 671	− 950	+ 279	0.714
(10)	735	147	290	160	100	− 1936	− 1232	− 2000	+ 768	0.636
C_2H_6	800	100	200	50	0	− 1926	− 1268	− 1987	+ 719	0.658
(2)	500	100	377	260	220	− 2330	− 1707	− 2392	+ 685	0.733
CH_3Cl	260	100	297	200	171	− 1167	− 818	− 1188	+ 370	0.701
(9)	200	26	344	160	89	− 2602	− 1851	− 2465	+ 614	0.711
CCl_2F_2	200	36	230	130	100	− 1919	− 1423	− 1998	+ 575	0.719
(3)	220	50	350	270	231	− 2131	− 1328	− 2179	+ 851	0.623
$CHClF_2$	240	55	202	100	60	− 1545	− 1007	− 1581	+ 574	0.652
(5)	260	50	366	250	200	− 2296	− 1542	− 2367	+ 825	0.672

P and V values from Tables 19.1 and 19.2 and the exponents from Tables 19.3 and 19.4.

From these ΔH and $\int VdP$ values, the values of Q are found by difference. As pointed out previously, this difference may be regarded as the value of $QW = Q + LW$, if there is enough information to break QW into these two components. Calling the difference Q is equivalent to assuming that $LW = 0$ and that the process is reversible. Calling the difference QW is equivalent to recognizing that the process is irreversible, even though the value of LW is not known.

Lastly, the adiabatic efficiencies are computed by the indicated ratios of enthalpy differences. These efficiency values are less than unity and can be considered as reflecting an adiabatic departure from reversibility, or a reversible departure from an adiabatic process, or a combination of these two departures.

It is hoped the information given in Tables 19.1 through 19.6 on the thermodynamic behavior of pure components in compression and expansion processes are of guidance. For example, one might use the exponent values given there as a means of approximating an exponent for a similar gas, such as a mixture, when making rough approximations.

A Computer Calculation Procedure

A more practical procedure for establishing isentropes for gaseous mixtures is via a series of iterative entropy calculations to determine the final temperature, followed by enthalpy calculations to determine the enthalpy difference for the isentropic process. Then the known, or assumed, adiabatic efficiency is applied to isentropic enthalpy difference to find the polytropic enthalpy difference and the final enthalpy of the polytropic path. In other words, the polytropic path ΔH value is found by dividing ΔH_s by ϵ_c for compression and multiplying ΔH_s by ϵ_e for expansion. The final temperature of the polytropic path can then be found by an iterative enthalpy balance calculation.

These calculations are made for gas mixtures by the computer programs that are included in EQUIL, in which ICOD = 10 for expansion and ICOD = 11 for compression. These programs are described in Chapters 16 and 17 and are the calculation procedures that are recommended and demonstrated in the following solutions.

Sample computer outputs are given for a gas expander problem in Tables 19.7 and 19.8, and for a compressor problem in Tables 19.9 and 19.10. The source of these problems is the typical turbo-expander gas processing plant used by Chappelear et. al. (11) in their 1977 evaluation of thermophysical property correlations. The gas compositions and conditions used in these computer runs are those for the Plant A of the referenced source.

For both expansion and compression examples the isentropic and the polytropic results are tabulated sepa-

rately, Tables 19.7 and 19.9 containing the isentropic results and Tables 19.8 and 19.10 containing the polytropic results. It will be noted that expanded effluents were part liquid, i.e., 18.98% for the isentropic path and 15.71% for the polytropic path. For compression the effluents were all vapor for both paths.

Enthalpy decrease during expansion and enthalpy increase during compression, both in Btu/lb mol, are shown for both isentropic and polytropic paths. These ΔH values may be considered as the theoretical work values for the processes, i.e., work output by the expansion and work input by the compressor.

Note that the isentropic paths, i.e., Case 1 for expansion and Case 1 for compression, were not necessary preliminaries for the polytropic path calculations. The Case 2 calculations were run independently of the Case 1 calculations. Both were run and are shown here for the benefit of the reader of this book and the user of the software being described.

A Manual Calculation Procedure

Equation 19.10 gave the theoretical work of compression, or expansion, as a function of compression ratio, the exponent n, and the initial conditions. Equation 19.10 is valid as long as $PV^n = $ constant defines the path, regardless of the value of the exponent n.

Figure 19.5 is a graphical solution of this equation for compression and expansion. As shown on the chart

$$\int_1^2 VdP = Z_1 T_1 F_w \tag{19.11}$$

$$F_w = R \left\{ n/(n-1) \right\} [(P_2/P_1)^{(n-1)/n} - 1] \tag{19.12}$$

where Z_1 = compressibility factor of gas at inlet conditions
T_1 = inlet temperature of the gas, °R
R = gas constant

The work factor F_w is given as a function of compression, or expansion ratio and exponent in Figure 19.5. Separate plots are given for expansion and compression. The units of work found this way are Btu/lbmole of gas. Division by 2,546 converts to horsepower/lbmole of gas. This work will be an indicator, or theoretical work, to which a mechanical efficiency must be applied to get brake horsepower.

Reciprocating Compressors

Volumetric efficiency and multistage compression are two items that are of interest in sizing reciprocating com-

(text continued on page 118)

Table 19.7
Case 1 of Computer Thermodynamic Calculations and Results for Gas Plant Expander—Isentropic (100% Efficiency)

Isentropic Expansion Calculations by Methods of
Soave Redlich-Kwong for K, H, and S Predictions

Material Balance Basis Component Name	Mole Percent			K-Value
	Feed	Vapor	Liquid	
Nitrogen	0.473	0.567	0.074	7.680E+00
Carbon dioxide	0.446	0.239	1.333	1.791E−01
Methane	95.386	98.654	81.436	1.211E+00
Ethane	2.953	0.532	13.287	4.004E−02
Propane	0.550	0.009	2.860	2.974E−03
2-Methylpropane (*I*-butane)	0.115	0.000	.605	4.422E−04
N-butane	0.051	0.000	.270	2.228E−04
2-Methylbutane (*I*-pentane)	0.013	0.000	.068	2.999E−05
N-pentane	0.013	0.000	.068	1.582E−05
Totals	100.000	100.000	100.000	
Temperature, °F	−80.00	−169.20	−169.20	
Pressure, psia	800.00	200.00	200.00	
Molecular weight	16.88	16.25	19.53	
Compressibility factor		0.8079	0.0485	
Density, lb-mols/cu-ft		0.0794	1.3225	
Molar volume, cu-ft/lb-mole		12.5930	0.7561	
Total flows, lb-moles/hour	3.260E+02	2.641E+02	6.187E+01	
Total flows, pounds/hour	5.501E+03	4.293E+03	1.209E+03	
Enthalpy, Btu/lb-mole*	1886.47	1994.91	−1483.23	
Enthalpy, Btu/hour*	.615E+06	.527E+06	−.918E+05	
Entropy, Btu/lb-mole-°F	32.38	33.88	25.99	
ΔH, Btu/lb-mole				−551.72
Mole percents		81.02	18.98	

* Zero enthalpy datum is ideal gas at 0°R.

Table 19.8
Case 2 of Computer Thermodynamic Calculations and Results for Gas Plant Expander—Polytropic (80% Efficiency)

Polytropic Expansion Calculation by Methods of
Soave Redlich-Kwong for K, H, and S Predictions

Material Balance Basis Component Name	Mole Percent			K-Value
	Feed	Vapor	Liquid	
Nitrogen	0.473	0.549	0.067	8.180E+00
Carbon dioxide	0.446	0.263	1.429	1.844E−01
Methane	95.386	98.547	78.421	1.257E+00
Ethane	2.953	0.630	15.418	4.088E−02
Propane	0.550	0.010	3.445	3.024E−03
2-Methylpropane (*I*-butane)	0.115	0.000	0.731	4.511E−04
N-butane	0.051	0.000	0.326	2.257E−04
2-Methylbutane (*I*-pentane)	0.013	0.000	0.082	3.045E−05
N-pentane	0.013	0.000	0.082	1.601E−05
Total	100.000	100.000	100.000	
Temperature, °F	−80.00	−167.57	−167.57	
Pressure, psia	800.00	200.00	200.00	
Molecular weight	16.88	16.27	20.12	
Compressibility factor		.8111	.0485	
Density, lb-mols/cu-ft		.0787	1.3149	
Molar volume, cu-ft/lb-mole		12.7140	.7605	
Total flows, lb-moles/hour	3.260E+02	2.748E+02	5.119E+01	
Total flows, pounds/hour	5.501E+03	4.471E+03	1.030E+03	
Enthalpy, Btu/lb-mole*	1886.47	2010.94	−1592.09	
Enthalpy, Btu/hour*	.615E+06	.553E+06	−.815E+05	
Entropy, Btu/lb-mole-°F	32.38	33.96	26.34	
ΔH, Btu/lb-mole				−441.41
Mole percents		84.29	15.71	

* Zero enthalpy datum is ideal gas at 0°R.

Table 19.9
Case 1 of Computer Thermodynamic Calculations and Results for Gas Plant Compressor—
Isentropic (100% Efficiency)

Isentropic Compression Calculations by Methods of
Soave Redlich-Kwong for K, H, and S Predictions

Material Balance Basis Component Name	Mole Percent			K-Value
	Feed	Vapor	Liquid	
Nitrogen	0.400	0.400	0.000	1.985E+01
Carbon dioxide	0.520	0.520	0.000	4.910E+00
Methane	90.166	90.166	0.000	8.656E+00
Ethane	4.690	4.690	0.000	2.353E+00
Propane	1.850	1.850	0.000	9.111E−01
2-Methylpropane (*I*-butane)	0.783	0.783	0.000	4.559E−01
N-butane	0.510	0.510	0.000	3.655E−01
2-Methylbutane (*I*-pentane)	0.270	0.270	0.000	1.864E−01
N-pentane	0.180	0.180	0.000	1.569E−01
N-heptane	0.630	0.630	0.000	4.205E−02
Total	100.000	100.000	100.000	
Temperature, °F	120.00	277.26	277.26	
Pressure, psia	250.00	810.00	810.00	
Molecular weight	18.74	18.74	0.00	
Compressibility factor		0.9753	0.0000	
Density, lb-mols/cu-ft		0.1050	0.0000	
Molar volume, cu-ft/lb-mole		9.5228	0.0000	
Total flows, lb-moles/hour	7.685E+03	7.685E+03	0.000E+00	
Total flows, pounds/hour	1.440E+05	1.440E+05	0.000E+00	
Enthalpy, Btu/lb-mole*	4787.29	6275.67	0.00	
Enthalpy, Btu/hour*	.368E+08	.482E+08	.000E+00	
Entropy, Btu/lb-mole-°F	42.19	42.19	0.00	
ΔH, Btu/lb-mole				1488.38
Mole percents		100.00	0.00	

* Zero enthalpy datum is ideal gas at 0°R.

Table 19.10
Case 2 of Computer Thermodynamic Calculations and Results for Gas Plant Compressor—
Polytropic (75% Efficiency)

Polytropic Compression Calculations by Methods of
Soave Redlich-Kwong for K, H, and S Predictions

Material Balance Basis Component Name	Mole Percent			K-Value
	Feed	Vapor	Liquid	
Nitrogen	0.400	0.400	0.000	1.985E+01
Carbon dioxide	0.520	0.520	0.000	4.910E+00
Methane	90.166	90.166	0.000	8.656E+00
Ethane	4.690	4.690	0.000	2.353E+00
Propane	1.850	1.850	0.000	9.111E−01
2-Methylpropane (*I*-butane)	0.783	0.783	0.000	4.559E−01
N-butane	0.510	0.510	0.000	3.655E−01
2-Methylbutane (*I*-pentane)	0.270	0.270	0.000	1.864E−01
N-pentane	0.180	0.180	0.000	1.569E−01
N-heptane	0.630	0.630	0.000	4.205E−02
Total	100.000	100.000	100.000	
Temperature, °F	120.00	318.86	318.86	
Pressure, psia	250.00	810.00	810.00	
Molecular weight	18.74	18.74	0.00	
Compressibility factor		0.9843	0.0000	
Density, lb-mols/cu-ft		0.0985	0.0000	
Molar volume, cu-ft/lb-mol		10.1532	0.0000	
Total flows, lb-moles/hour	7.685E+03	7.685E+03	0.000E+00	
Total flows, pounds/hour	1.440E+05	1.440E+05	0.000E+00	
Enthalpy, Btu/lb-mole*	4787.29	6771.79	0.00	
Enthalpy, Btu/hour*	0.368E+08	0.520E+08	0.000E+00	
Entropy, Btu/lb-mole-°F	42.19	42.84	0.00	
ΔH, Btu/lb-mole				1984.51
Mole percents		100.00	0.00	

* Zero enthalpy datum is ideal gas at 0°R.

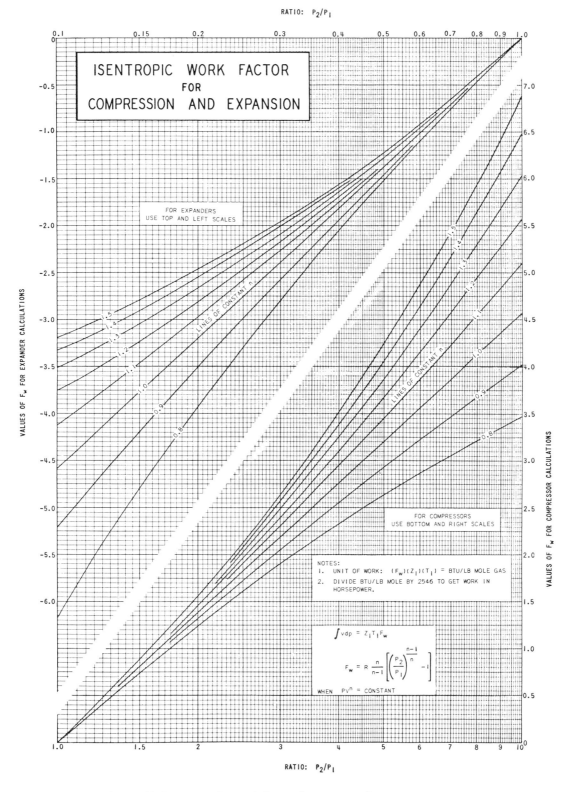

Figure 19.5. Isentropic work factor for compression and expansion.

pressors, but have no applications for centrifugal compressors, nor for turbo-expanders. The simple equations used in these calculations are presented here.

Volumetric Efficiency

Reciprocating compressors have clearance spaces between the pistons and the cylinder heads. This clearance volume remains full of high-pressure gases at the end of the compression stroke. These clearance volume gases expand during the intake stroke and occupy some of the cylinder volume, thus reducing the quantity of gas that can be taken in during the suction stroke.

This compressor capacity reduction may be expressed as the ratio of the gas intake to the displacement and called the "volumetric efficiency." An equation for the volumetric efficiency in terms of the clearance volume, the compression ratio, and the pressure-volume relationship exponent, n, is derived, using the following terms:

V_d = displacement volume
V_c = clearance volume
V_e = volume to which clearance gases expand
$C = V_c/V_d$ = ratio of clearance to displacement volumes
$V_{net} = V_d + V_c - V_e$ = net volume of intake gases

The complete compression cycle is shown in Figure 19.6 where numbers 1 through 6 designate the points of change. Eight parts of the cycle are described as follows for two cases, one with clearance and one without clearance volume, the latter being of theoretical interest only:

Points on Figure 19.6

Without Clearance	With Clearance	Operation
3	3	Intake valve closes
3–4	3–4	Compression stroke
4	4	Outlet valve opens
4–6	4–5	Discharge stroke
6	5	Outlet valve closes
6–1	5–2	Re-expansion stroke
1	2	Inlet valve opens
1–3	2–3	Intake stroke

In practice, reciprocating compressors must have some clearance volume to avoid metal-to-metal contact at the end of each piston stroke.

Assuming that the re-expansion of clearance volume gases follows the same PV_n = constant path as compression, it is possible by using Equation 19.8 to write: $V_e = V_c (P_2/P_1)^{1/n}$. Combining these relationships gives the following expression for the volumetric efficiency:

$$\epsilon_v = (V_{net}/V_d) [1 + C - C(P_2/P_1)^{1/n}] \qquad (19.13)$$

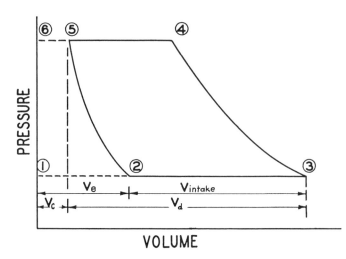

Figure 19.6. Diagram of reciprocating compressor cycle.

Multistage Compression

Two or more stages, with intercooling between stages, are often used in compression because of the lower volumetric efficiency at higher pressure ratios. There is a value of P_2/P_1 for each clearance volume percentage and gas exponent at which the re-expansion gases will fill the cylinder completely and exclude taking in any new gas.

This ratio is found from Equation 19.13 by setting $\epsilon_v = 0$, which leads to

$$[(P_2/P_1)_{\epsilon_v=0}] = [(1 + C)/C]^n \qquad (19.14)$$

For a 10% clearance volume, the P_2/P_1 at zero capacity is found to be 28.6 when $n = 1.4$; 17.7 for $n = 1.2$; and 11.0 for $n = 1.0$. This effect of clearance volume, pressure ratio, and gas exponent on compressor capacity provides an incentive to use multistages.

The intermediate pressure for two-stage compression may be chosen so that the total work is a minimum. This criterion can be expressed mathematically when the interstage cooling takes the gas back to the inlet gas temperature. This case is illustrated by the PV diagram in Figure 19.7, where an intermediate pressure of P_i is indicated and the two compression paths are reversible and adiabatic, i.e., isentropic.

The total work for two stages, with P_i the intermediate pressure, is

$$(-W_s)_{total} = (-W_s)_1 + (-W_s)_2 = \{n/(n - 1)\}\ P_1V_1$$
$$[(P_i/P_1)^{(n-1)/n} + (P_2/P_i)^{(n-1)/n} - 2]\quad (19.15)$$

The total work is a minimum when the term inside the large bracket of the right side of Equation 19.15 is a min-

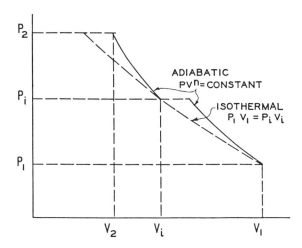

Figure 19.7. Diagram showing *PV* path for two-stage compression with intercooling.

imum. Differentiating with respect to P_i, which is the only variable, and setting the derivative equal to zero gives

$$P_i = (P_1P_2)^{1/2} \qquad (19.16)$$

Combining Equations 19.15 and 19.6 gives

$$(-W_s)_{total} = 2n/(n-1) \, P_1V_1 \, [(P_2/P_1)^{(n-1)/2n} - 1] \qquad (19.17)$$

For *S* stages

$$\text{T.H.P.} = (144/33{,}000) \, Sn/(n-1) \, P_1V_1 \\ [(P_2/P_1)^{(n-1)/Sn} - 1] \qquad (19.18)$$

An analogous expression gives the intermediate temperature

$$T_2 = T_1 \, (P_2/P_1)^{(n'-1)/Sn'} \qquad (19.19)$$

Typical Indicator Diagrams

Figures 19.8 and 19.9 are replots of indicator diagrams taken from the head end of a reciprocating compressor. These two diagrams (12), are for the same cylinder with different clearance volumes, the clearance pockets being closed in one case and open in the other.

The compressor indicator diagrams in Figure 19.8 were replotted on log-log scales in Figure 19.9 to illustrate the paths of the compression and expansion strokes. The straight lines on log-log scales proves that the $PV^n = $ constant relationship is applicable. The values of the exponents are shown on the plots. Note that the exponents for the expansion strokes are of lower magnitude than the exponents for the compression strokes. This was true for most of the indicator diagrams I examined.

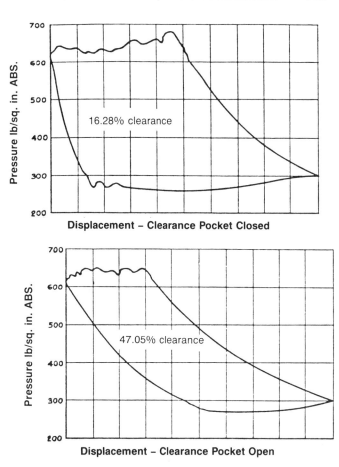

Figure 19.8. Typical indicator diagrams from reciprocating compressor.

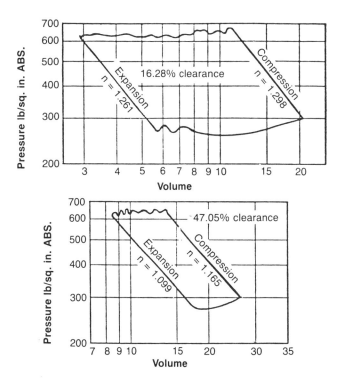

Figure 19.9. Log-log plot of compressor indicator diagrams.

It will also be noted, by examination of Figure 19.9, that the exponents are smaller on the diagram for the cylinder with the clearance pockets open. Although the values of these exponents are different, the straightness indicates that the values for each path are constant for the entire path. The data shown on these plots plus similar indicator card data justify using the $PV^n = $ constant for polytropic compression and expansion paths.

References

1. Anonymous, Nat. Bur. Standards (U.S.), Circular No. 142, (1923).
2. Barkeley, C. H., J. L. Valentine, and C. O. Hurd, *Chem. Engr. Progress*, 43, 25, (1947).
3. Buffington, R. M. and W. K. Gilkey, *Refrig. Engr.*, Circular No. 12, (1931).
4. Edmister, W. C., *Chem. Engr. Progress*, 47, 191, (1951).
5. Graham, D. P. and R. C. McHarness, "Thermodynamic Properties of Freon-22," Kinetic Chemicals, Inc. (1945).
6. Keenan, J. H. and F. G. Keyes, *Thermodynamic Properties of Steam,* John Wiley and Sons, New York, NY (1936).
7. Matthews, C. S. and C. O. Hurd, *Trans. Am. Inst. Chem. Engrs.*, 42, 55, (1946).
8. Rynning, D. F. and C. O. Hurd, *Trans. Am. Inst. Chem. Engrs.*, 41, 265, (1945).
9. Tanner, H. G., A. F. Benning, and W. F. Mathewson, *Ind. Engr. Chem.*, 31, 878, (1939).
10. York, R., Jr. and E. F. White, *Trans. Am. Inst. Chem. Engrs.*, 40, 227, (1944).
11. Chappelear, P. S., R. J. J. Chen, and D. G. Elliot, Proceedings of Fifty-Sixth Annual Convention, Gas Processors Association, (1977).
12. Private Communication from B. C. Thiel, Cooper-Bessemer Corporation (1949).

20

Compressible Fluid Flow Through Orifices and Pipes

The flow of gases and vapors, for which the density and temperature change appreciably with pressure, is a process operation that can be calculated by the application of thermodynamics. Properties involved in this application include the compressibility factor, enthalpy and entropy changes, and the exponents for isentropic and isenthalpic changes of temperature and volume with pressure. These items were presented in Chapter 19.

With these thermodynamic properties, three types of compressible fluid flow can be predicted for real gases, namely:

Nozzle Flow—Pressure energy is converted into kinetic energy by flowing through a nozzle.

Pipe Flow—Fluid flows in a channel of uniform cross section, with friction and, in some cases, heat transfer, causing pressure drop.

Diffuser Flow—Kinetic energy of a high velocity gas is converted into pressure energy; the opposite of nozzle flow.

The first two of these flow processes will be considered after deriving the basic equations for a steady flow system.

Energy Balances

Consider the steady flow system shown schematically in Figure 20.1 and equate the sum of the *input* energy terms to the sum of the *output* terms.

$$E_1 + P_1V_1 + U_1^2/2g_c + X_1(g/g_c) + Q =$$
$$E_2 + P_2V_2 + U_2^2/2g_c + X_2(g/g_c) + W_s \quad (20.1)$$

Combining terms and using the symbol Δ to designate a difference, i.e., $\Delta X = (X_2 - X_1)$, gives

$$\Delta E + \Delta(PV) + \Delta(U^2/2g_c)$$
$$+ \Delta X(g/g_c) = Q - W_s \quad (20.2)$$

Note: An inconsistency in terminology between Equations 20.1 and 20.2 and Chapter 2 exists. In Equation 2.1, U is energy. In Equations 20.1 and 20.2, E is energy, U is velocity, and V is specific volume. The latter is the terminology that was used in the first edition of this book. It will be retained here to avoid assigning new symbols to velocity and specific volume.

Combining Equation 20.2 with the definition of enthalpy, $\Delta H = \Delta E + \Delta(PV)$, gives the following very useful expression:

$$\Delta H + \Delta(U^2/2g_c) + \Delta X(g/g_c) = Q - W_s \quad (20.3)$$

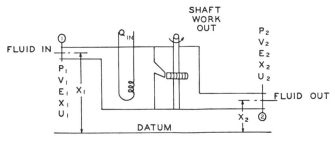

Figure 20.1. Schematic diagram of steady flow system for energy balance.

121

The energy units of Equations 20.1, 20.2, and 20.3 are "ft lb-force per lb-mass," sometimes called "feet of fluid flowing." The g/g_c ratio, following the elevation variable, X, is the local acceleration of gravity, g, divided by a conversion constant, g_c, which is defined as $g_c = (m/F)g = $ (lb-mass/lb-force)(ft/sec^2). Q is heat added to the system and W_s is shaft work done by the system.

For a reversible nozzle the potential energy change is zero and so are the Q and W_s terms, giving

$$\Delta H = -\Delta U^2/2g_c \qquad (20.4)$$

Equation 20.4 is used with P-H, or Mollier, chart readings to find the nozzle exit velocity, which is usually done by assuming that the initial velocity is zero, giving

$$U = (2g_c(H_1 - H_2))^{0.5} \qquad (20.5)$$

Units of enthalpy in the previous equations are "ft lb-force/lb-mass." Another nozzle flow formula will be derived later.

It is sometimes more convenient to have Equation 20.3 in another form, without the enthalpy term. The H term can be eliminated by combining with the first law expression from Chapter 2:

$$dH = TdS + VdP \qquad (2.9)$$

Where only reversible work is involved, $Q = TdS$. In real processes there will be friction so we will write $TdS = Q + F$. Combining with Equation 2.9 gives

$$\Delta H = Q + \int VdP + F \qquad (20.6)$$

where F = friction in the consistent units

Combining Equations 20.3 and 20.6, eliminating Q and differentiating gives

$$VdP + dF + UdU/g_c + (g/g_c)dX = -d'W_m \qquad (20.7)$$

where the operator d' indicates an inexact differential.

Equation 20.7 is the differential form of the Bernoulli Theorem. When the potential energy and the shaft work terms are zero, as in a nozzle, horizontal pipe, and diffuser flows, Equation 20.7 becomes

$$VdP + UdU/g_c + dF = 0 \qquad (20.8)$$

Equation 20.8 shows the relationship between *pressure* energy, *kinetic* energy, and *friction* energy. This relationship will be used in solving compressible fluid flow problems.

When velocity change is negligible and shaft work is zero, as in the case of gas-lift oil well flow, Equation 20.7 becomes

$$VdP + dF + (g/g_c)dX = 0 \qquad (20.9)$$

This application will not be discussed here.

Nozzle Flow

Nozzles are usually designed to give a nearly perfect frictionless flow so an ideal nozzle is the only case of interest here. From the nozzle inlet up to the throat, flow is generally assumed to be reversible and adiabatic even though acoustic velocity may be reached at this point.

Maximum flow will be obtained through a frictionless nozzle when the change in kinetic energy equals the change in the pressure energy. This occurs at a critical flow velocity, sometimes called acoustic or sonic velocity. At this point the pressure will be as low as it will go. The pressure will not drop below P_{CF}, the critical flow pressure, even if the nozzle discharged into a complete vacuum.

The critical flow pressure for an ideal adiabatic nozzle can be found from the exponent defining the volume change with pressure for an isentropic path, which has been expressed as

$$P_1V^n_1 = P_2V^n_2 \qquad (19.8)$$

from which it has been shown that

$$\int_1^2 VdP = P_1V_1 \{n/(n-1)\} [(P_2/P_1)^{(n-1)/n} - 1] \qquad (19.10)$$

For the frictionless case Equation 20.8 is

$$VdP + UdU/g_c = 0 \qquad (20.10)$$

Integrating Equation 20.10 gives

$$\int_1^2 VdP = -(U^2_2 - U^2_1)/2g_c \qquad (20.11)$$

Combining Equations 19.10 and 20.11, assuming that the initial velocity is negligible and rearranging gives

$$(U^2_2)/g_c nP_1V_1 = 2/(n-1) \\ [1 - (P_2/P_1)^{(n-1)/n}] \qquad (20.12)$$

In mass flow units Equation 20.12 becomes

$$G^2 = 2g_cP_1/V_1(n/(n-1))(P_2/P_1)^{2/n}$$
$$[1 - (P_2/P_1)^{(n-1)/n})] \qquad (20.12A)$$

Leaving Equations 20.12 and 20.12A until later and returning to Equations 19.8 and 20.10, from Equation 19.8

$$dP/dV = -n(P/V) \qquad (20.13)$$

From Equation 20.10

$$VdP = -G^2VdV/g_c \qquad (20.14)$$

where $U = GV$
$\qquad dU = GdV$
$\qquad G$ = mass velocity
(when G = lb/sec-ft^2, V = ft^3/lb, and U = ft/sec)

Combining Equations 20.13 and 20.14 gives the following two relationships for acoustic mass and acoustic linear velocities:

$$G^2 = g_cnP/V \qquad (20.15)$$

and

$$U^2 = g_cnPV \qquad (20.16)$$

Equations 20.15 and 20.16 give the velocity at the end of any line when the pressure beyond the end point is below the critical pressure, whether the flow is frictionless or not, as long as critical flow is reached. From Equation 20.16 it follows that

$$U^2_2 = g_cnP_2V_2$$

Combining with Equation 20.12 gives

$$(P_2V_2)/(P_1V_1) = 2/(n-1) [1 - (P_2/P_1)^{(n-1)/n}] \quad (20.17)$$

The left side of Equation 20.17 is equal to $(P_2/P_1)^{(n-1)/n}$. Substitution, rearranging and identifying P_2 as P_{CF} gives the following expression for the critical flow pressure

$$P_{CF}/P_1 = (2/(n+1))^{n/(n-1)} \qquad (20.18)$$

Figure 20.2 is a graphical solution of Equation 20.18, giving the critical flow pressure ratio for ideal nozzles as a function of the n values for the gas.

The maximum rate of flow through the nozzle is a function of the critical flow velocity and pressure. Writing Equation 20.15 for the mass velocity at the nozzle throat and eliminating $(P/V)_{CF}$ by means of Equations

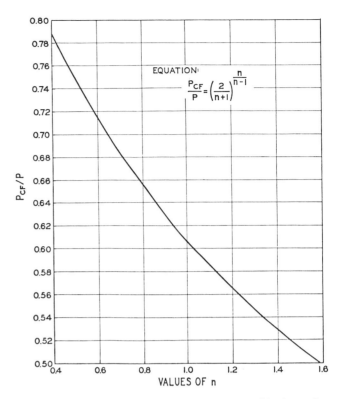

Figure 20.2. Critical flow pressure ratio for ideal nozzle.

20.18 and 19.8 gives the mass velocity in terms of upstream or nozzle inlet conditions.

$$V_{CF} = V_1(P_1/P_{CF})^{1/n} = V_1((n+1)/2)^{1/(n-1)}$$

$$P_{CF} = P_1(2/(n+1))^{n/(n-1)}$$

$$(P/V)_{CF} = (P/V)_1 (2/(n+1))^{(n+1)/(n-1)} \qquad (20.19)$$

Combining Equations 20.15 and 20.19 gives

$$G_{max} = [g_cP_1/V_1 (2/(n+1))^{(n+1)/(n-1)}]^{1/2} \qquad (20.20)$$

where $1/V_1$ = density of gas at inlet
$\qquad G_{max}$ = mass velocity, lb/sec-ft^2

When $n = 1.0$ Equations 20.12A and 20.20 become:

$$G^2 = 2g_cP_1/V_1(P_2/P_1)^2 \ln(P_1/P_2) \qquad (20.12A)$$

and

$$G_{max} = 0.605(g_cP_1/V_1)^{1/2} \qquad (20.20A)$$

Equation 20.20A can be used to make a solution when $n = 1.0$, at which point a direct solution of Equation 20.20 is not possible.

The density term in Equation 20.20 will now be replaced by

$$\rho_1 = PM/ZRT \qquad (20.21)$$

where
ρ = density, lb mol/ft^3
P = pressure, lb/ft^2
M = molecular weight
Z = compressibility factor
R = gas constant = 1,544
T = temperature, °R
0.605 = value of P_{CF}/P at n = 1.0

Equation 20.21 gives the density in lb/ft^3. Combining Equations 20.20 and 20.21, introducing a nozzle discharge coefficient, K, and rearranging gives

$$W = 3,600(32.2/1,544)^{1/2}$$
$$[n(2/(n + 1))^{(n+1)/(n-1)}]^{1/2} \, KAP(M/ZT)^{1/2} \quad (20.22)$$

where W = rate of flow in lb/hr

Equation 20.22 has been applied to the calculation of flow through nozzles to pressure relieving systems, for which application a gas coefficient has been plotted against n and Equation 20.22 has been put in the following form:

$$W = C K A P (M/ZT)^{1/2} \qquad (20.23)$$

where $C = 520 \, [n \, (2/(n + 1))^{(n+1)/(n-1)}]^{1/2}$ (20.24)
K = discharge coefficient (about 0.95 for a well designed nozzle)
A and P must be in consistent units. When P is in pounds/in.^{2}, A should be in in.^{2} units.

Figure 20.3 is a plot of Equation 20.24, giving C as a function of n. This chart and equation apply to an ideal frictionless nozzle, the same as does Figure 20.2 for the critical flow pressure. The temperature and pressure conditions, the compressibility factor, and the exponent of the gas are those at the nozzle inlet.

These nozzle flow equations were developed previously. One application has been to the problem of sizing pressure-relieving systems for pressure vessels in oil refinery service. (See API RP 520, September 1955.)

In this application, the objective is to size the nozzle for a flow rate equal to the rate of hydrocarbon vaporization in the vessel while exposed to fire. The rate of heat absorption, an important factor in the design calculations, is beyond the scope of this chapter. When properly sized, the safety valve will open at the design pressure and the vapors will flow through the nozzle, reaching acoustic velocity at the outlet; the pressure in the vessel will not exceed the design value.

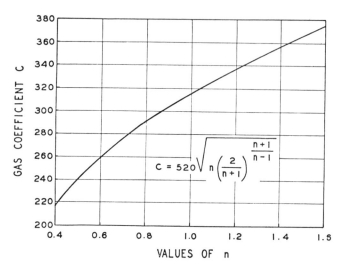

Figure 20.3. Gas coefficient, C, for nozzle flow formula.

Vapors leaving the vessel via the nozzle will be saturated, i.e., at the dew point. During the vaporization of a wide boiling hydrocarbon liquid mixture at constant pressure, i.e., the safety valve setting, the dew-point temperature will increase as the composition of the vapors being generated changes to higher concentrations of the heavier components. With sufficient heat input to keep the vapors at their dew point, the temperature will increase.

The value of n for a given hydrocarbon increases with increasing temperature. Increasing molecular weight, for the same temperature and pressure, reduces the n value. Thus, these effects are in opposite directions and off-setting so the initial conditions of hydrocarbon composition, temperature, and pressure may be used with reasonable accuracy to find the n value for safety valve calculations.

A value of the exponent n can be found for a gas mixture by making an isentropic, or polytropic, expansion calculation to find the values of P and V at terminal points 1 and 2 in Equation 19.8, and then solve this equation for n, i.e.,

$$n = [(\ln(P_1/P_2))/(\ln(V_2/V_1))] \qquad (19.8A)$$

A satisfactory procedure is to find n by computer and then use the n value to make nozzle calculations. A more direct method of calculating nozzle flow is the enthalpy balance method, which follows.

Enthalpy Difference Method

This method is based on Equation 20.5, where point 1 is the plenum and point 2 is the nozzle throat, giving the throat velocity U. With enthalpies in Btu/lb, the ΔH must be divided by 778; $g_c = 32.2$ and U is velocity in ft/sec. The n need not be evaluated when using this method.

Gas Flow in Pipe Line

When the kinetic energy term is negligible, the equation for gas flow in a pipe of uniform cross section is obtained by starting with the following form of Equation 20.8:

$$VdP + dF = 0 \qquad (20.25)$$

The friction term, dF, is a function of the friction factor, f, as follows:

$$dF = (f/(2g_cD))U^2\,dL \qquad (20.26)$$

In terms of the friction factor, f, and the gas density, $1/V$, Equation 20.25 becomes

$$dP/dL = -(fU^2(1/V))/(2g_cD) \qquad (20.27)$$

where f = Darcy-Weisbach friction factor, defined by this equation and four times the magnitude of the Fanning friction factor, which is defined in terms of hydraulic radius instead of diameter
$1/V$ = gas density in ft^3/lb
D = pipe diameter, ft
$g_c = 32.2$
U = velocity in ft/sec

Equation 20.27 is put in a more convenient form by replacing $1/V$ and U by their equivalents:

$$1/V = PM/ZRT$$

and

$$U = G/(1/V) = GZRT/PM$$

and by

$$PdP = -[(fZRTG^2)/(2g_cDM)]dL \qquad (20.28)$$

Integrating from P_1, at $L = 0$, and rearranging

$$G^2 = (g_cDM)/(ZRTf)(P^2_1 - P^2_2)/L \qquad (20.29)$$

where M = molecular weight and other terms as already defined

A general equation that includes a term for kinetic energy change is obtained for gas flow in a pipe of uniform

diameter by starting with Equation 20.8, as shown in detail in the following.

Combining Equations 20.8 and 20.26 gives

$$VdP + UdU/g_c + fU^2dL/(2g_cD) = 0 \qquad (20.30)$$

Make the following substitutions: $U = GV$, $dU = GdV$, divide by V and solve for $-dP$, giving

$$-dP = (G^2/g_c)[(Vf/2D)dL + dV] \qquad (20.31)$$

Integrating and solving for G^2 gives

$$G^2 = 2g_c \int_{P_2}^{P_1}(1/V)dP\,[\,fL/D + 2\,\ln(V_2/V_1)]^{-1} \qquad (20.32)$$

Equation 20.32 is a general solution of Equation 20.8. It will next be put into more convenient form for the calculation of isothermal flow of gas in a pipe line. Two density terms will be replaced, i.e., $(1/V) = PM/ZRT$ and $V_2/V_1 = P_1/P_2$, for isothermal flow conditions:

$$G^2 = \frac{\dfrac{2g_cM}{ZRT}\displaystyle\int_{P_2}^{P_1}PdP}{\left[\dfrac{fL}{D} + 2\ln\left(\dfrac{P_1}{P_2}\right)\right]} \qquad (20.33)$$

Integrating and rearranging gives the following desired expression:

$$G^2 = \frac{\dfrac{2g_cM}{ZRT}(P_1^2 - P_2^2)}{\left[\dfrac{fL}{D} + 2\ln\left(\dfrac{P_1}{P_2}\right)\right]} \qquad (20.34)$$

Another arrangement of Equation 20.34 is

$$f/LD = 2g_cM/(ZRTG^2)\,[P^2_1 - P^2_2] \\ -2\,\ln(P_1/P_2) \qquad (20.35)$$

Equation 20.34 was used in preparing a computer program for making flow calculations for gases in pipelines. In this program the gas amount, pipe length, and pipe diameter are given, or assumed. Initial and final conditions of temperature and pressure are given or assumed and gas properties computed. With this information, Equation 20.34 is solved to find a correct exit pressure for the given flow quantity, diameter, and length.

Transfer Line Flow

In furnace-to-column transfer lines, the flow velocity of gas or vapor may approach or equal critical velocity so that it is necessary to allow for the kinetic energy change in the calculations for sizing the line. Flow will not occur at constant temperature so the simplifying assumption of isothermal conditions cannot be made.

The principle objective of transfer line flow calculations is to size the transfer line so that its inlet pressure does not exceed the desired furnace outlet pressure. Of course, it is not necessary that the outlet pressure be as low as the column pressure, which, in many cases, operates at a substantial vacuum. The first step in the design of a transfer line is to determine the critical outlet pressure for the tentatively chosen line size.

It will be assumed that a gas-vapor-liquid mixture behaves essentially the same as a vapor of the same density so that the same flow equations can be used. This is in accord with actual experience when the flow is turbulent and the velocity is high enough to keep the liquid and gas moving at substantially the same velocity. These conditions are fulfilled in the usual transfer line from a crude oil vaporization furnace to a fractionating column. The friction factor will vary from 0.020 to 0.025 and will be taken as 0.025 for design calculations.

An expression for the critical flow pressure as a function of mass velocity, fluid density and expansion is obtained by applying Equation 20.15 at the critical flow point, giving

$$P_{CF} = G^2/g_c n \rho \qquad (20.35)$$

The value of n, which is defined by $PV^n = $ constant for the flowing fluid and the flow path, can be found by applying Equation 19.8A and the same procedure described previously in this chapter. The value of the specific volume V that is used in Equation 19.8A is the value at the critical flow conditions, making solution an iterative calculation.

In the iterative calculation required to find P_{CF}, V, and n at the critical flow point, enthalpy balances must be made, and the assumption is made that the flowing fluid is at constant enthalpy, i.e., irreversible and adiabatic. This will be illustrated later.

An equation relating pressures, expansion coefficient, and a fluid friction parameter is obtained from Equation 20.32, as follows: The two density terms in Equation 20.32 are transformed and combined with $PV^n = C$ giving

$$\int_2^1 (1/V)dP = (1/C^{1/n}) \int_2^1 P^{1/n}dP$$
$$= (1/C^{1/n})(n/(n=1))[P^{(n+1)/n}]^{1/2}$$

Replacing $C^{1/n}$ by its equivalent $VP^{1/n}$ and writing the value of the integral for the limits gives

$$\int_2^1 (1/V)dP = n/(n+1)[P_1/V_1 - P_2/V_2]$$

$$= n/(n + 1)(P_2/V_2)[(P_1/P_2)^{(n+1)/n} - 1] \qquad (20.36)$$

In like manner

$$\ln(V_2/V_1) = \ln(P_1/P_1)^{1/n} = (1/n)\ln(P_1/P_2) \qquad (20.37)$$

Combining Equations 20.32, 20.36, and 20.37 gives

$$G_2 = \frac{(2ng_cP_2/V_2)/(n+1)[(P_1/P_2)^{(n+1)/n} - 1]}{fL/D + 2/n + \ln(P_1/P_2)} \qquad (20.38)$$

where subscript 1 refers to the furnace end of transfer line; subscript 2 refers to the column end of the transfer line, i.e., critical flow velocity point

Combining Equations 20.35 and 20.38 to eliminate G^2 and rearranging gives the following important and useful expression:

$$(fL/D) = 2/(n + 1)[(P/P_{CF})^{(n+1)/n} - 1]$$
$$- (2/n)\ln(P/P_{CF}) \qquad (20.39)$$

Equation 20.39 gives (fL/D) as a function of (P/P_{CF}) and n. This relationship is represented graphically by Figure 20.4, where the coordinates are the values at the flow condition where friction losses and changes in pressure and kinetic energy are in balance to give flow at critical velocity at the transfer line outlet.

Expansion Exponents

Finding values of the expansion exponent for a flowing fluid that is a mixture, whether it's all gas or a gas-liquid combination, is a large part of the work required in making orifice or pipe line flow calculations for such fluids. A manual procedure for a petroleum fraction will be presented with examples to illustrate this application. Afterwards a computer procedure for evaluating exponents and making flow calculations is presented.

Chapter 19 showed plots of isentropic and isenthalpic exponents to demonstrate that these exponents may be assumed constant over rather large ranges of conditions for pure components in the vapor phase. The petroleum naphtha and natural gas fluids that will be used later in two examples differs from these pure component fluids in two respects: (a) the naphtha and natural gas are complex mixtures rather than single components; and (b) the phase condition of the naphtha is very complex along the flow path, starting at the critical point and going through the liquid/vapor two-phase region.

Treating the complex mixture as a homogeneous system in the vapor/liquid region is obviously questionable, but making this simplification for the vapor only region should be acceptable. This mixed-phase flaw in this application is believed to be trivial compared to the large

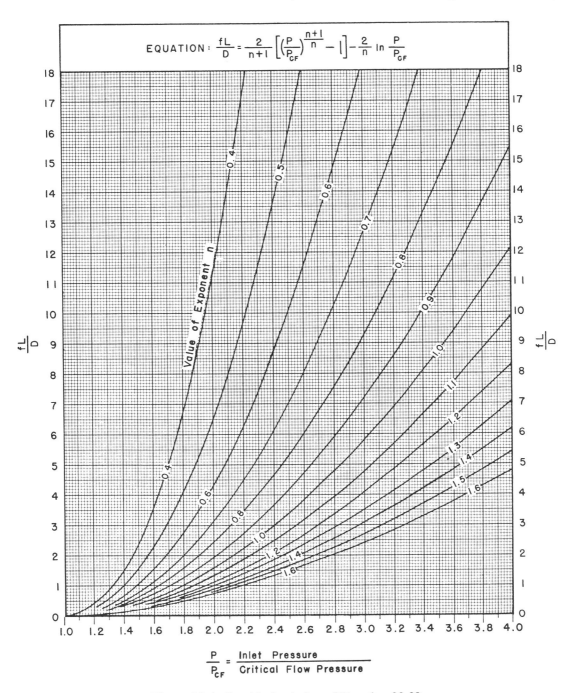

Figure 20.4. Graphical solution of Equation 20.39.

errors that can, and do, result from grossly improper methods of transfer line calculations, however.

Another uncertainty is the definition of the path, which is assumed to be at, or near to, isentropic for nozzles, and at, or near to, isenthalpic for transfer lines. Non-adiabatic conditions in the nozzle or pipe would cause departure, resulting in a polytropic flow. Also an unsmooth orifice or nozzle would have a less than 100% efficiency and a polytropic flow.

Naphtha Diagrams

Figures 20.5–20.8 were prepared for and used in the first edition of this book. Although the manual methods used in making and using these charts is now obsolete, repeating these items, and Tables 20.1 and 20.2, will be useful in illustrating procedures. The objective of the chart preparation was Figure 20.8, a pressure-enthalpy diagram for the co-existing vapor and liquid region of a

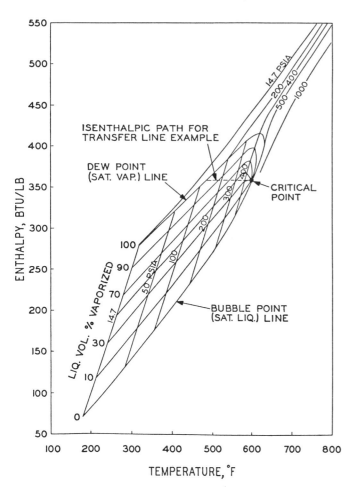

Figure 20.5. Enthalpy-temperature diagram for petroleum naphtha.

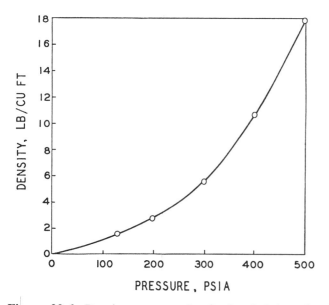

Figure 20.6. Density-pressure plot for isenthalpic path of naphtha in transfer line problems.

66°API petroleum naphtha, defined by the following ASTM Distillation Assay:

Point:	IBP	10%	30%	50%	70%	90%	FBP
T°F:	130	190	240	200	320	360	400

The temperature-enthalpy chart in Figure 20.5 was prepared previously for a design problem and was taken as the basis for making Figure 20.8, which was needed for the transfer line calculations. Only the two-phase (vapor and liquid) region was drawn on this pressure-enthalpy diagram. The enthalpies and phase conditions were taken from Figure 20.5 and densities were calculated in Tables 20.1 and 20.2. Values of the densities and the exponents along the isenthalpy path chosen for the two examples are shown in Figures 20.6 and 20.7.

Using the enthalpy-temperature plot in Figure 20.5 and other data on liquid and vapor densities, compressibility factors, molecular weights, etc., densities of the vapor-liquid mixtures were calculated at five points on the constant enthalpy line shown dashed on Figure 20.5.

From these densities and the pressures, the values of the exponent *n* were computed for the four intervals between these five points. These calculations are shown in Table 20.1. Densities are plotted against pressure in Figure 20.6 and the exponent *n* is plotted against pressure in Figure 20.7. Readings of the exponent and the density from these plots can be used in making flow calculations for sections of the pipe length.

Flow Calculation Examples

Two transfer line examples will be solved using Equation 20.39 (Figure 20.4) and manual calculations to il-

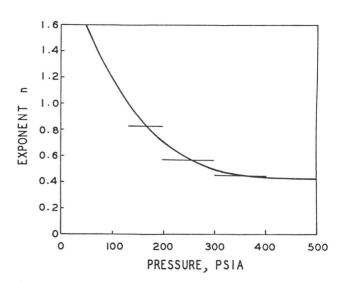

Figure 20.7. Exponent *n* vs. pressure for isenthalpic path of naphtha in transfer line problem.

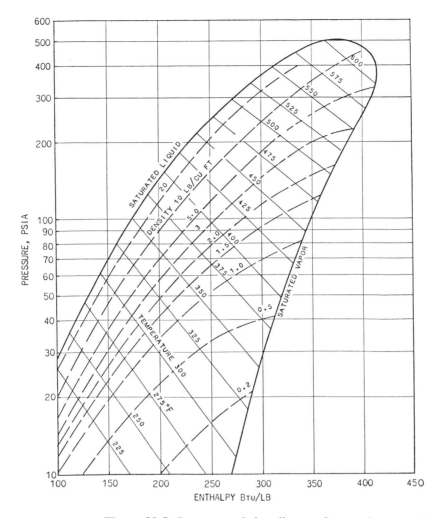

Figure 20.8. Pressure-enthalpy diagram for petroleum naphtha.

lustrate the method. Both examples are for the 66°API naphtha. In the first example the critical pressure is above the column pressure, while in the second P_{CF} is below the column pressure. Following are the flow conditions for these two examples:

Flow rate: 50,000 bbl/day
Furnace outlet: 600°F and 500 psia
Tower pressure: 200 psia
Assume: Friction factor $f = 0.025$; isenthalpic (irreversible and adiabatic) flow.

Example 1. Assume a 4.026-in. I.D. (4-in. schedule 40 pipe) transfer line with an equivalent length of 100 ft.

Rate of flow = 145.2 lb/sec
Flow area = 0.0894 ft²
Flow velocity $G = 145.2/0.0884 = 1,640$ lb. mass/sec. ft²
$P_{CF} = G^2/g_c n \rho = 83,500/n \rho$

P_{CF} is next determined by trial and error with Figures 20.6 and 20.7.

Pressure psia	Density lb/ft³ Fig. 20.6	Exponent n Fig. 20.7	Value of P_{CF} lb/ft²	Value of P_{CF} lb/in.²
200	2.4	0.70	49,700	345
250	4.0	0.57	36,600	255

From the previous calculations, the critical outlet pressure is obviously between 250 and 255 psia and is assumed to be 252 psia.

The friction-length-diameter parameter of Equation 20.39, and ordinate of Figure 20.4, is calculated next:

$$fL/D = (0.025)(100)/(4.026/12) = 7.47$$

Solution for the ratio P/P_{CF}, by Equation 20.39 or by Figure 20.4 requires an average value of n, which is assumed to be 0.57, to be conservative. With these two parameters, i.e., 7.47 and 0.57, the value of P/P_{CF} is found to be 2.22.

Then $P_1 = (2.22)(252) = 560$ psia. Comparing this to the 500-psia furnace outlet pressure, it can be seen that the 4-in. line is a bit small for this flow.

Example 2. In Example 1 it was found that a 4-in. pipe line requires a higher inlet pressure than is available for the required flow conditions. So now a 6-in. schedule 40 transfer line, with a cross-sectional flow area of 0.201 ft², will be tried.

$$G = 145.2/0.201 = 722 \text{ lb-mass/sec ft}^2$$

The critical outlet pressure is obviously below the column pressure of 200 psia. Nevertheless, its determination will aid in the calculation of the furnace outlet pressure for this column pressure.

Readings of Figures 20.6 and 20.7 at 200 psia give an exponent value of 0.70 and a fluid density value of 2.4 lb/ft³. Putting these values into Equation 19.8 and rearranging gives

$$P_{CF} = ((200)(144))/((2.4)(V_{CF}))^n$$

Putting these same values, i.e., $n = 0.70$ and $1/V = 2.4$ at 200 psia, into Equation 20.35 gives

$$P_{CF} = 16,200 \ V_{CF}/0.70$$

Solving these two expressions simultaneously gives the following critical flow pressure for a 6-in. transfer line in this service:

$$P_{CF} = 18,200 \text{ lb/ft}^2 \text{ abs} = 127 \text{ psia}$$

Table 20.1
Calculation of Densities and Exponents of 66°API Naphtha for an Isenthalpic Expansion

Point Interval	1	1–2	2	2–3	3	3–4	4	4–5	5
Readings From Figure 20.5									
Pressure, psia	500	...	400	...	300	...	200	...	130
Vaporization, liq. vol. %	0	...	32	...	50	...	84	...	100
Temperature, °F	600	...	580	...	556	...	520	...	483
Enthalpy, Btu/lb	365	...	365	...	365	...	365	...	365
Vapor Portion									
Gallons	0	...	32	...	50	...	84	...	100
°API	...	...	76	...	74	...	68	...	66
Lb/gal	...	...	5.676	...	5.731	...	5.904	...	5.964
Lb	...	...	181.5	...	286.8	...	496.0	...	596.4
Molecular weight	...	...	65	...	70	...	88	...	104
Z	...	...	0.53	...	0.65	...	0.73	...	0.80
ρ, lb/ft³	...	...	4.40	...	2.97	...	2.29	...	1.67
Ft³ vapor	...	...	41.2	...	96.6	...	216.8	...	357.3
Liquid Portion									
Gallons	100	...	68	...	50	...	16	...	0
°API	66	...	62	...	60	...	58	...	...
Lb/gal	5.964	...	6.087	...	6.151	...	6.216	...	...
Lb	596.4	...	414.0	...	307.5	...	99.4	...	...
$\rho_{T,P}/\rho_{60.0}$	0.40	...	0.60	...	0.64	...	0.70	...	...
ρ, lb/ft³	17.85	...	27.3	...	29.3	...	42.5	...	...
Ft³ liquid	33.4	...	15.16	...	10.45	...	2.33	...	...
Vapor-Liquid Mixture									
Total lb	596.4	...	596.4	...	596.4	...	596.4	...	596.4
Total ft³	33.4	...	56.4	...	107.1	...	219.1	...	357.3
ρ_{mix} lb/ft³	17.85	...	10.59	...	5.57	...	2.73	...	1.67
P_i/P_{i+1}	...	1.250	...	1.333	...	1.500	...	1.539	...
ρ_i/ρ_{i+1}	...	1.687	...	1.900	...	2.045	...	1.633	...
Exponent n-intervals	...	0.427	...	0.448	...	0.567	...	0.826	...
Exponent n-points*	0.42	...	0.44	...	0.50	...	0.70	...	1.00

* From Figure 20.7

Table 20.2
Sample Calculations of Densities for Pressure-Enthalpy Diagram of 66°API Naphtha

Readings From Figure 20.5								
Pressure, psia	50	50	100	100	200	200	300	300
Vaporization, liq. vol. %	10	70	10	70	10	70	10	70
Temperature, °F	310	362	382	438	467	510	528	565
Enthalpy, Btu/lb	177	270	220	310	273	344	313	372
Vapor Portion								
Gallons	10	70	10	70	10	70	10	70
Lb/gal	5.6	5.8	5.6	5.8	5.6	5.8	5.6	5.8
Lb	56	406	56.0	406	56.0	406	56.0	406
Molecular weight	60	83	60	83	60	83	60	83
Lb moles	0.934	4.90	0.934	4.90	0.934	4.90	0.934	4.90
Z	0.85	0.88	0.75	0.80	0.625	0.685	0.475	0.565
ZRT/P	140.4	155.1	67.8	77.0	33.1	35.6	16.76	20.72
Ft^3 vapor	131.2	760.0	63.3	377.3	29.0	174.5	15.7	101.6
Liquid Portion								
Gallons	90	30	90	30	90	30	90	30
Lb/gal	6.0	6.19	6.0	6.19	6.0	6.19	6.0	6.19
Lb	540	185.7	540	185.7	540	185.7	540	185.7
Lb/ft^3 at 60.0	44.9	46.3	44.9	46.3	44.9	46.3	44.9	46.3
$\rho_{T,P}/\rho_{60.0}$	0.820	0.775	0.760	0.715	0.657	0.625	0.518	0.500
Lb/ft^3	36.8	35.9	34.1	33.1	29.5	28.5	23.28	23.15
Ft^3 liquid	14.66	5.18	15.84	5.62	18.32	6.52	23.2	8.03
Vapor-Liquid Mixture								
Total lb	596.4	596.4	596.4	596.4	596.4	596.4	596.4	596.4
Total ft^3	145.86	765.18	79.14	382.92	47.32	181.02	38.9	109.63
ρ_{mix}, lb/ft^3	4.09	0.78	7.54	1.56	12.60	3.29	15.33	5.44

With these conditions, the furnace outlet pressure would be lower than the permissible maximum of 500 psia. It is calculated as follows:

$$fL/D = (0.025)(100)/(6.065/12) = 4.94$$

From Figure 20.4 or Equation 20.39, with this value of fL/D and $n = 0.7$, the ratio of inlet pressure to the critical flow pressure is found to be 2.3, giving a furnace outlet pressure of $127 \times 2.3 = 292$ psia, which is well below the furnace outlet of 500 psia.

From the previous calculations, it is clear that an intermediate size pipe would provide the desired flow conditions, i.e., an inlet pressure of 500 psia and an outlet pressure of 200 psia for a transfer line of 100 feet. Such a pipe would have to be made special at added expense.

A 6-in. line could be longer than 100 feet and still handle this flow with these pressures. Substituting the pressure ratio of 3.94 and the exponent of 0.7 into Equation 20.39 gives a value of $fL/D = 19.76$, from which $L = (19.76/0.025) \times (6.065/12) = 400$ feet.

Another option is to operate with the critical outlet pressure above the pressure of the column, a usual practice for vacuum columns because this makes use of a smaller size line. Transfer lines in vacuum units are large

and expensive, often being fabricated from costly alloy steel. Proper sizing is of great importance in such cases.

Computer Program

Subroutine GASFLO was prepared for compressible fluid flow calculations of three types:

1. Orifice design for nozzle flow, with plenum pressure given and critical velocity at the throat.
2. Pipe line design for isothermal flow, with pressures given.
3. Transfer-line design, with pressures given and a critical velocity discharge.

Values of orifice or pipe diameter, pipe length, and flow rate are specified by the user, according to the nature of the flow.

In the case of the nozzle, the orifice diameter or the flow rate can be fixed and the other calculated, which is the procedure followed in the program. Thus there are two calculations for the orifice, one for diameter and one for flow rate. For the isothermal pipe line, two of the three variables (diameter, length, and flow rate) are fixed and one calculated for each of the three calculations made by the program.

Although there are the same three variables (diameter, length, and flow rate) for the transfer-line problem, the value of one can be fixed and two calculated for each solution made by the program, making three cases also. The following tabulation summarizes these run conditions.

For these fluid flow calculations, preparatory thermodynamic calculations are made at the initial and final conditions, the latter being guessed if not known. In these preliminary calculations, pressure, temperature, density, molecular weight, enthalpy, and flow quantity information is provided in a convenient form for the gas flow calculations. The ICOD's of the preliminary calculations required for the three types of gas flow calculations are shown in the following. Also shown are the conditions that are specified and those that are calculated.

Prep. Runs

Flow Type	Nozzle		Isothermal Pipeline		Transfer Line	
Case	1	2	1	2	1	2
ICOD	1	9	1	1	1	8

Flow Runs (ICOD = 13)

Flow Type	Nozzle		Isothermal			Transfer Line		
Case	3	4	3	4	5	3	4	5
Flow Rate	SP	CA	SP	SP	CA	SP	SP	CA
Diameter	CA	SP	CA	SP	SP	CA	SP	SP
Length	NA	NA	SP	CA	SP	CA	CA	SP
Outlet T	CA	CA	SP	SP	SP	SP	SP	SP
Outlet P	CA	CA	CA	SP	SP	CA	SP	SP

where SP = specified; CA = calculated; NA = not applicable

ICOD = 1 is for finding the starting properties of the fluid, usually all vapor at the given T and P conditions. If the fluid is not all vapor, ICOD = 1 will give properties for the mixed phase system, as well as each phase. ICOD = 8 is an isenthalpic flash calculation, for which final temperature is found for a given pressure. The output may be vapor or vapor/liquid. ICOD = 9 is an isentropic flash calculation, for which final temperature is found for a given pressure. The output may be vapor or vapor/liquid.

Flow rates are required in making the first three types of calculations. Normal operating rates can be used for nozzle and isothermal pipe line flow rates. All three types of gas flow calculations are direct, i.e., without iterations to find the desired result. This is not convenient for isothermal pipe line flow where the usual objective of the calculations is the pipe diameter or pressure drop as the flow rate and length are usually known.

With the present form of the program, it is necessary to solve for diameter and pressure drops by trials.

Transfer-line flow calculation requirements may be determined by normal design specifications or by emergency conditions. A transfer-line from an oil heating furnace to a fractionating column is an example of the design application. An emergency caused by a fire in a hydrocarbon processing plant might make it necessary to empty a vessel quickly to avoid an explosion and more damage. Heat from the fire would cause the pressure to rise high enough to rupture the safety disc and send the hot vapors through the transfer line of the blow-down system.

The size of the line is determined by the amount of material that must be removed in a given length of time. Factors are rates of heat and vapor generation, inventory of the vessel, insulation protection, etc., items that are beyond the scope of this book. A computer subroutine for making the thermodynamic calculations is illustrated in the following examples.

Compressible Fluid Flow Solutions by Computer

Four problems will illustrate the computer program for compressible fluid flow calculations. The first is for nozzle flow, the second for pipe line flow and the last two are for transfer-line flow. Except for the fourth problem, which is for a petroleum naphtha, all problems are for a gas from a high-pressure separator of the same turbo-expander gas plant for which compression and expansion calculations were made in Chapter 19.

Two tabulations are used to present the input and output data in each example, the first containing the control codes and the feed mixture component information plus the results of the two preliminary runs made to get thermodynamic data; while the second tabulation contains the results from the second preliminary run plus the fluid flow calculations.

Each computer run is identified by a case number and a name. Cases 1 and 2 are the preparatory runs. Cases 3, 4, and 5 are the fluid flow calculation runs. Case 1 is for the starting point properties and is at the same ICOD for all three types of flow. The other cases differ for each flow type.

Example 3. The nozzle flow option is illustrated by calculations that appear in Tables 20.3 and 20.4 for the gas from a high-pressure separator of a gas processing plant. Case 2, in this example, is an isentropic flash made to provide P, T, and V values for calculating the isentropic exponents. Cases 3 and 4 give the flow calculation results, i.e., orifice diameter for a given flow rate in Case 3 and the flow rate for a given orifice diameter in Case 4. Values of the isentropic exponents, critical flow pressure, and mass velocity are also given.

Controls IKV	IHS	IHD	IOUT	MULT	XTOL	TTOL	PTOL
1	1	1	1	0	0.00001	0.10000	0.00010

Feed Mixture

Comp. ID	Feed mols/hr	Boil Pt °F	Specific Gravity	Component Name
603	1.542	− 320.40	.0000	Nitrogen
606	1.458	− 109.30	.0000	Carbon dioxide
101	310.917	− 258.70	.3000	Methane
102	9.625	− 127.50	.3564	Ethane
103	1.792	− 43.70	.5077	Propane
105	.375	10.90	.5631	2-Methylpropane (*i*-Butane)
104	.167	31.10	.5844	*n*-Butane
107	.042	82.10	.6247	2-Methylbutane (*i*-Pentane)
106	.042	96.90	.6310	*n*-Pentane

Problem Conditions for Calculations that Follow:

ICOD	KHF	KYF	KTL	TI	PI	VF	TO	PO
1	0	0	1	− 50.00	800.00	0.00	0.00	0.00

Case 1 of Gas Plant High-Pressure Nozzle Design Vapor Properties at Given Temperature and Pressure Using *K*-Values and *H* & *S* Values by Soave Redlich-Kwong Methods

Material Balance Basis Component Name	Mole Percent Feed	Mole Percent Vapor	Mole Percent Liquid	K-Value
Nitrogen	0.473	0.473	0.000	1.268E+00
Carbon dioxide	0.447	0.447	0.000	9.620E−01
Methane	95.385	95.385	0.000	1.039E+00
Ethane	2.953	2.953	0.000	9.419E−01
Propane	0.550	0.550	0.000	9.133E−01
2-Methylpropane (*i*-Butane)	0.115	0.115	0.000	8.935E−01
n-Butane	0.051	0.051	0.000	8.889E−01
2-Methylbutane (*i*-Pentane)	0.013	0.013	0.000	8.702E−01
n-Pentane	0.013	0.013	0.000	8.657E−01
Total	100.000	100.000	00.000	
Temperature, °F	−50.00	−50.00	−50.00	
Pressure, psia	800.00	800.00	800.00	
Molecular weight	16.88	16.88	.00	
Compressibility factor		0.7039	0.0000	
Density, lb-mols/ft^3		0.2585	0.0000	
Molar volume, ft^3/lb-mole		3.8684	0.0000	
Total flows, lb-moles/hr	3.260E+02	3.260E+02	.000E+00	
Total flows, lb/hr	5.501E+03	5.501E+03	.000E+00	
Enthalpy, Btu/lb-mole*	2435.17	2435.17	.00	
Enthalpy, Btu/hr*	.794E+06	.794E+06	.000E+00	
Mole percents		100.00	.00	

* Zero enthalpy datum is ideal gas at 0°R.

Example 4. The isothermal pipe line option is illustrated by calculations that appear in Tables 20.5 and 20.6 for the same gas used in Example 3. Case 2, in this example, is for output conditions only. Exponents are not needed in making isothermal pipe line flow calculations. Both of the preliminary runs, i.e., Case 1 in Table 20.5 and Case 2 in Table 20.6, are ICOD = 1 VLE calcula-tions to find beginning and ending properties. Pressure drop is known or assumed, but not calculated, for all three cases. Case 3 calculations find the pipe line diameter for a given flow rate and pipe length. Case 4 calculations find the pipe length and Case 5 calculations find the flow rate, other variables being fixed. The numerical results in Table 20.6 should be self-explanatory.

(text continued on page 137)

Table 20.4
Gas Plant High-Pressure Nozzle Design Thermodynamic Properties at Output and Flow Calculations

Problem Conditions for Calculations that Follow:

ICOD	KHF	KYF	KTL	TI	PI	QF	TO	PO
9	0	0	2	−50.00	800.00	0.00	−100.00	500.00

Case 2 of Gas Plant High-Pressure Nozzle Design
   Isentropic Flash Temperature at Given Pressure Using K-Values and
   H & S Values by Soave Redlich-Kwong Methods

| Material Balance Basis | Mole Percent | | | |
Component Name	Feed	Vapor	Liquid	K-Value
Nitrogen	0.473	0.478	0.073	6.542E+00
Carbon dioxide	0.447	0.441	0.907	4.861E−01
Methane	95.385	95.870	59.428	1.613E+00
Ethane	2.953	2.763	16.999	1.626E−01
Propane	0.550	0.384	12.873	2.979E−02
2-Methylpropane (i-Butane)	0.115	0.046	5.214	8.873E−03
n-Butane	0.051	0.015	2.735	5.497E−03
2-Methylbutane (i-Pentane)	0.013	0.001	0.870	1.529E−03
n-Pentane	0.013	0.001	0.901	1.007E−03
Total	100.000	100.000	100.000	
Temperature, °F	−50.00	−98.35	−98.35	
Pressure, psia	800.00	500.00	500.00	
Molecular weight	16.88	16.75	26.64	
Compressibility factor		0.7145	0.1197	
Density, lb-mols/ft^3		0.1805	1.0774	
Molar volume, ft^3/lb-mole		5.5416	0.9282	
Total flows, lb-moles/hr	3.260E+02	3.216E+02	4.338E+00	
Total flows, lb/hr	5.501E+03	5.386E+03	1.156E+02	
Enthalpy, Btu/lb-mole*	2435.17	2236.40	−1626.62	
Enthalpy, Btu/hr*	0.794E+06	0.719E+06	−.706E+04	
Entropy, Btu/lb-mole-°F	33.78	33.80	32.81	
ΔH, Btu/lb-mole				−250.18
Mole percents		98.67	1.33	

* Zero enthalpy datum is ideal gas at 0°R.

Problem Conditions for Calculations that Follow:

ICOD	KHF	KYF	KTL	TI	PI	FDX	TO	PO	DIAM	LENGTH
13	0	1	2	−50.00	800.00	1.00	−100.00	500.00	1.00	

Case 3 Nozzle Diameter Calculation Results
Adiabatic exponents: P(V)=1.5912; T(P)=1.3647
Given total flow=5443.6270 lb/hr
Critical flow pressure=398.4550 psia
Velocity in nozzle throat=60.8011 ft/sec
Diameter of nozzle throat=1.1913 in.

Case 4 Nozzle Flow Rate Calculation Results
Adiabatic exponents: P(V)=1.5912; T(P)=1.3647
Diameter of nozzle throat=1.0000 in.
Critical flow pressure=398.4550 psia
Velocity in nozzle throat=60.8011 ft/sec
Calculated flow rate=4494.7820 lb/hr

Table 20.5
Gas Plant High-Pressure Pipe Line Design Input Data and Preparatory Calculations

Controls

IKV	IHS	IHD	IOUT	MULT	XTOL	TTOL	PTOL
1	1	1	1	0	0.00001	0.10000	0.00010

Feed Mixture

Comp. ID	Feed mols/hr	Boil Pt °F	Specific Gravity	Component Name
603	1.542	−320.40	.0000	Nitrogen
606	1.458	−109.30	.0000	Carbon dioxide
101	310.917	−258.70	.3000	Methane
102	9.625	−127.50	.3564	Ethane
103	1.792	−43.70	.5077	Propane
105	.375	10.90	.5631	2-Methylpropane (*i*-Butane)
104	.167	31.10	.5844	*n*-Butane
107	.042	82.10	.6247	2-Methylbutane (*i*-Pentane)
106	.042	96.90	.6310	*n*-Pentane

Problem Conditions for Calculations that Follow:

ICOD	KHF	KYF	KTL	TI	PI	VF	TO	PO
1	0	0	1	−50.00	800.00	0.00	0.00	0.00

Case 1 of Gas Plant High-Pressure Pipe Line Design
    Vapor Properties at Given Temperature and Pressure Using
    K-Values and *H* & *S* Values by Soave Redlich-Kwong Methods

Material Balance Basis Component Name	Mole Percent			K-Value
	Feed	Vapor	Liquid	
Nitrogen	0.473	0.473	0.000	1.268E+00
Carbon dioxide	0.447	0.447	0.000	9.620E−01
Methane	95.385	95.385	0.000	1.039E+00
Ethane	2.953	2.953	0.000	9.419E−01
Propane	0.550	0.550	0.000	9.133E−01
2-Methylpropane (*i*-Butane)	0.115	0.115	0.000	8.935E−01
n-Butane	0.051	0.051	0.000	8.889E−01
2-Methylbutane (*i*-Pentane)	0.013	0.013	0.000	8.702E−01
n-Pentane	0.013	0.013	0.000	8.657E−01
Total	100.000	100.000	00.000	
Temperature, °F	−50.00	−50.00	−50.00	
Pressure, psia	800.00	800.00	800.00	
Molecular weight	16.88	16.88	.00	
Compressibility factor		0.7039	0.0000	
Density, lb-mols/ft³		0.2585	0.0000	
Molar volume, ft³/lb-mole		3.8684	0.0000	
Total flows, lb-moles/hr	3.260E+02	3.260E+02	.000E+00	
Total flows, lb/hr	5.501E+03	5.501E+03	.000E+00	
Enthalpy, Btu/lb-mole*	2435.17	2435.17	.00	
Enthalpy, Btu/hr*	0.794E+06	0.794E+06	0.000E+00	
Mole percents		100.00	0.00	

* Zero enthalpy datum is ideal gas at 0°R.

Table 20.6
Gas Plant High-Pressure Pipe Line Design Thermodynamic Properties at Output and Flow Calculations

Problem Conditions for Calculations that Follow:

ICOD	KHF	KYF	KTL	TI	PI	QF	TO	PO
1	0	0	2	−50.00	700.00	0.00	0.00	0.00

Case 2 of Gas Plant High-Pressure Pipe Line Design
   Vapor Properties at Given Temperature and Pressure Using
   K-Values and *H* & *S* Values by Soave Redlich-Kwong Methods

Material Balance Basis Component Name	Mole Percent			*K*-Value
	Feed	Vapor	Liquid	
Nitrogen	0.473	0.473	0.000	1.302E+00
Carbon dioxide	0.447	0.447	0.000	9.660E−01
Methane	95.385	95.385	0.000	1.048E+00
Ethane	2.953	2.953	0.000	9.505E−01
Propane	0.550	0.550	0.000	9.276E−01
2-Methylpropane (*i*-Butane)	0.115	0.115	0.000	9.122E−01
n-Butane	0.051	0.051	0.000	9.087E−01
2-Methylbutane (*i*-Pentane)	0.013	0.013	0.000	8.941E−01
n-Pentane	0.013	0.013	0.000	8.906E−01
Total	100.000	100.000	00.000	
Temperature, °F	−50.00	−50.00	−50.00	
Pressure, psia	700.00	700.00	700.00	
Molecular weight	16.88	16.88	.00	
Compressibility factor		0.7450	0.0000	
Density, lb-mols/ft^3		0.2137	0.0000	
Molar volume, ft^3/lb-mole		4.6797	0.0000	
Total flows, lb-moles/hr	3.260E+02	3.260E+02	0.000E+00	
Total flows, lb/hr	5.501E+03	5.501E+03	0.000E+00	
Enthalpy, Btu/lb-mole*	2567.10	2567.10	0.00	
Enthalpy, Btu/hr*	0.837E+06	0.837E+06	0.000E+00	
Mole percents		100.00	0.00	

* Zero enthalpy datum is ideal gas at 0°R.

Problem Conditions for Calculations that Follow:

ICOD	KHF	KYF	KTL	TI	PI	FDX	TO	PO	DIAM	LENGTH
13	0	2	2	−50.00	800.00	1.00	−50.00	700.00	2.00	1000.0

Case 3 Pipe Line Diameter Calculation Results
Given line length = 1000.0000 ft
Given pressure drop = 100.0000 lb/in.^{2}
Given total flow = 5501.4130 lb/hr
Mass velocity = 221.5689 lb/ft^2 sec
Calculated pipe diameter (ID) = 1.1245 in.

Case 4 Pipe Line Length Calculation Results
Given pipe diameter (ID) = 2.0000 in.
Given pressure drop = 100.0000 lb/in.^{2}
Given total flow = 5501.4130 lb/hr
Mass velocity = 70.0462 lb/ft^2 sec
Calculated line length = 10021.7700 ft

Case 5 Pipe Line Flow Rate Calculation Results
Given pipe diameter (ID) = 2.0000 in.
Given line length = 1000.0000 ft
Given pressure drop = 100.0000 lb/in.^{2}
Mass velocity = 221.5689 lb/ft^2 sec
Calculated total flow = 17401.9800 lb/hr

Note that the given flow rates for Cases 3 and 4 are equal at 5,501.4 lb/hr and that the calculated flow rate for Case 5 is 17,402 lb/hr, which is 3.16 times the size of the given flow rate for the first two cases. The value of *FDX*, the flow rate multiplier, was changed from 1.0 to 3.16 and the problem rerun with no other data changes. The results for all three cases were the same, i.e., a 2-in. diameter pipe line 1,000 feet long with a flow rate of 17,402 lb/hr and a pressure drop of 100 psi.

Example 5. The transfer-line flow option is illustrated in this example by calculations that appear in Tables 20.7 and 20.8. For this problem it is assumed that an emergency blow-down system is required for the 800 psia gas separator. It is also assumed that the separator will be at $-50°F$ when blow-down starts. The line is 2.0-in. ID, the outlet pressure is set at 100 psia, and it is assumed that "critical flow" conditions will exist. Flow rate and pipe length are the objectives of the calculations. These conditions will serve to illustrate this computer procedure.

Table 20.7 gives the data, which include the control codes, the gas feed (lb moles/hr) and the problem conditions for the first run. For this problem, the gas feed quantities are for defining composition only and do not specify the flow rate of the gas. That and other flow conditions, such as pipe diameter and length, are given in the lower part of Table 20.8, just before the results of the flow calculations. Preliminary run results labeled Case 1 in Table 20.7 give thermodynamic properties at the inlet of the transfer line.

Case 2 run for this problem is an isenthalpic expansion (ICOD = 8) of the feed gas from inlet to outlet (critical flow) point. Conditions for this preliminary run and the results thereof are given in Table 20.8. Problem conditions for the actual flow calculations (ICOD = 13) are given next and these are followed by the results of Cases 3, 4, and 5.

The values of *FDX* = the feed rate multiplier, pipe diameter, and pipe length are all required for making all three calculation options, i.e., fixing any one of these three conditions and calculating the other two. *FDX* affects Case 3 option only; DIAM affects Case 4 only; and LENGTH affects Case 5 only. For this example these three input values were selected to produce almost identical results from the three cases. This required a trial run, starting with a 2-in. pipe and guesses for length and flow rate. From this start, Case 4 gives length and flow rate values that were used for the final runs shown in Table 20.8.

The high rate of flow, 127,916.2 lb/hr is 23.4 times the normal production rate of the gas in the plant from which the data were taken. This ratio does not mean much however. A more meaningful number would be the time it would take to blow-down the entire inventory of the 800 separator. This problem is outside the scope of this book.

Example 6. The transfer-line option is also illustrated by calculations that appear in Tables 20.9 and 20.10 for the same petroleum naphtha problem that was used previously in the manual solutions in Examples 1 and 2. The present computer solutions, and comparisons with the manual solutions should be of interest. As in Example 5, the first tabulation contains the input data on the feed and the first preparatory run, Case 1; and the second tabulation contains the second preparatory run, Case 2, and the actual flow calculations, Cases 3, 4, and 5.

The breakdown procedure for representing the $66°$API petroleum naphtha as an eleven component (pseudo) mixture was presented in Chapter 15, where this oil was an example. The lb/hr equivalent of the 50,000 bbl/day feed rate that was used in Examples 1 and 2 is 514,000. In the first calculations made on this problem by the computer program, this flow rate was used with a 4-in. pipe 100 feet long, and the same pressure and temperature conditions. The resulting pipe length and flow rates were 148 feet and 699,479 lb/hr, respectively, i.e., Case 4 in Table 20.10.

Since both flow rate and pipe length were greater than the previous manual solution, the value of *FDX* was changed from 1.0 to 1.35 and the length was changed from 100 to 150 feet. These input data changes give the results shown in Table 20.10 for Cases 4 and 5. It can be seen that the three results, i.e., Cases 3, 4, and 5, are in close agreement.

These computer results are not in agreement with the manual results of Examples 1 and 2. The differences might be due to the different values of the isenthalpic exponent and density values, which were 0.5764 and 3.13 lb/ft^3 by the computer calculations, these being in agreement with the values used in Examples 1 and 2. The critical flow pressure in Example 1 was found to be 252 psia. This required a higher inlet pressure, i.e., 560 psia instead of 500. By increasing the flow rate by 35% and the pipe length by 50% the calculations led to a critical flow pressure of 200 psia at the end of the transfer line.

Discussion

Both Figures 20.2 and 20.4, which are graphical representations of Equations 20.18 and 20.39, respectively, are exponent versus pressure relationships. They are for different processes, however. Figure 20.2 is for ideal nozzles, for which friction is zero and so is the initial velocity. Figure 20.4 is for transfer line flow, in which there is friction and a finite initial velocity. Also, Figure 20.4 contains the combination of friction, pipe diameter,

(text continued on page 142)

Table 20.7
Gas Plant High-Pressure Blowdown Transfer Line Diameter Input Data and Preparatory Calculations

Controls							
IKV	IHS	IHD	IOUT	MULT	XTOL	TTOL	PTOL
1	1	1	1	0	0.00001	0.10000	0.00010

Feed Mixture				
Comp. ID	Feed mols/hr	Boil Pt °F	Specific Gravity	Component Name
603	1.542	−320.40	.0000	Nitrogen
606	1.458	−109.30	.0000	Carbon dioxide
101	310.917	−258.70	.3000	Methane
102	9.625	−127.50	.3564	Ethane
103	1.792	−43.70	.5077	Propane
105	0.375	10.90	.5631	2-Methylpropane (*i*-Butane)
104	0.167	31.10	.5844	*n*-Butane
107	0.042	82.10	.6247	2-Methylbutane (*i*-Pentane)
106	0.042	96.90	.6310	*n*-Pentane

Problem Conditions for Calculations that Follow:

ICOD	KHF	KYF	KTL	TI	PI	VF	TO	PO
1	0	0	1	−50.00	800.00	0.00	0.00	0.00

Case 1 of Gas Plant High-Pressure Blowdown Transfer Line Diameter
      Vapor Properties at Given Temperature and
      Pressure Using *K*-Values and *H* & *S* Values by Soave Redlich-Kwong Methods

Material Balance Basis Component Name	Mole Percent			K-Value
	Feed	Vapor	Liquid	
Nitrogen	0.473	0.473	0.000	1.268E+00
Carbon dioxide	0.447	0.447	0.000	9.620E−01
Methane	95.385	95.385	0.000	1.039E+00
Ethane	2.953	2.953	0.000	9.419E−01
Propane	0.550	0.550	0.000	9.133E−01
2-Methylpropane (*i*-Butane)	0.115	0.115	0.000	8.935E−01
n-Butane	0.051	0.051	0.000	8.889E−01
2-Methylbutane (*i*-Pentane)	0.013	0.013	0.000	8.702E−01
n-Pentane	0.013	0.013	0.000	8.657E−01
Total	100.000	100.000	00.000	
Temperature, °F	−50.00	−50.00	−50.00	
Pressure, psia	800.00	800.00	800.00	
Molecular weight	16.88	16.88	0.00	
Compressibility factor		0.7039	0.0000	
Density, lb-mols/ft³		0.2585	0.0000	
Molar volume, ft³/lb-mole		3.8684	0.0000	
Total flows, lb-moles/hr	3.260E+02	3.260E+02	0.000E+00	
Total flows, lb/hr	5.501E+03	5.501E+03	0.000E+00	
Enthalpy, Btu/lb-mole*	2435.17	2435.17	0.00	
Enthalpy, Btu/hr*	0.794E+06	0.794E+06	0.000E+00	
Mole percents		100.00	.00	

* Zero enthalpy datum is ideal gas at 0°R.

Table 20.8
Gas Plant High-Pressure Blowdown Transfer Line Diameter Thermodynamic Properties at Output and Flow Calculations

Problem Conditions for Calculations that Follow:

ICOD	KHF	KYF	KTL	TI	PI	QF	TO	PO
8	0	0	2	−50.00	800.00	0.00	−100.00	200.00

Case 2 of Gas Plant High-Pressure Blowdown Transfer Line Diameter
  Isenthalpic Flash Temperature at Given Pressure
  Using *K*-Values and *H* & *S* Values by Soave Redlich-Kwong Methods

Material Balance Basis	Mole Percent			
Component Name	Feed	Vapor	Liquid	K-Value
Nitrogen	0.473	0.476	0.021	2.281E+01
Carbon dioxide	0.447	0.445	0.734	6.063E−01
Methane	95.385	95.847	30.435	3.143E+00
Ethane	2.953	2.826	20.802	1.358E−01
Propane	0.550	0.368	26.623	1.369E−02
2-Methylpropane (*i*-Butane)	0.115	0.032	11.767	2.736E−03
n-Butane	0.051	0.008	6.113	1.396E−03
2-Methylbutane (*i*-Pentane)	0.013	0.000	1.762	2.523E−04
n-Pentane	0.013	0.000	1.789	1.413E−04
Total	100.000	100.000	100.000	
Temperature, °F	−50.00	−118.14	−118.14	
Pressure, psia	800.00	200.00	200.00	
Molecular weight	16.88	16.74	36.13	
Compressibility factor		0.8805	0.0550	
Density, lb-mols/ft^3		0.0620	0.9927	
Molar volume, ft^3/lb-mole		16.1368	1.0073	
Total flows, lb-moles/hr	3.260E+02	3.237E+02	2.301E+00	
Total flows, lb/hr	5.501E+03	5.418E+03	8.316E+01	
Enthalpy, Btu/lb-mole*	2435.17	2478.80	−3700.88	
Enthalpy, Btu/hr*	0.794E+06	0.802E+06	−0.852E+04	
Mole percents		99.46	0.54	

* Zero enthalpy datum is ideal gas at 0°R.

Problem Conditions for Calculations that Follow:

ICOD	KHF	KYF	KTL	TI	PI	FDX	TO	PO	PDIAM	PLENG
13	0	3	2	−50.00	800.00	23.40	−100.00	200.00	2.00	74.0

Case 3 Transfer Line Flow Diameter and Length Results
Given total flow = 127761.1 lb/hr
Isenthalpic exponents: P(V) = 1.0443; T(P) = 1.1510
Mass velocity = 1628.680 lb/ft^2 sec
Critical flow pressure = 200.000 psia
Calculated pipe diameter (ID) = 1.9988 in.
Calculated pipe length = 74.1255 ft

Case 4 Transfer Line Flow Rate and Length Results
Given pipe diameter (ID) = 2.000 in.
Isenthalpic exponents: P(V) = 1.0443; T(P) = 1.1510
Mass velocity = 1628.680 lb/ft^2 sec
Critical flow pressure = 200.000 psia
Calculated total flow = 127916.200 lb/hr
Calculated line length = 74.171 ft

Case 5 Transfer Line Flow Rate and Diameter Results
Given transfer line length = 74.000 ft
Isenthalpic exponents: P(V) = 1.0443; T(P) = 1.1510
Mass velocity = 1628.680 lb/ft^2 sec
Critical flow pressure = 200.000 psia
Calculated total flow = 127327.900 lb/hr
Calculated line diameter (ID) = 1.9954 in.

Table 20.9
Petroleum Naphtha Transfer Line Design Input Data and Preparatory Calculations

Controls IKV	IHS	IHD	IOUT	MULT	XTOL	TTOL	PTOL
1	1	1	1	0	.00001	.10000	.00010

Feed Mixture

Comp. ID	Feed mols/hr	Boil PT °F	Specific Gravity	Component Name
1	136.206	50.24	.6372	Petroleum Naphtha Comp 01
2	544.824	119.17	.6645	Petroleum Naphtha Comp 02
3	272.412	165.75	.6823	Petroleum Naphtha Comp 03
4	544.824	197.70	.6938	Petroleum Naphtha Comp 04
5	272.412	223.87	.7026	Petroleum Naphtha Comp 05
6	544.824	252.65	.7123	Petroleum Naphtha Comp 06
7	272.412	281.26	.7217	Petroleum Naphtha Comp 07
8	544.824	310.76	.7312	Petroleum Naphtha Comp 08
9	272.412	338.05	.7398	Petroleum Naphtha Comp 09
10	544.824	368.56	.7491	Petroleum Naphtha Comp 10
11	136.206	422.87	.7651	Petroleum Naphtha Comp 11

Problem Conditions for Calculations that Follow:

ICOD	KHF	KYF	KTL	TI	PI	VF	TO	PO
1	0	0	1	600.00	500.00	.00	.00	.00

Case 1 of Petroleum Naphtha Transfer Line Design
   Vapor Properties at Given Temperature and Pressure Using
   K-Values and H & S Values by Soave Redlich-Kwong Methods

Material Balance Basis	Mole Percent			
Component Name	Feed	Vapor	Liquid	K-Value
Petroleum Naphtha Comp 01	3.333	3.333	.000	1.289E+00
Petroleum Naphtha Comp 02	13.333	13.333	.000	1.159E+00
Petroleum Naphtha Comp 03	6.667	6.667	.000	1.095E+00
Petroleum Naphtha Comp 04	13.333	13.333	.000	1.058E+00
Petroleum Naphtha Comp 05	6.667	6.667	.000	1.031E+00
Petroleum Naphtha Comp 06	13.333	13.333	.000	1.003E+00
Petroleum Naphtha Comp 07	6.667	6.667	.000	9.777E−01
Petroleum Naphtha Comp 08	13.333	13.333	.000	9.526E−01
Petroleum Naphtha Comp 09	6.667	6.667	.000	9.303E−01
Petroleum Naphtha Comp 10	13.333	13.333	.000	9.062E−01
Petroleum Naphtha Comp 11	3.333	3.333	.000	8.646E−01
Total	100.000	100.000	00.000	

Temperature, °F	600.00	600.00	600.00
Pressure, psia	500.00	500.00	500.00
Molecular weight	125.79	125.79	.00
Compressibility factor		.3590	.0000
Density, lb-mols/ft³		.1225	.0000
Molar volume, ft³/lb-mole		8.1648	.0000
Total flows, lb-moles/hr	4.086E+03	4.086E+03	.000E+00
Total flows, lb/hr	5.140E+05	5.140E+05	.000E+00
Enthalpy, Btu/lb-mole*	41787.05	41787.05	.00
Enthalpy, Btu/hr*	.171E+09	.171E+09	.000E+00
Mole percents		100.00	.00

* Zero enthalpy datum is ideal gas at 0°R.

Problem Conditions for Calculations that Follow:

ICOD	KHF	KYF	KTL	TI	PI	QF	TO	PO
8	0	0	2	600.00	500.00	.00	400.00	200.00

Case 2 of Petroleum Naphtha Transfer Line Design Calculations
Isenthalpic Flash Temperature at Given Pressure Using
K-Values and H & S Values by Soave Redlich-Kwong Methods

Material Balance Basis Component Name	Mole Percent			K-Value
	Feed	Vapor	Liquid	
Petroleum Naphtha Comp 01	3.333	3.333	.000	2.883E+00
Petroleum Naphtha Comp 02	13.333	13.333	.000	1.960E+00
Petroleum Naphtha Comp 03	6.667	6.667	.000	1.483E+00
Petroleum Naphtha Comp 04	13.333	13.333	.000	1.213E+00
Petroleum Naphtha Comp 05	6.667	6.667	.000	1.022E+00
Petroleum Naphtha Comp 06	13.333	13.333	.000	8.394E−01
Petroleum Naphtha Comp 07	6.667	6.667	.000	6.846E−01
Petroleum Naphtha Comp 08	13.333	13.333	.000	5.496E−01
Petroleum Naphtha Comp 09	6.667	6.667	.000	4.444E−01
Petroleum Naphtha Comp 10	13.333	13.333	.000	3.466E−01
Petroleum Naphtha Comp 11	3.333	3.333	.000	2.160E−01
Total	100.000	100.000	00.000	
Temperature, °F	600.00	536.76	536.76	
Pressure, psia	500.00	200.00	200.00	
Molecular weight	125.79	125.79	.00	
Compressibility factor		.7486	.0000	
Density, lb-mols/ft^3		.0250	.0000	
Molar volume, ft^3/lb-mole		40.0271	.0000	
Total flows, lb-moles/hr	4.086E+03	4.086E+03	.000E+00	
Total flows, lb/hr	5.140E+05	5.140E+05	.000E+00	
Enthalpy, Btu/lb-mole*	41787.05	41787.05	.00	
Enthalpy, Btu/hr*	.171E+09	.171E+09	.000E+00	
Mole percents		100.00	.00	

* Zero enthalpy datum is ideal gas at 0°R.

Problem Conditions for Calculations that Follow:

ICOD	KHF	KYF	KTL	TI	PI	FDX	TO	PO	PDIAM	PLENG
13	0	3	2	600.00	500.00	1.35	400.00	200.00	4.00	150.00

Case 3 Transfer Line Diameter and Length Results
Given total flow = 693898.600 lb/hr
Isenthalpic exponents P(V) = 0.5764; T(P) = 1.0720
Mass velocity = 2226.510 lb/ft^2 sec
Critical flow pressure = 200.0000 psia
Calculated pipe diameter (ID) = 3.9840 in.
Calculated pipe length = 147.4256 ft

Case 4 Transfer Line Flow Rate and Length Results
Given line diameter (ID) = 4.000 in.
Isenthalpic exponents P(V) = 0.5764; T(P) = 1.0720
Mass velocity = 2226.510 lb/ft^2 sec
Critical flow pressure = 200.0000 psia
Calculated total flow = 699478.700 lb/hr
Calculated line length = 148.0172 ft

Case 5 Transfer Line Flow Rate and Diameter Results
Given transfer line length = 150.0000 ft
Isenthalpic exponents P(V) = 0.5764; T(P) = 1.0720
Mass velocity = 2226.510 lb/ft^2 sec
Critical flow pressure = 200.0000 psia
Calculated total flow = 718344.400 lb/hr
Calculated line diameter (ID) = 4.0536 in.

and pipe length, fL/D, in which each variable is taken as a constant for the process.

In addition, the effect of a finite initial velocity is included in Figure 20.4. Because of the initial velocity difference between the two cases, Figure 20.2 cannot be regarded as a special case of Figure 20.4 with fL/D equal to zero.

Although the same symbol n has been used for the exponent defining the changes of volume with pressure in both the nozzle and transfer line applications, the paths are different. The expansion of the gases in the ideal nozzle is isentropic, i.e., reversible and adiabatic. For the transfer line, on the other hand, the expansion is isenthalpic, i.e., irreversible and adiabatic. As shown in Chapter 19, the isenthalpic value of n is always lower than the isentropic value.

Actually, the expansion of a gas in a transfer line is somewhere between isenthalpic and isentropic, i.e., polytropic, as represented by Path B or C in Figure 19.4, as there is substantial change in kinetic energy, especially for short lines. It has been observed, however, that the density-pressure relationship for a mixture is not materially different for isentropic and isenthalpic expansions, and assuming isenthalpic expansion always leads to lower calculated densities as the pressure drops so that the calculated line capacity is always conservative, or low. The actual difference is trivial.

Equation 20.39 (Figure 20.4) was derived for a homogeneous compressible fluid, which is usually considered to be a gas or vapor and not a two-phase vapor/liquid mixture. In applying Figure 20.4 to the calculation of transfer line flow of the complex mixture, petroleum naphtha, from the critical point along an irreversible and adiabatic path through the vapor/liquid two-phase region, as was done in Examples 1 and 2, a homogeneous mixture was assumed.

The computer solution of the petroleum naphtha transfer-line design problem showed that the phase conditions at both ends of the line were all vapor. In other words, the flow path did not start or end in the vapor/liquid region. This disagreement between the older graphical analysis and the newer analytical analysis is clearly due to the reliability of the two methods. Experimental phase measurements would answer this question.

Notation

A = area, ft^2

D = diameter of pipe or orifice

L = length of pipe line or transfer line

P = pressure, lb-force/ft^2

T = temperature, °R

R = gas constant, 1,544

M = molecular weight

g = local acceleration due to gravity

g_c = 32.2, a conversion constant, (lb-mass/lb-force)(ft/sec^2)

X = elevation, feet

V = specific volume, ft^3/lb-mass

$\rho = 1/V$ = density, lb-mass/ft^3

U = linear velocity, ft/sec

G = mass velocity, lb-mass/sec-ft^2

H = enthalpy, ft-lb/lb-mass in equations
Btu/lb-mass in Figures 20.5 and 20.8

$E = H - PV$ = energy, same units as H

Q = heat added to fluid, same units as H

W_s = shaft work done by the fluid, same units as H

F = friction, same units as H

f = friction factor, Darcy-Weisbach friction factor is defined in terms of diameter and is four times the magnitude of the Fanning friction factor, which is defined in terms of hydraulic radius instead of diameter

n = exponent in PV^n = constant

n' = exponent in $TP^{(1-n')/n'}$ = constant

Z = compressibility factor, dimensionless
$= PV/RT$ in consistent units; $= MPV/RT$ in above units

d = differential operator

21

Heats of Chemical Reactions

Heat and material balances are the two most important calculations in the design of plants and equipment for processing hydrocarbons, petroleum, gases, and petrochemicals. Heat quantities are sometimes referred to as "sensible" or "latent," distinguishing between heat changes that can be measured by a thermometer and heat that does not change the temperature. The heat of vaporization is an example of the "latent" heat of a physical process. In a process reactor, where molecular structures of process fluid components change, the heat effect is the "latent" heat of a chemical process.

The 1st edition treatment of this subject included charts and numerical illustrations, most of which are still relevant and will be included here. Chapter 8 of Volume 1 contains 11 pressure-enthalpy charts prepared by Starling (4) that include the heat of formation in the enthalpy scales.

Tables and pressure-enthalpy charts by Canjar and Manning (2) are also included for 12 hydrocarbons and 8 other gases. The H scales for these 20 substances were absolute enthalpy. After a review of this background, a new computer program for calculating heats of formation is described.

Basic Concepts

Heats of Reaction

Following are the symbols for seven different heats of reaction, some for physical and some for chemical reactions:

ΔH_c = heat of combustion to CO_2, H_2O, etc

ΔH_f = heat of formation from the elements; carbon, hydrogen, etc

ΔH_r = heat of reaction of one or two molecules to form others

ΔH_v = heat of vaporization of liquid to vapor

$- \Delta H_v$ = heat of condensation of vapor to liquid

$- \Delta H_s$ = heat of fusion of liquid to solid

ΔH_s = heat of sublimation of solid to vapor

Heats of nuclear, i.e., atomic splitting or fusion, reactions have been omitted from these definitions as being beyond the scope of this book. In this chapter, the first three ΔH's are the ones of interest.

Heats of reaction and phase change can be measured in a bomb or a flow calorimeter, which is usually operated adiabatically for reactions and isothermally for phase changes. The results of such measurements for chemical reactions are corrected to isothermal conditions by using heat capacities of the reactants and the products. The value of the Q from a flow calorimeter is equal to ΔH, the change in enthalpy. Q from a bomb calorimeter is equal to ΔU, the internal energy change. Applying Equation 2.6 to the results from a constant volume bomb, in which changes in pressure, volume, and Q are measured, gives: $\Delta H_r = Q + \Delta(PV)$.

Experimental measurement of the heat of combustion of a hydrocarbon to CO_2 and H_2O, plus calculations, leads to the value of the heat of formation. This kind of calorimetry is the source of the standard heats of formation that are given in Table 7.5, which will be further discussed later.

Heat of Reaction Equation

Three types of chemical reaction: combustion, formation and molecular reactions to form other compounds, are of interest. All of these can be represented by the same symbolic equation, as follows:

$$aA + bB + \ldots = mM + nN + \ldots, \Delta H = \pm \qquad (21.1)$$

$$\Delta H = (mH_M + nH_N + \ldots) - (aH_A + bH_B + \ldots) \quad (21.2)$$

when ΔH is positive, the reaction is endothermic, i.e., heat must be added to keep the system at constant temperature. When ΔH is negative, the reaction is exothermic, i.e., heat must be removed to keep the system at constant temperature.

The individual enthalpies, i.e., H_A, H_B, etc., may be the enthalpies of formation, or they may be the absolute enthalpies that are referred to the same zero enthalpy datum, which is generally the elements in the ideal gas state at absolute zero temperature.

Corrections of the heat of reaction for changes or differences in temperature and pressure are discussed next.

Effect of Temperature

From Equation 21.2, at constant pressure,

$$\left(\frac{\partial(\Delta H_r)}{\partial T}\right)_P = m\left(\frac{\partial H_M}{\partial T}\right)_P + n\left(\frac{\partial H_N}{\partial T}\right)_P + \ldots$$

$$- a\left(\frac{\partial H_A}{\partial T}\right)_P - b\left(\frac{\partial H_B}{\partial T}\right)_P - \ldots \qquad (21.3)$$

The right side of Equation 21.3 is a combination of heat capacities of the products and the reactants. The constant pressure heat capacities can usually be represented by a simple polynomial function of the temperature as follows:

$$C_{pi} = \alpha_i + \beta_i T + \gamma_i T^2 \qquad (21.4)$$

where α_i, β_i and γ_i values are known or can be estimated for each substance, designated by subscript i.

In terms of this heat capacity expression, the effect of temperature on the reaction is obtained from Equation 21.3, as follows:

$$\left(\frac{\partial(\Delta H_r)}{\partial T}\right)_P = \Delta\alpha + \Delta\beta T + \Delta\gamma T^2 \qquad (21.5)$$

where $\Delta\alpha = m\alpha_M + n\alpha_N + \ldots - a\alpha_A - b\alpha_B$ (similar for $\Delta\beta$ and $\Delta\gamma$)

Integrating Equation 21.5 between the limits T_1 and T_2 gives

$$\Delta H_2 - \Delta H_1 = \Delta\alpha(T_2 - T_1) + \frac{\Delta\beta}{2}(T_2^2 - T_1^2)$$

$$+ \frac{\Delta\gamma}{3}(T_2^3 - T_1^3) \qquad (21.6)$$

Equation 21.5 is put into the following more practical general form

$$\Delta H = \Delta H_o + \Delta\alpha T + \frac{\Delta\beta}{2}T^2 + \frac{\Delta\gamma}{3}T^3 \qquad (21.7)$$

where ΔH_o = integration constant, which is usually obtained from the heat of reaction at 25°C.

An alternate method for correcting the heat of reaction for the effect of temperature at constant pressure is by an enthalpy balance technique, using the enthalpy values of the reactants and products. The following diagram illustrates this heat balance method.

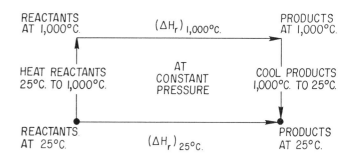

From the above heat balance diagram it is obvious that

$$(\Delta H_r)_{25°C} = (\Delta H)_{\text{Heat Reactants}} + (\Delta H_r)_{1,000°C}$$
$$+ (\Delta H_r)_{\text{Cool Products}} \qquad (21.8)$$

It is a simple matter to find the heat of reaction at 1000°C, using Equation 21.8 with the value of the standard heat of reaction at 25°C and tables or plots of enthalpies vs. temperature for the reactants and products.

Effect of Pressure

The effect of pressure on the heat of reaction is obtained from the value at a standard reference pressure by making a similar heat balance calculation, which refers to the following diagram:

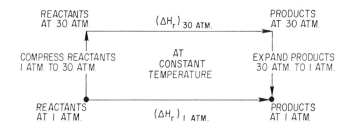

From this heat balance diagram it is obvious that

$$(\Delta H_r)_{1 \text{ Atm}} = (\Delta H)_{\text{compress reactants}} + (\Delta H_r)_{30 \text{ Atm}}$$
$$+ (\Delta H)_{\text{expand products}} \qquad (21.9)$$

Using Equation 21.9, with the values of the heat of reaction at a standard reference pressure and tables, or plots, or equations for the isothermal pressure effects on the enthalpies of the reactants and products, it is easy to calculate the heat of reaction at some other pressure, 30 atmospheres in this case.

Combining Reactions

ΔH may be found for any reaction by combining the values of ΔH for a series of replacement reactions, the combination of which reduces to the original reaction. These replacement reactions do not necessarily represent steps of the process. This procedure is illustrated by the following example for the production of sulfuric acid from steam and sulfur trioxide. The calculations are made by a combination of three oxidation reactions.

Reaction	ΔH, Heat of Reaction at 18°C and one atm
1. $H_{2(g)} + 2.0\ O_{2(g)}$ $+ S_s \rightarrow H_2SO_{4(1)}$	− 193,750 cal/gm mole
2. $H_{2(g)} + 0.5\ O_{2(g)} \rightarrow H_2O_{(1)}$	− 68,370 cal/gm mole
3. $S_{(s)} + 1.5\ O_{2(g)} \rightarrow SO_{(3)}$	− 104,200 cal/gm mole

Subtracting the equations for Reactions 2 and 3 from the equation for Reaction 1, and doing the same for the ΔH values, gives

4. $H_2O_{(1)} + SO_{3(1)} \rightarrow H_2SO_{4(1)}$ − 21,180 cal/gm mole

This technique is used in combining heats of formation to obtain a heat of reaction, for which the general relationship is:

$$\Delta H_r = \underset{\text{products}}{\Sigma \Delta H_f} - \underset{\text{reactants}}{\Sigma \Delta H_f} \qquad (21.10)$$

where ΔH_f = the ΔH for the reaction in which the compound is formed from the elements

Heats of reaction are usually derived in this way from experimental heats of combustion for the substance and its elements because it is easier to get reproducible heats of combustion data experimentally than it is to measure the heat effect of a reaction.

Examples

Typical Heats of Reaction

Heats of reaction are given in Table 21.1 for nine reactions between compounds of carbon, hydrogen, and oxygen. These nine reactions are typical for the hydrocarbon process industries, although the industrial applications of these processes are much more complicated than the reactions shown. Stoichiometric formulas and the ΔH values are given.

Table 21.1
Reactions Between Compounds of Carbon, Hydrogen, and Oxygen

Type	Reaction	ΔH at 77°F M Btu
Oxidation	$C_2H_4 + 3O_2 \rightarrow 2CO_2 + 2H_2O$	− 596.2
Hydrogenation	$C_2H_4 + H_2 \rightarrow C_2H_6$	− 58.9
Cracking	$C_2H_4 \rightarrow C_2H_4 + H_2$	+ 58.9
Cracking	$2C_2H_6 \rightarrow 2CH_4 + C_2H_4$	+ 29.3
Dehydrocyclization	$C_6H_{14} \rightarrow C_6H_6 + 4H_2$	+ 106.5
Polymerization	$2C_2H_4 \rightarrow C_4H_8$	− 99.3
Hydrolysis	$C_2H_4 + H_2O \rightarrow C_2H_5OH$	− 12.5
Fisher-Tropsch	$2CO + 4H_2 \rightarrow C_2H_4 + 2H_2O$	− 138.0
Oxo	$C_2H_4 + CO + H_2 \rightarrow C_2H_5CHO$	− 63.4

It will be noted that some of the ΔH values are negative and some are positive. The negative sign indicates that the reaction is exothermic, i.e., heat is liberated during the reaction. The plus sign indicates that the reaction is endothermic, i.e., heat is absorbed during the reaction. With no heat addition or removal, temperatures would rise in exothermic reactions and decrease in endothermic reactions.

The ΔH values are given at 77°F because this is a widely accepted standard state for reporting such data. ΔH values at this standard temperature may be applied to reactions occurring at other temperatures by including the sensible heat changes for the reactants and the products, as given by Equation 21.8. Pressure effects are calculated by Equation 21.9. The heat effects of phase changes are also considered.

Heats of Reaction from Heats of Combustion

Heats of reaction can be measured in a calorimeter but this is too complex for routine use and is not sufficiently precise for some reactions. The most widely accepted and used method is to obtain the heats of chemical reaction from the heats of formation of the reactants and the products of the reaction, as in Equation 21.10. These heats of formation are obtained from the heats of combustion of the substances and the elements appearing in them.

This method is illustrated in Table 21.2 for the formation of ethylene from graphite and hydrogen. This technique is identical to that used for the sulfuric acid example discussed previously. In both of these illustrations, the phase condition of the elements and compounds are indicated by the subscripts (g), (l), (s), or (graphite) following the formula. Phase change effects may be included by adding another "reaction" and its ΔH value.

Table 21.2
Heat of Formation for Ethylene from Heats of Combustion

Reaction	ΔH at 77°F M Btu/lb mole
1. $C_2H_4(g) + 3O_2(g) \rightarrow 2CO_2(g) + 2H_2O(g)$	− 569.2
2. $C(graphite) + O_2(g) \rightarrow CO_2$	− 169.3
3. $H_2(g) + \frac{1}{2} O_2(g) \rightarrow H_2O(g)$	− 104.0
Combine Equations 1, 2 and 3 and their ΔH values	
4. $2C(graphite) + 2H_2(g) \rightarrow C_2H_4(g)$	+ 22.5

Heats of combustion data, for this purpose, must be obtained with a high degree of precision because the heat of formation calculation applied with the experimental heat of combustion data involves small differences in large numbers. Small errors in heats of combustion become large errors in heats of formation.

Calorimetric research of this kind has been conducted at the Petroleum Research Laboratory of the U.S. Bureau of Mines in Bartlesville, Okla.; at the National Bureau of Standards, Washington, D.C.; the University of California, Berkeley; Stanford University, Palo Alto; and others.

Absolute Enthalpy

As pointed out in Chapter 18, a zero enthalpy datum state of the elements at 0°R and in the ideal gas state results in absolute enthalpies that are referred to the elements at 0°R and include the heat of formation of the compound. Absolute enthalpy of a compound is then a combination of the enthalpies of the elements, the heat of formation, and the enthalpy of the compound.

Absolute enthalpies of the reactants and products of a chemical reaction can be combined directly in making a heat balance on a reactor, without a separate evaluation of the heat of reaction and the effects of temperature and pressure, because these heat effects are all incorporated into the absolute enthalpies.

As an illustration of this absolute enthalpy concept, Figure 21.1 shows several points on an absolute enthalpy chart for ethylene. ΔH for the elements above the 0°R datum of ideal H_2 gas and graphite C is first found to be 8.18 MBtu/lbmole at 536.7°R (see Figure 21.1). To this is added 22.5, the standard heat of formation of ethylene, in the same units, to obtain 30.68, the H value for the vapor at 536.7°R and 0 psia. Next, three other points are calculated for gaseous ethylene, using ideal gas state heat capacities or enthalpies, obtaining values at 0, 1,000, and 1,500°R. The value at 0°R is the heat of formation of ethylene at absolute zero, i.e., ΔH_f^o.

The absolute enthalpy technique is also illustrated by Figure 21.2 for the substances that enter into and come from the synthesis of benzene, i.e., hydrogen, methylcyclopentane, n-hexane, and benzene. Absolute H values are shown on the same plot for all substances for convenient reference and comparison. Note that the heats of formation of the two paraffin hydrocarbons are negative, while the value is positive for benzene, as it is for ethylene on Figure 21.1.

The two phase (liquid-vapor) boundary curve and one vapor phase isobar are included in the benzene part of the diagram. Similar additions could be made for the other hydrocarbons as well, if needed.

Figure 21.3 is a similar absolute enthalpy diagram for the substances that appear in the synthesis of toluene. Note that curves on Figure 21.3 for the toluene synthesis are similar to those on Figure 21.2 for the benzene synthesis.

Evaluating heats of formation and absolute enthalpy is a straightforward procedure for discrete components when the necessary data are available. Fortunately, the data are complete for most of the components in Tables 7.1 and 7.5, but not all.

Pseudo components of petroleum fractions are characterized by normal boiling points and specific gravities of the pseudo components. Methods for predicting the standard heats of combustion and formation are given in Chapter 24.

Petroleum Reactions

Catalytic Cracking Heat of Reaction

Catalytic cracking of a petroleum fraction to produce gas, gasoline, and cycle gas oil is a complex endothermic

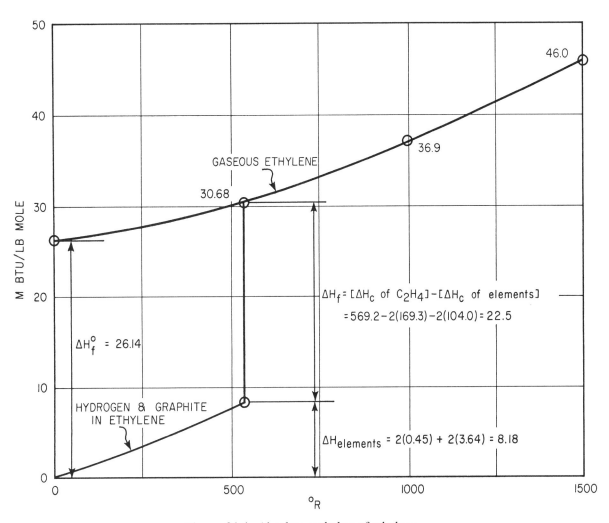

Figure 21.1. Absolute enthalpy of ethylene.

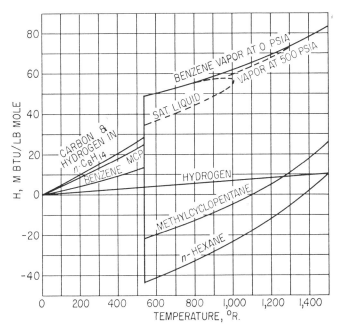

Figure 21.2. Absolute enthalpies of hydrocarbons in benzene synthesis: H = 0 for graphite and H_2 gas at 0°R and 0 psia.

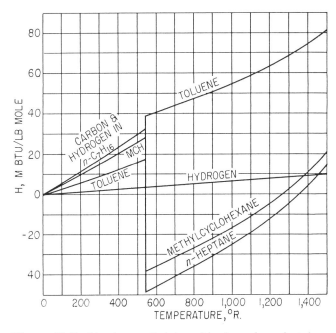

Figure 21.3. Absolute enthalpies of hydrocarbons in toluene synthesis: H = 0 for graphite and H_2 gas at 0°R and 0 psia.

process. There are an infinite number of hydrocarbons in the feed and the products. Four ways of obtaining the heat of reaction for catalytic cracking are:

1. Calculate the heat effects from measurements taken during a test in a reactor-calorimeter.
2. Calculate from plant test data taken from an operating unit.
3. Calculations based on thermodynamic data and analyses of the products.
4. Calculate by combining the heats of combustion data obtained for the feed stock and each of the products with product yields.

To each of these four methods could be added ". . . stoichiometric quantities," which should be obvious to chemical and process engineers.

In all four procedures it is necessary to have experimental yields, conditions, and product compositions. Methods 1 and 2 are similar in that heat balances on an operating reactor are used. Methods 3 and 4 are similar in that the heat of cracking is found by the combination of the heats of several other reactions, as illustrated in Table 21.2.

In 1949, Dart and Oblad (3) presented results of heat of cracking determinations made on an East Texas heavy gas oil. Seven experimental runs covered conversions ranging from 36.0 to 81.5 wt% gas and gasoline, on a once-through basis. Method 4 (above) was used in finding the heat of cracking. Figure 21.4 gives the reactor conditions and yields for one of the experiments. More details of the experiments and the calculations may be found in the published report. Tables 21.3, 21.4, and 21.5 summarize the data and calculations for the run shown in Figure 21.4.

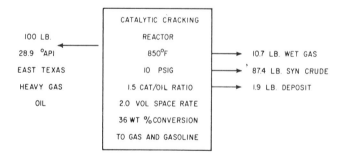

Figure 21.4. Catalytic cracker for gas oil example.

Heats of combustion for the four streams (one feed and three products) are given in Table 21.3. Heats of combustion of the heavy gas oil charge and the synthetic crude (gasoline and cycle gas oil) were determined by experimental measurements, using a constant volume ca-

Table 21.3
Heats of Combustion at 77°F of Materials in Gas Oil Cracking Example

Stream	ΔH Btu/lb	Source
Gas oil charge	− 19,479	experimental
Wet gas	− 21,392	calculated
Synthetic crude	− 19,445	experimental
Catalyst deposit	− 16,740	estimated

Table 21.4
Heat of Cracking at 77°F for Gas Oil Example

		ΔH Btu/ lb charge
Wet gas	(0.107)(− 21,392) =	− 2,289
Synthetic crude	(0.874)(− 19,445) =	− 16,995
Catalyst deposit	(0.019)(− 16,740) =	− 318
Total for reactor effluent		− 19,602
Charge		− 19,479
ΔH_c Btu/lb charge ...		123

Table 21.5
Heat of Cracking at 850°F for Gas Oil Example

Correction from 77°F to 850°F	
ΔH for reactor effluent	+ 551 Btu/lb charge
ΔH for charge	− 572 Btu/lb charge
ΔH correction	− 21 Btu/lb charge
ΔH cracking at 850°F = 123 − 21	102 Btu/lb charge
ΔH cracking at 850°F = 102 ÷ 0.36 ...	283 Btu/lb product
ΔH cracking at 850°F = (283)(77)	21,840 Btu/lb mole product

lorimeter with results corrected to 77°F, and knowing the initial and final temperatures and the heat capacity of the bomb. It is interesting to note that the experimental heat of combustion values for the gas oil charge and the synthetic crude differ by only 17 parts in 10,000, even though the two stocks are of widely different compositions.

The heat of combustion of the wet gas was calculated and that of the coke deposited on the catalyst was estimated. The wet gas heat of combustion calculations were based upon component analyses. Catalyst deposit heats of combustion were estimated from the carbon/hydrogen ratios of the coke plus a plot of ΔH_c vs. C/H ratio for solid hydrocarbons. It is of interest to point out that the catalyst deposit accounts for only 1.9% of the charge, which makes its heat of combustion of minor significance, compared to the other heats of combustion values.

Using the yields given in Figure 21.4 and the heats of combustion values in Table 21.3, the heat of cracking at 77°F was calculated in Table 21.4. Next, the heat of cracking was corrected from 77 to 850°F. The correction calculation is given in Table 21.5, where these final results are given on three bases, i.e., per pound of charge, per pound of product, and per pound mole of product, using an average molecular weight of 77 for the products.

An examination of the last three lines in Table 21.5 shows how sensitive this method is to the accuracy of the heat of combustion determinations. The difference of 123 two numbers that are over 19,000 illustrates the importance of having accurate heats of combustion.

Heat of Cracking Equations

In some other previous work on petroleum feed stocks by Benedict (1), experimental measurements in calorimetric reactors plus accurate mass and energy balances were used to derive heats of thermal and catalytic cracking. From such data on petroleum fractions and cracking conditions of interest, analytical equations were derived. Benedict's equations are included here with permission of The M. W. Kellogg Company.

Thermal cracking
Heat of reaction
   Btu = −1.38 (total lb millimoles of gas and liquid
        made)
        +38.87 (lb millimoles mono olefins made)
        +26.49 (lb millimoles aromatics
        made)   (21.11)

Catalytic cracking
Heat of reaction
   Btu = −0.091 (total lb millimoles of gas and liquid
        made)
        +17.21 (lb millimoles hydrogen made)
        −8.10 (lb millimoles methane made)
        +37.25 (lb millimoles mono olefins made)
        +29.46 (lb millimoles aromatics made)
        +(k + 8,992 w − 433)
        (lb carbon-hydrogen complex made)   (21.12)

where k = heat of formation of C-H complex in Btu/lb
          (between −586 and +10 Btu/lb)
     w = wt% hydrogen in C-H complex

The coefficients of hydrogen and methane in Equation 21.12 were based upon heats of formation. All the other

coefficients were based upon experimental data. Thus Equations 21.11 and 21.12 are largely empirical, although their forms are theoretical. These equations apply at conversion levels of 30 to 40% and at average cracking temperatures of 900°F.

Heat of Reaction Chart

Complete composition and rate information is required for the charge stock and the products of cracking for the application of Equations 21.11 and 21.12. All of this information is not always available so an approximation method for predicting heats of reaction was developed. These Benedict equations were used in the development. Average molecular weights of the charge and the products were suggested by the previous expressions as parameters for a new correlation.

Another background item that was useful in developing the new correlation is a plot of heats of formation at 25°C for different types of hydrocarbons against molecular weight. This plot is given in Figure 21.5 where it can be noted that, except for the component at the start of each class, straight and parallel lines are formed. The parallel lines suggest that molecular weight differences as an important parameter.

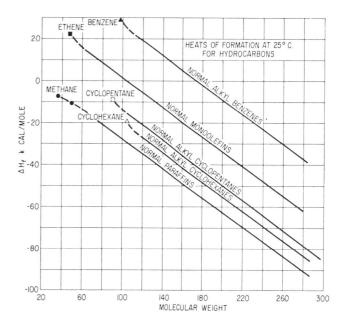

Figure 21.5. Heats of formation at 25°C for hydrocarbons.

For the development of a heat of reaction correlation, a knowledge of charge stocks and the proportions of different types of hydrocarbons in the products is needed for each petroleum refining process of interest. The PONA (Paraffin, Olefin, Naphthene, Aromatic) analyses of the gasoline from six different processes are given in Table 21.6.

Table 21.6
Typical PONA Compositions of Gasolines from Six Cracking Processes

Percent	P	O	N	A
Thermal Cracking				
Mixed phase lt. gas oil	40	39	14	7
Mixed phase hvy. gas oil	42	44	8	6
Vapor phase, De Florez	10	49	25	16
Vapor phase, Gyro	19	45	14	22
Catalytic Cracking				
Houdry	44	18	23	15
Destructive Hydrogenation				
Gas Oil	63	10	22	5

These data were used in determining the variations of the constant in the correlation for different processes. The following formulation was found to be a good approximation for the endothermic heat of reaction in petroleum refining processes.

$$\text{Heat of reaction, Btu/lb} = C_R(M_c - M_P)/M_c M_P \quad (21.13)$$

where C_R = constant for each process and feedstock type

M_c = average molecular weight of charge or feed

M_P = average molecular weight of products, i.e., gas + gasoline + cycle oil + fuel oil.

Values of the constant C_R in Equation 21.13 are given in Table 21.7 for ten refining processes.

A two-parameter chart was prepared from Equation 21.13 and the constants in Table 21.7. This chart is given as Figure 21.6, on which the 10 processes are identified by the alphabetic key in Figure 21.6. An example (dashed lines with arrows) illustrates the method of reading the chart. This chart and equation are applicable only to endothermic reactions and should not be applied to exothermic processes. A similar chart might be developed along similar lines for exothermic reactions, however.

Chemical Processes

Two non-hydrocarbon examples will illustrate the absolute enthalpy technique of making reactor heat balances. These are (a) synthesis of ammonia at 400 atmospheres, and (b) recovery of sulfur from H_2S at approximately atmospheric pressure. Both of these reactors operate at high temperatures, i.e., 1,020 and 2,200°F, respectively. A pressure correction to the en-

Table 21.7
Approximate Endothermic Heat of Reaction Equation Constant for 10 Petroleum Refining Processes

Curve in Figure 21.6	Process description	Values of C_R in Equation 21.13
A	Low pressure vapor phase thermal cracking gas oil .	73,200
B	Vapor phase thermal cracking gas oil	67,500
C	Mixed phase thermal cracking heavy gas oil . .	53,600
D	Mixed phase thermal cracking light gas oil	48,800
E	Thermal reforming naphtha	43,700
F	Catalytic cracking heavy gas oil	38,200
G	Catalytic cracking light gas oil	33,700
H	Catalytic reforming naphtha	30,100
I	Destructive hydrogenation gas oil	23,600
J	Destructive hydrogenation naphtha	19,350

thalpy is required in the ammonia process heat balance but no such correction is required in the sulfur calculations.

Ammonia Synthesis

The heat balance calculations for the ammonia synthesis process are given in Tables 21.8 and 21.9. Table 21.8 gives the calculation of the absolute enthalpy and Table 21.9 gives the actual heat balance. Note that the feed gas to the reactor contains 5% NH_3 with the 3:1 syn gas. Also note that the temperatures are 850°F inlet and 1,020°F outlet. These conditions are from other process

Table 21.8
Absolute Enthalpies for Ammonia Synthesis Calculation

	Ideal gas state enthalpy above −459.6°F Btu/lb mole		Btu/lb mole		Absolute enthalpy at 6,000 psia Btu/lb mole	
			Heat of Formation	Pressure correction to 6,000 lb		
	850°F	1,020°F	ΔH^o_f	ΔH_P	850°F	1,020°F
Hydrogen . . .	8,300	10,137	0	+ 300	8,600	10,437
Nitrogen	8,480	10,496	0	+ 300	8,780	10,796
Ammonia . . .	12,000	. . . . .	− 16,657	− 1,600	− 6,257	. . . . .
Ammonia . . .	. . . . .	14,020	− 16,657	− 1,110	. . . . .	− 3,747

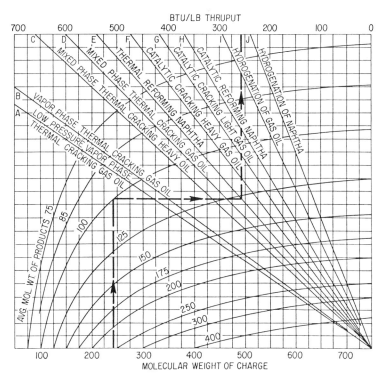

Figure 21.6. Approximate endothermic heats of reaction.

Table 21.9
Ammonia Synthesis Reactor Heat Balance
Pressure = 6,000 psia

	Moles/hr	Btu/mole	Btu/hr
Inlet (850°F)			
Hydrogen	71.25	8,600	612,500
Nitrogen	23.75	8,780	208,600
Ammonia	5.00	− 6,257	− 31,290
Total	100.00		789,810
Outlet (1,020°F)			
Hydrogen	58.2	10,437	607,000
Nitrogen	19.4	10,796	209,300
Ammonia	13.7	− 3,747	− 51,300
Total	91.3		755,000
Net exothermic heat to be removed from reactor ...			34,810

considerations rather than the enthalpy and heat balance techniques.

Using reactor material balance information, from another source, and the enthalpy values in Table 21.8, the heat balance shown in Table 21.9 was obtained. Note that the input total is higher than the output by 34,810 Btu/hr, which must be removed from the reactor during the synthesis operation in order to maintain the outlet temperature at 1,020°F.

Sulfur Recovery

The sulfur calculations are given in Tables 21.10 and 21.11. In this process sulfur is produced from hydrogen sulfide by burning the H_2S in air and then condensing the water and recovering the sulfur in the condensation operation. The reaction occurs at 2,200°F and low pressure. An 85% conversion of the hydrogen sulfide is assumed in these calculations, i.e., 85% of the H_2S goes to S_2 and SO_2. The recovery and recycle of the SO_2 is not included

Table 21.10
Sulfur from H₂S Reactor Material Balance

Reactor conditions: Temperature = 2,200°F			
H₂S conversion = 85%			
	Moles	MW	lb
Charge gas: H_2S	33.0	34	1,122
O_2	16.5	32	528
N_2	78.5	28	2,200
Total input	128.0		3,850
Product gas: H_2S (33)(0.15)	4.95	34	168.4
H_2O (33)(0.85)	28.05	18	505.0
SO_2 (16.5 − 28.05/2)	2.475	64	158.4
S_2 (28.05 − 2.475)/2	12.85	64	818.2
N_2	78.50	28	2,200.0
Total output	126.825		3,850.0

Table 21.11
Sulfur from H₂S Reactor Heat Balance

Gas	Moles	$H° - H_o°$	$\Delta H°_f$	Btu
Input 100°F				
H₂S	33.0	4,479	− 35,754	− 1,012,000
O₂	16.5	3,884	0	64,100
N₂	78.5	3,887	0	305,000
Total Input	128.0			− 642,900
Output 2,200°F				
H₂S	4.95	26,451	− 35,451	− 44,550
H₂O	28.05	24,454	− 102,793	− 2,198,000
SO₂	2.475	30,840	− 154,343	− 305,500
S₂	12.85	24,200	0	+ 311,000
N₂	78.50	19,907	0	+ 1,563,000
Total Output ..	126.825			− 674,050
Net heat to be removed from reactor				31,150

in these calculations. The material balance is given in Table 21.10.

Two types of enthalpy values are required, i.e., the ideal gas state enthalpies and the heats of formation. These are given in Table 21.11, along with the heat balance calculations. The numbers should be self explanatory. With an inlet temperature of 100°F, it is necessary to remove 31,150 Btu from the reactor in order to maintain the outlet temperature at 2,200°F.

These results, and those for the NH₃ reactor, illustrate an important point about the use of this technique in making reactor heat balances. In both cases the net exothermic heats are differences between large numbers so that small errors in the enthalpies, especially the heats of formation, have a large effect on the overall results.

Computer Program for Heat of Formation

Heats of formation for discrete components, such as the hydrocarbons and associated substances that appear in BLOCK DATA of the EQUIL program, can be calculated at any desired temperature, providing all of the necessary coefficients and constants are available. This calculation is made by Equation 7.32 (from Volume 1):

$$\Delta H^*_{f,T} = \Delta H^*_{f,536.7} + \left(\Delta H^* - \sum_i^N n_i \Delta H^*_i \right) \qquad (7.32)$$

where $\Delta H^* = H^* - H^*_{536.7}$ for the compound
n_i = number of moles of element i per mole of compound
$\Delta H^*_i = H^*_i - H^*_{i,536.7}$ for element i

Equation 7.32 can be solved to find the heat of formation for each component in the reaction, i.e., both reactants and products, and these individual heats of formation combined to give the heat of reaction. This is illustrated by Equations 21.1 and 21.2. For the solution of Equation 7.32, it is necessary to have the following data for each component: (a) amount of each element, i.e., C, H, etc., in the component molecule; (b) standard heat of formation at 536.7°R (25°C or 77°F); and (c) coefficients of the ideal gas state enthalpy equation. The first of these items comes from the component formula. The enthalpy equation coefficients are given in Table 7.5 for the compound of interest and all of the elements except sulfur. All of these data, and sulfur, have been included in BLOCK DATA of the EQUIL computer program.

The element composition information in BLOCK DATA is given by a term, called CHONS, which is made up of seven integers, two of two digits each and three of one digit each. These seven integers represent the number of atoms of carbon, hydrogen, oxygen, nitrogen and sulfur in the formulas as given by two, two, one, one, and one digits, respectively, reading from left to right. The computer program reads the value of CHONS from Block Data, and then interprets that reading to get the number of atoms of each element in the molecule of the compound, i.e., the n_i values in Equation 7.32.

The remainder of the programming for the solution of Equation 7.32, in the calculation of the heat of formation of each component is straightforward. After finding the heats of formation for each compound involved in the reaction, the heat of reaction can be calculated by combining these heats of formation, as has been previously illustrated. This combination is made in the computer program by using the amounts and algebraic signs of the substances as they appear in Equation 21.2. Note that m and n are positive, while a and b are negative. In this way the summation is very easy.

The FORTRAN program was prepared, and tested, for calculating enthalpy changes for chemical reactions. Later another program was prepared for free energy calculations in chemical reactions. This program, which is presented in Chapter 22, and named EQRK3, includes the enthalpy change calculations, as well as the free energy changes and the reaction equilibrium constant.

For illustrating this heat of reaction program, two reactions of methane, ethene and propane have been calculated: one for propane formation from methane and ethene, and the other for propane cracking. Both are at the same temperature of 1,000°F. The results of these calculations are given in Table 21.12. Note that the component heats of formation are the same for both directions, but that the signs of the number of moles are opposite according to direction.

Table 21.12
Results of Computer Calculations of Enthalpy Changes for Reactions of Methane, Ethene, and Propane

Methane + Ethene = Propane Reaction (polymerization)

Heats of Formation at Temperature of 1,000.00°F = 1,459.70°R

Component	Enthalpy Btu/lb mole	Moles
Methane	− 37,509.41	− 1.00
Ethene	17,313.48	− 1.00
Propane	− 54,358.33	1.00
Heat of reaction	− 34,162.40 Btu/lb mole propane	

Propane = Methane + Ethene Reaction (cracking)

Heats of formation at temperature of 1,000.00°F = 1,459.70°R

Component	Enthalpy Btu/lb mole	Moles
Methane	− 37,509.41	1.00
Ethene	17,313.48	1.00
Propane	− 54,358.33	− 1.00
Heat of reaction	34,162.40 Btu/lb mole propane	

The first reaction is exothermic, i.e., heat must be removed to keep temperature constant during the reaction. The second reaction is endothermic, i.e., heat must be added during the reaction to keep temperature constant. Other heat effects, such as sensible heat, would be included with these heats of reaction in making a heat balance on the reactors.

"Moles" in Table 21.12 designate the number of moles of each component in the reaction and its position relative to the equal sign in the equilibrium reaction equation. The minus sign means "feed or left side" and the plus sign means "product or right side" of the equation. These signs control the computer summation of the component enthalpies.

EQRK3, the program for reaction equilibrium and heat effects calculations that is presented in Chapter 22, has been made into Subroutine REACK and integrated into EQUIL.

Alternate Method

An alternate method for calculating the enthalpy of formation, and the heat of combustion of a petroleum fraction was recently presented (5). This method is given in Chapter 24.

References

1. Benedict, M. The M. W. Kellogg Company, Jersey City Laboratory, (1940).
2. Canjar, L. N. and F. S. Manning, *Thermodynamic Properties and Reduced Correlations for Gases,* Gulf Publishing Company, Houston, (1967).
3. Dart, J. C. and A. G. Oblad, "Heat of Cracking and Regeneration in Catalytic Cracking," *Chem. Engr. Prog.* 45, 110 (1949).
4. Starling, K. E., *Fluid Thermodynamic Properties for Lighter Hydrocarbons,* Gulf Publishing Company, Houston, (1973).
5. Montgomery, R. L., "Estimation of Thermochemical Properties of Undefined Hydrocarbon Mixtures: Heats of Combustion of Petroleum Fractions," presented at the American Chemical Society National Meeting, Kansas City, Missouri, Sept. 15, 1982.

22

Chemical Reaction Equilibria

The equilibrium distribution of the components leaving the chemical reactor, i.e., product and unreacted feed components, is another important and interesting application of thermodynamics. This chapter presents theoretical relationships for predicting reaction equilibrium constants, charts of equilibrium constants for twelve reactions, computer calculation of reaction equilibrium K-values, and examples of reactor equilibrium calculations.

This material is covered in two parts, corresponding to the two phases of the work reported. First, there will be theory, charts, and illustrations of the original first edition material. Then, a new computer method is described that can be applied to all of the 213 components of Tables 7.1 and 7.5 (Volume 1), i.e., BLOCK DATA of the EQUIL program, and calculates reaction equilibrium.

Theoretical Relationships

Equilibrium Criterion

Equilibrium is expressed symbolically for the same isothermal homogeneous reaction that was used previously in the heat of reaction presentation, i.e.,

$$aA + bB + \ldots = mM + nN + \ldots \tag{21.1}$$

where lower case letters designate the number of moles, and upper case letters designate the identity of components.

The criterion of equilibrium, $\Sigma \mu_i N_i = 0$, is applied by writing Equation 21.1 in the following form:

$$\mu_A dN_A + \mu_B dN_B + \ldots$$
$$+ \mu_M dN_M + \mu_N dN_N + \ldots = 0 \tag{22.1}$$

Dividing by dN_A gives

$$\mu_A + \mu_B \frac{dN_B}{dN_A} + \ldots$$

$$+ \mu_M \frac{dN_M}{dN_A} + \mu_N \frac{dN_N}{dN_A} + \ldots = 0 \tag{22.2}$$

From stoichiometric considerations

$$\frac{dN_B}{dN_A} = \frac{b}{a}; \ \frac{dN_M}{dN_A} = -\frac{m}{a}; \ \& \ \frac{dN_N}{dN_A} = -\frac{n}{a},$$

which are combined with Equation 22.2 to give

$$a\mu_A + b\mu_B + \ldots - n\mu_M - n\mu_N - \ldots = 0 \tag{22.3}$$

The chemical potential, μ_i, of each component in Equation 22.3 may be replaced by the following function of activity, a_i:

$$\mu_i = \mu^o_i + RT \ln a_i \tag{22.4}$$

where a_i = activity of component "i"

Substituting Equation 22.4 into Equation 22.3 and changing signs,

$$-a(\mu^o_A + RT \ln a_A) - b(\mu^o_B + RT \ln a_B) - \ldots$$
$$+ m(\mu^o_M + RT \ln a_M) +$$
$$n(\mu^o_N + RT \ln a_N) + \ldots = 0 \qquad (22.5)$$

Collecting terms gives the following equilibrium relationship

$$\Delta\mu^o = -RT \ln \frac{a_M^m \, a_N^n}{a_A^a \, a_B^b} \qquad (22.6)$$

where $\Delta\mu^o = (m\mu^o_M + n\mu^o_N + \ldots$
$$= a\mu^o_A - b\mu^o_B - \ldots) \qquad (22.7)$$

$\Delta\mu^o$ is a constant that is independent of everything but temperature. The equilibrium constant is defined as follows:

$$\Delta\mu^o = -RT \ln K_a \qquad (22.8)$$

where K_a = reaction equilibrium constant, which is a function of temperature and independent of pressure.

Reaction Equilibrium Constants

There are four different reaction equilibrium constants of interest. These are defined for convenient use later. Two are identical and all are interrelated, as will be shown.

K_a = activity equilibrium constant
K_f = fugacity equilibrium constant
K_p = pressure equilibrium constant
K_ϕ = fugacity-coefficient equilibrium constant

The interrelations between these equilbrium constants is described next. By definition, the activity is $a_i = \bar{f}_i/f^o_i$, which is combined with the definition of K_a that is given by Equations 22.6, 22.7 and 22.8 to give the following:

$$K_a = \left(\frac{a_M^m \, a_N^n}{a_A^a \, a_B^b}\right) = \frac{(\bar{f}_M/f^o_M)^m (\bar{f}_N/f^o_N)^n}{(\bar{f}_A/f^o_A)^a (f_B/f^o_B)^b}$$

$$= \left(\frac{\bar{f}_M^m \bar{f}_N^n}{\bar{f}_A^a \bar{f}_B^b}\right)\left(\frac{f^{oa}_A f^{ob}_B}{f^{om}_M f^{on}_N}\right) \qquad (22.9)$$

By definition,

$$K_f = \frac{\bar{f}_M^m \bar{f}_N^n}{\bar{f}_A^a \bar{f}_B^b} \qquad (22.10)$$

By choosing unit fugacity as the reference state

$$\frac{f^{oa}_A f^{ob}_B}{f^{om}_M f^{on}_N} = 1.0$$

These lead to the simplification: $K_a = K_f$.

It is convenient to define an equilibrium constant in terms of partial pressures, as follows:

$$K_P = \frac{(Y_M P)^m (Y_N P)^n}{(Y_A P)^a (Y_B P)^b} \qquad (22.11)$$

where Y_i = mol fraction of "i" in equilibrium products from the reactor*
P = reactor (system) pressure

K_P is the equilibrium constant that is used in the stoichiometric calculations, so it is important to know the relationship between K_P and K_f (or K_a). For a real gas mixture

$$\bar{f}_i = Y_i P \phi_i \qquad (22.12)$$

Combining Equations 22.10 and 22.12 gives

$$K_f = \frac{(Y_M P)^m (Y_N P)^n}{(Y_A P)^a (Y_B P)^b} \frac{\phi_M^m \, \phi_N^n}{\phi_A^a \, \phi_B^b} \qquad (22.13)$$

from which

$$K_f = K_p K\phi = K_a \qquad (22.14)$$

where

$$K\phi = \frac{\phi_M^m \, \phi_N^n}{\phi_A^a \, \phi_B^b} \qquad (22.15)$$

In many cases the fugacity coefficient ϕ_i may be approximated by f_i/P. The fugacity coefficient terms are evaluated from P-V-T data, equations of state, or generalized charts.

Evaluation of Equilibrium Constants

Reaction equilibrium constants may be obtained from experimental measurements made in a laboratory reactor in which the reaction can be studied under the desired temperature (usually isothermal), pressure, and equilibrium conditions. In this experimental method, the value of K_p is found from Y_i values and the pressure. Then K_p is

---

* Capital Y is used here to designate vapor mol fraction so as to be consistent with the charts, which had already been prepared, using this symbol rather than y_i.

combined with the fugacity corrections to obtain K_f or K_a, where needed, or converted to other pressures, where this is required. An example of the experimental method of evaluating the reaction equilibrium constant is given later.

It is frequently more convenient to compute the reaction equilibrium constants by thermodynamic methods. The procedure is to combine three types of data for the reactants and the products of the reaction, namely, (a) heat of formation at a reference temperature, (b) heat capacities, and (c) free energy of formation at a reference temperature.

Tabulations have been prepared giving values of these three thermodynamic data for most hydrocarbons and chemicals of interest to petroleum and process engineers. These property values are given as functions of temperature over wide ranges.

The equilibrium constant for a chemical reaction is defined in Equation 22.8 in terms of the change in chemical potential. When pure components are the reference states, $\mu^o_i = G^o_i$, and this definition may be written

$$\Delta(G^o_i) = -RT \ln K_a \qquad (22.16)$$

Equation 22.16 is the basic relationship for the thermodynamic method of calculating chemical reaction equilibrium constants.

Free Energy Relationships

Differentiating G/T with respect to temperature, at constant pressure, and combining with $dG/dT = -S$, give

$$\frac{d(G/T)}{dT} = -\frac{G}{T^2} + \frac{1}{T}\frac{dG}{dT} = -\frac{G + T\dfrac{dG}{dT}}{T^2}$$

$$= -\left(\frac{G + TS}{T^2}\right) = -\frac{H}{T^2} \qquad (22.17)$$

For a chemical reaction this expression may be written

$$d\left(\frac{\Delta G^o}{T}\right) = -\frac{\Delta H^o}{T^2}\, dT \qquad (22.18)$$

From Equation 21.7 we can write

$$\Delta H^o = \Delta H^o_o + (\Delta\alpha)T + \frac{1}{2}(\Delta\beta)T^2 + \frac{1}{3}(\Delta\gamma)T^3 \qquad (22.19)$$

where ΔH^o_o = value of ΔH^o at $T = 0$, which integration constant is determined by using the ΔH^o value at some given temperature.

Combining Equations 22.18 and 22.19, integrating and solving for ΔG^o, gives

$$\Delta G^o = \Delta H^o_o - (\Delta\alpha)T \ln T$$

$$- \frac{1}{2}(\Delta\beta)T^2 - \frac{1}{6}(\Delta\gamma)T^3 + IT \qquad (22.20)$$

where I = integration constant

The value of ΔG^o at one temperature is required for the evaluation of the integration constant, I.

After the integration constants, ΔH^o_o and I, have been determined, Equations 22.16 and 22.20 may be used to find values of the equilibrium constant, K_a. From Equations 22.16 and 22.18 if follows that

$$\frac{d \ln K_a}{d\dfrac{1}{T}} = -\frac{\Delta H^o}{R} \qquad (22.21)$$

Third Law Application

As previously stated, it is necessary to have a value of G^o for the given reaction at some temperature in order to calculate the value of the integration constant I in Equation 22.20. This information is more difficult to find, or determine, than is the value of H^o required in the evaluation of the integration constant H^o_o.

Changes in absolute entropy, as well as enthalpy, for the reaction are involved, as can be seen in the following basic thermodynamic relationship that is used in the computation of G^o;

$$G^o = H^o - T S^o \qquad (22.22)$$

where S^o = entropy change for reaction at T

All the terms in Equation 22.22 must be in absolute units and referred to correct datum states. Absolute enthalpies were discussed in Chapter 21.

Absolute entropies are evaluated relative to a special reference state, using the Third Law of Thermodynamics, which states that "the entropy of a perfect crystalline substance is zero at absolute zero temperature." This is the zero entropy datum. Absolute entropy for each substance is obtained by evaluating the integral

$$S = \int_0^T (C_P/T)\, dT \qquad (22.23)$$

Heat capacities at temperatures down to $T = 0$ are required for solving Equation 22.23. Special calorimetric research, i.e., measurements and data processing, must be employed to get these low temperature heat capacities. Such research, which is done in very few, properly equipped laboratories, is outside the scope of this book.

Free Energy Tabulations

Tabulations of $((G_o - H^o_o)/T)$, the free energy function for the formation reactions of various substances, including hydrocarbons and chemicals, have been prepared and are available in the API Project 44 publications (1). Values of this function can be used in finding G_o for any reaction of interest, by using the expression derived as follows:

For any reaction

$$\Delta \left(\frac{G^o - H^o_o}{T}\right) = \sum \left(\frac{G^o - H^o_o}{T}\right)_{products}$$

$$- \sum \left(\frac{G^o - H^o_o}{T}\right)_{reactants} \qquad (22.24)$$

The left side of Equation 22.24 may be written

$$\Delta \left(\frac{G^o - H^o_o}{T}\right) = \frac{\Delta G^o - \Delta H^o_o}{T} = \frac{\Delta G^o}{T} - \frac{\Delta H^o_o}{T}$$

from which

$$\frac{\Delta G^o}{T} = \frac{\Delta H^o_o}{T} + \Delta \left(\frac{G^o - H^o_o}{T}\right) \qquad (22.25)$$

Combining Equations 22.16 and 22.25 gives

$$\ln K_a = -\frac{1}{R} \left[\frac{\Delta H^o_o}{T} + \Delta \left(\frac{G^o - H^o_o}{T}\right)\right] \qquad (22.26)$$

where ΔH^o_o = heat of reaction at 0°K, calculated via H^o_o values from tabulations,

$$\left(\frac{G^o - H^o_o}{T}\right) = \text{free energy function, from tabulations.}$$

Values of H^o_o and of $(G^o - H^o_o)/T$ are given in standard tables for hydrocarbons and chemicals of interest, the free energy function being a function of temperature. These may be read from the tabulations for the reactants and the products of any given reaction and then combined in Equation 22.26 to give the value of K_a.

Manual Applications

Charts for Reaction Equilibrium Constants

Using free energy function tabulations from another source, standard heats of formation and the previous equations, values of equilibrium constants were calculated and plotted for twelve reactions of C, H_2, O_2, H_2O, CH_4, CO_2, CO, H_2S, and SO_2. These charts are at the end of this chapter. The charts for the first ten of these reactions were calculated and plotted in 1947 by Hydrocarbon Research Inc. using thermodynamic information originally published by API Project 44 (1). The charts were reproduced in the first edition by permission of HRI (2). That source is again acknowledged in this reprinting of them.

Reaction	Figure
$C + 2H_2 = CH_4$	22.1
$C + H_2O = H_2 + CO$	22.2-1 and 22.2-2
$2 H_2O + C = 2 H_2 + CO_2$	22.3-1 and 22.3-2
$CO_2 + C = 2 CO$	22.4-1 and 22.4-2
$CO + H_2O = CO_2 + H_2$	22.5
$3 H_2 + CO = CH_4 + H_2O$	22.6
$H_2 + \frac{1}{2}O_2 = H_2O$	22.7-1 and 22.7-2
$CO + \frac{1}{2}O_2 = CO_2$	22.8-1 and 22.8-2
$CO + H_2S = COS + H_2$	22.9
$H_2S + CO_2 = H_2O + COS$	22.10
$2 H_2 + S_2 = 2 H_2S$	22.11-1 and 22.11-2
$SO_2 + 3 H_2 = H_2S + 2 H_2O$	22.12-1 and 22.12-2

The charts for the last two reactions were prepared from similar information (6,7) using the same techniques.

The coordinates of all these charts are temperature in °F and equilibrium constant K_p, which assume that $K_\phi = 1.0$, a low pressure condition. If necessary, these charts can be applied at high pressures by using the chart reading as K_a and combining with K_ϕ to obtain a corrected K_p.

As equilibrium constant values were calculated by Equation 22.16, it would have been more general to label the equilibrium constant scales as K_a, rather than K_p. Since most applications of these charts are at low pressures where $K_p = K_a$, it was more convenient to label the scales K_p, as shown. These K_p values are used in calculating reactor effluent compositions.

(text continued on page 177)

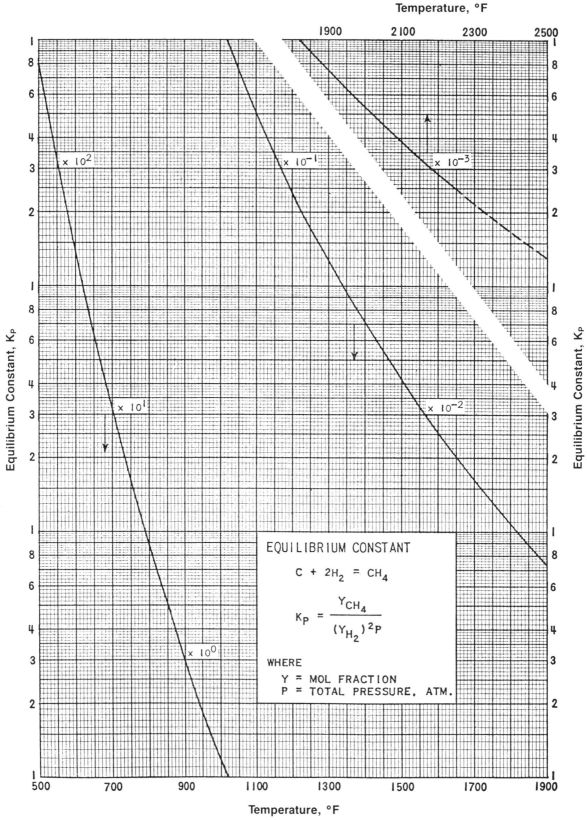

EQUILIBRIUM CONSTANT

$$C + 2H_2 = CH_4$$

$$K_P = \frac{Y_{CH_4}}{(Y_{H_2})^2 P}$$

WHERE

Y = MOL FRACTION
P = TOTAL PRESSURE, ATM.

Figures 22.1–22.10 courtesy Hydrocarbon Research, Inc.

Figure 22.1

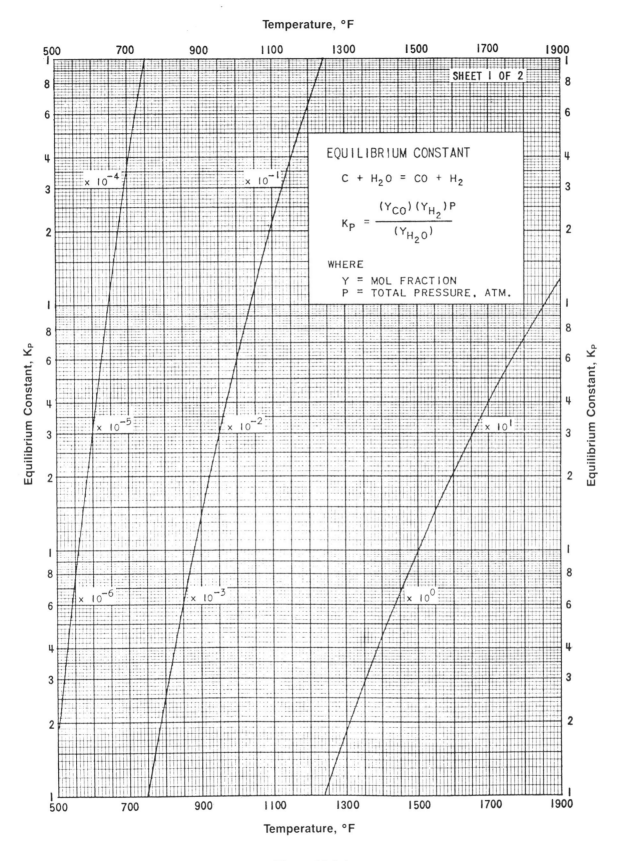

Temperature, °F

EQUILIBRIUM CONSTANT

$$C + H_2O = CO + H_2$$

$$K_P = \frac{(Y_{CO})(Y_{H_2})P}{(Y_{H_2O})}$$

WHERE

Y = MOL FRACTION
P = TOTAL PRESSURE, ATM.

SHEET 1 OF 2

Equilibrium Constant, K_P

Temperature, °F

Figure 22.2-1

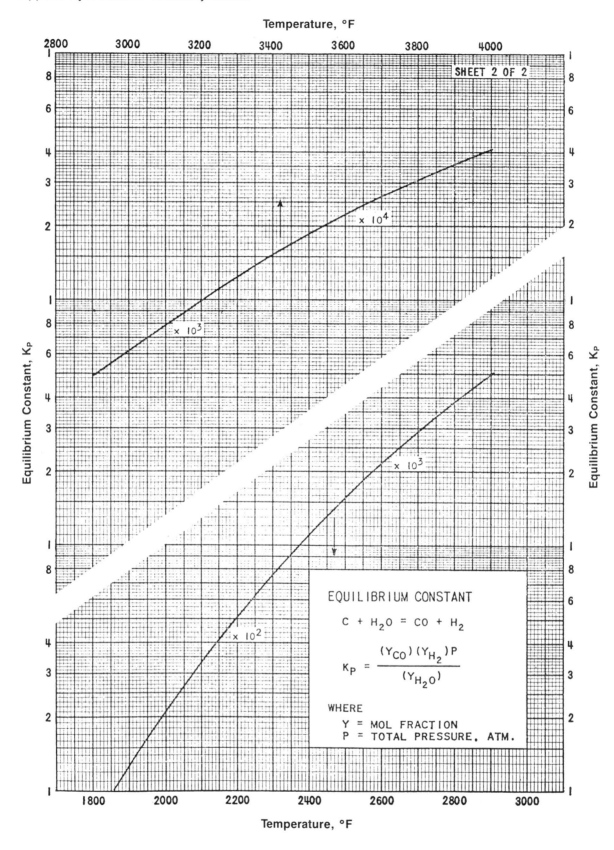

Figure 22.2-2

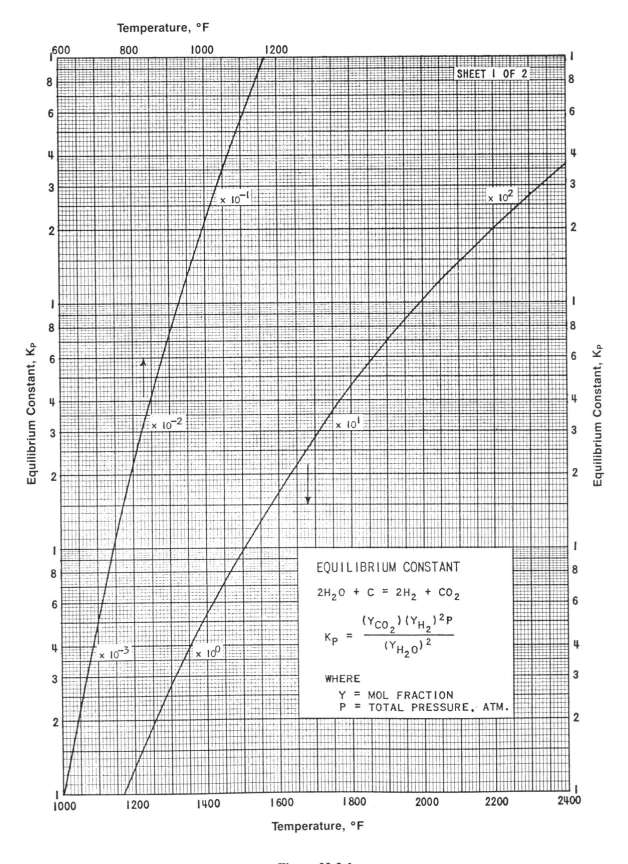

Figure 22.3-1

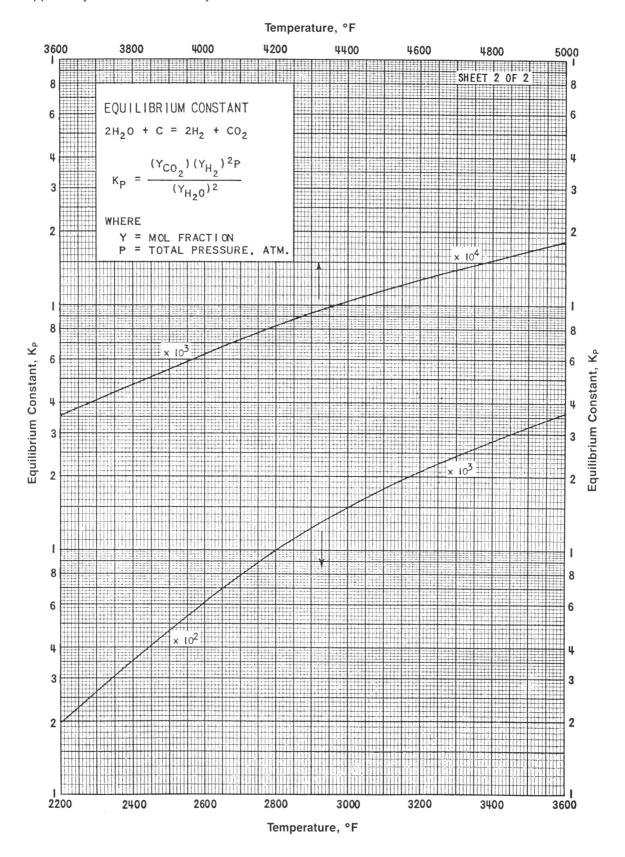

Figure 22.3-2

Temperature, °F

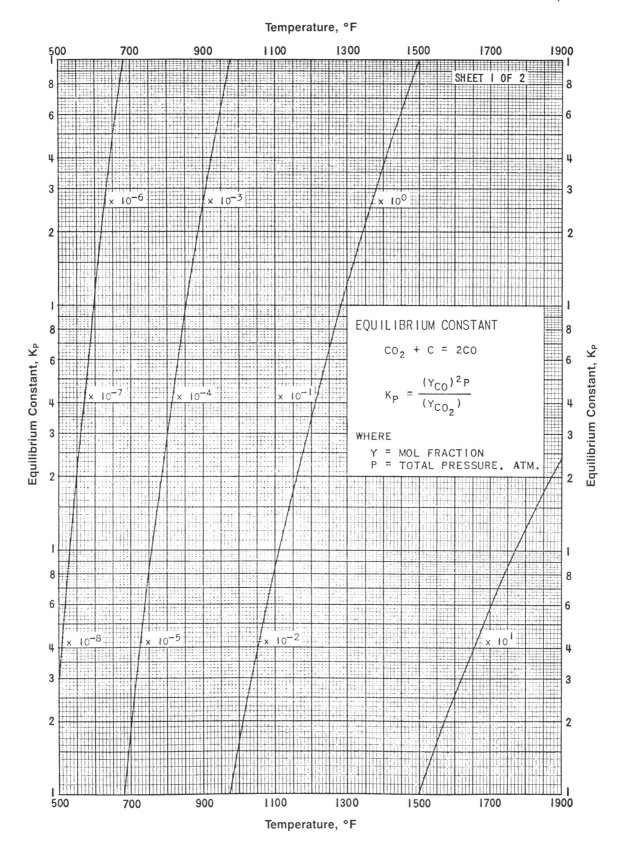

EQUILIBRIUM CONSTANT

$$CO_2 + C = 2CO$$

$$K_P = \frac{(Y_{CO})^2 P}{(Y_{CO_2})}$$

WHERE

Y = MOL FRACTION
P = TOTAL PRESSURE, ATM.

SHEET 1 OF 2

Temperature, °F

Figure 22.4-1

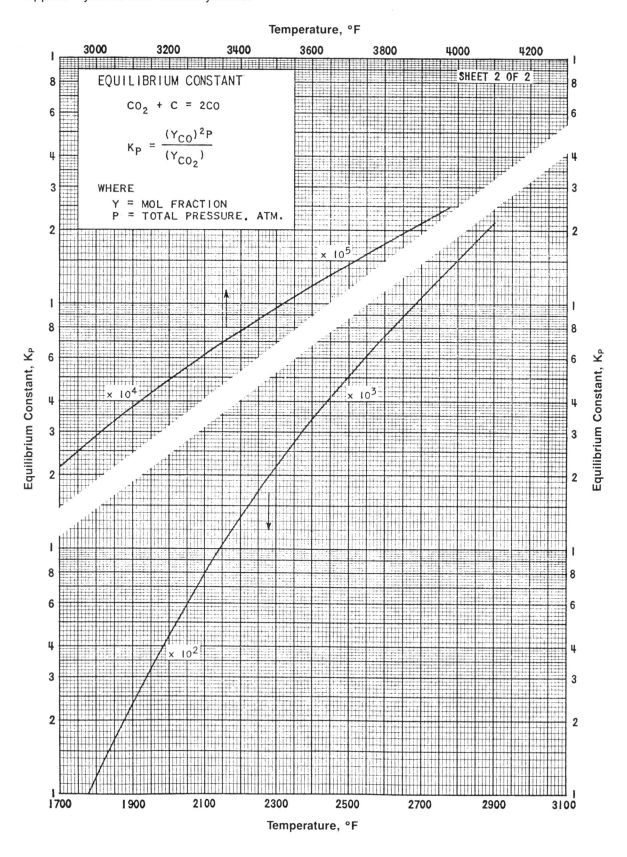

Figure 22.4-2

Temperature, °F

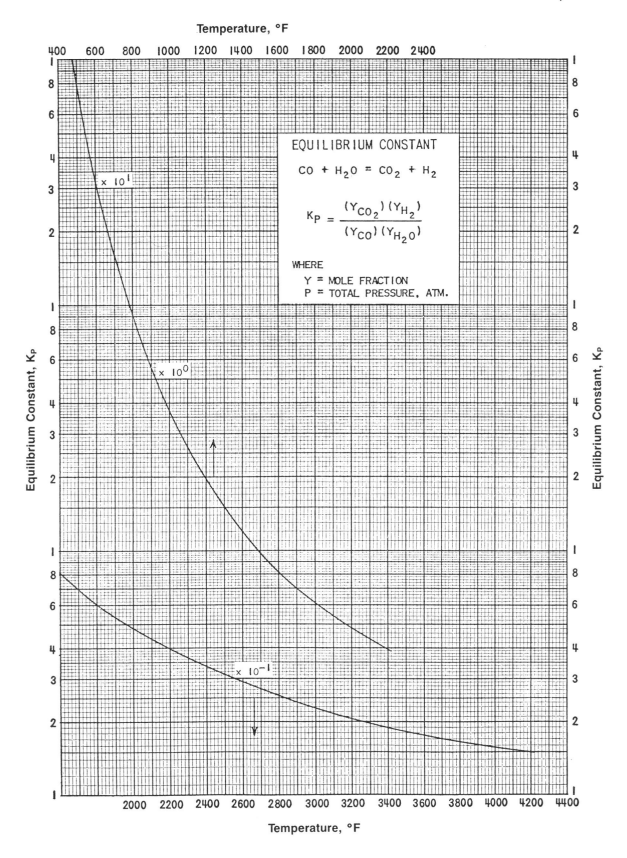

EQUILIBRIUM CONSTANT

$CO + H_2O = CO_2 + H_2$

$$K_P = \frac{(Y_{CO_2})(Y_{H_2})}{(Y_{CO})(Y_{H_2O})}$$

WHERE
Y = MOLE FRACTION
P = TOTAL PRESSURE, ATM.

Temperature, °F

Figure 22.5

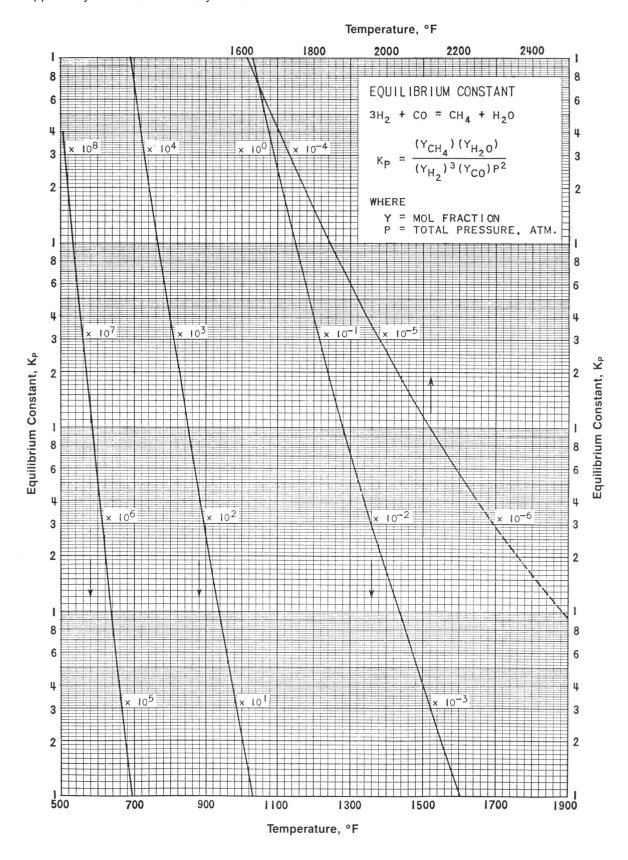

Temperature, °F

EQUILIBRIUM CONSTANT

$$3H_2 + CO = CH_4 + H_2O$$

$$K_P = \frac{(Y_{CH_4})(Y_{H_2O})}{(Y_{H_2})^3(Y_{CO})P^2}$$

WHERE

Y = MOL FRACTION
P = TOTAL PRESSURE, ATM.

Equilibrium Constant, K_P

Temperature, °F

Figure 22.6

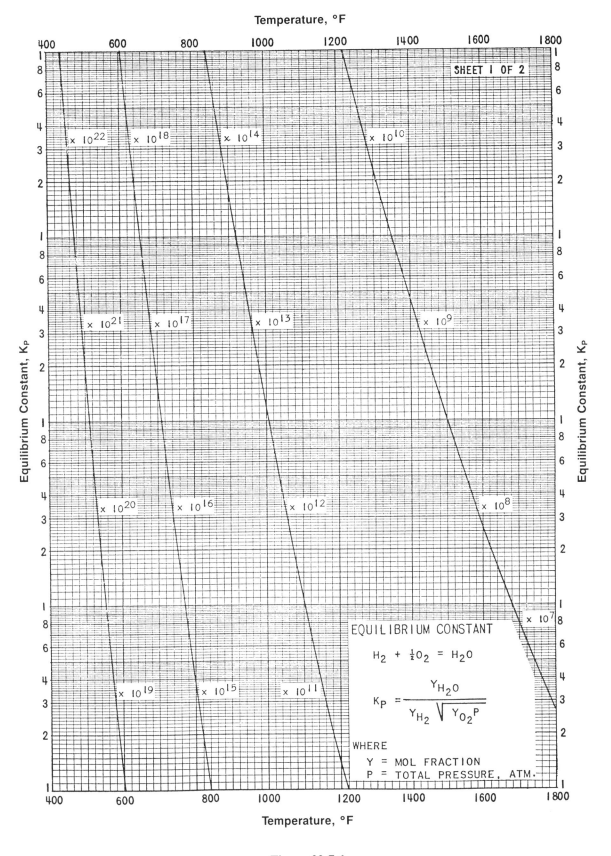

Figure 22.7-1

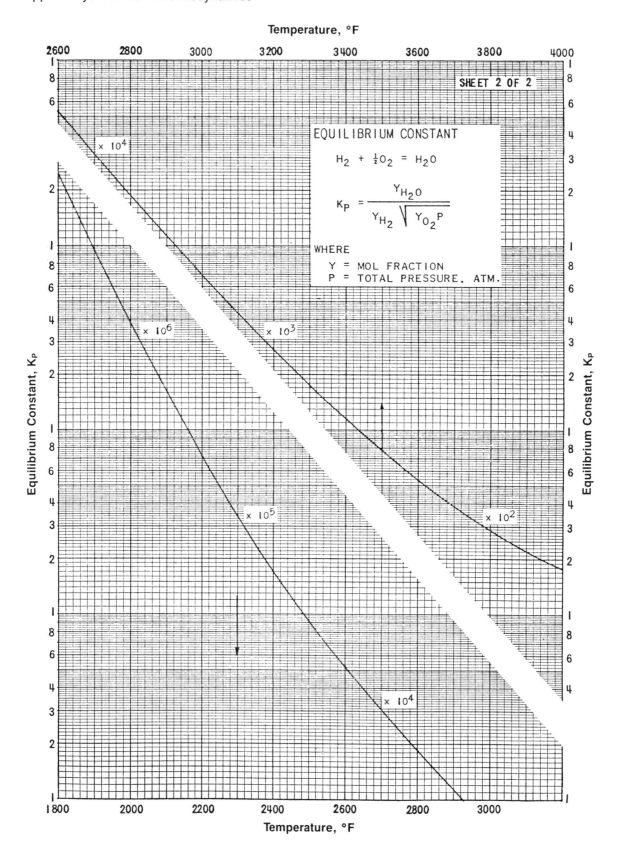

Figure 22.7-2

Temperature, °F

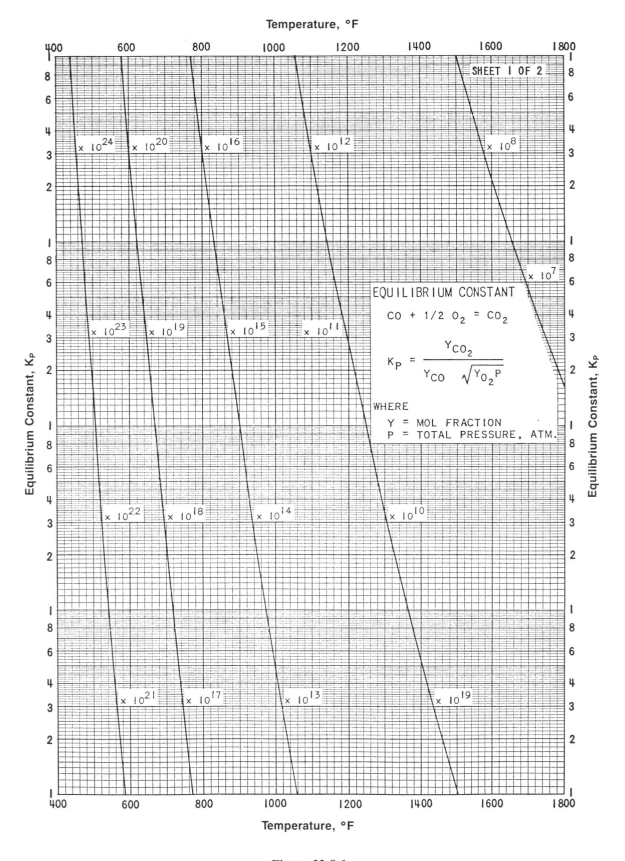

Figure 22.8-1

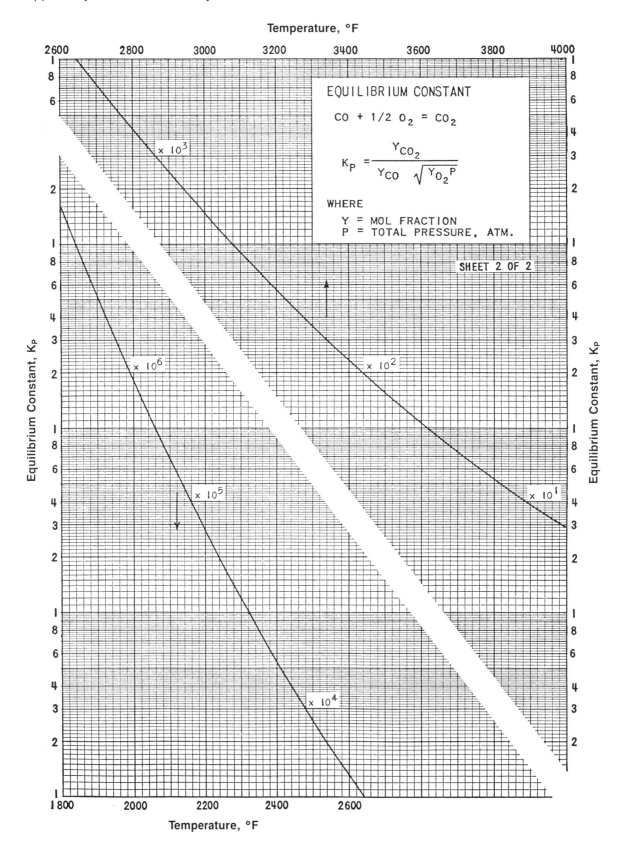

Figure 22.8-2

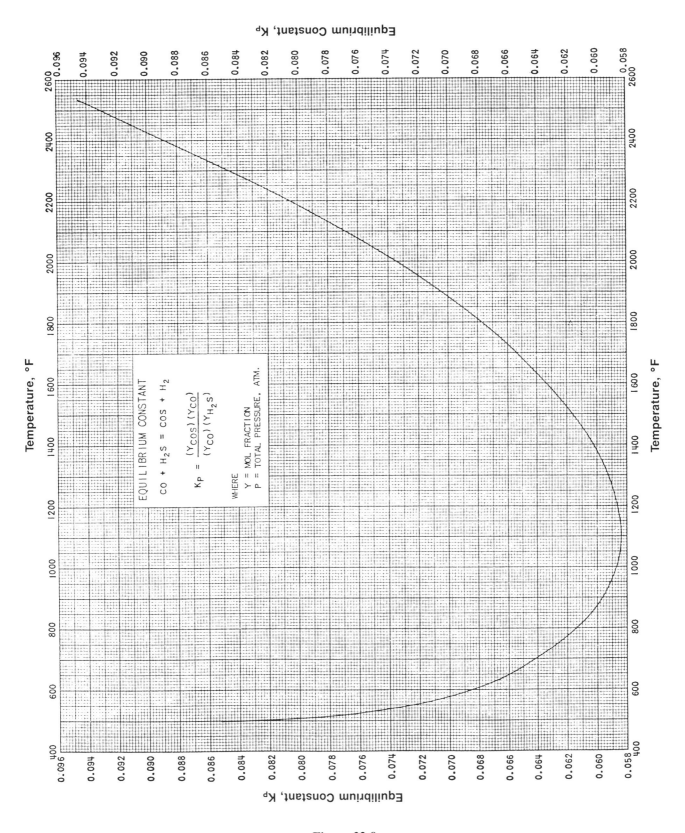

Figure 22.9

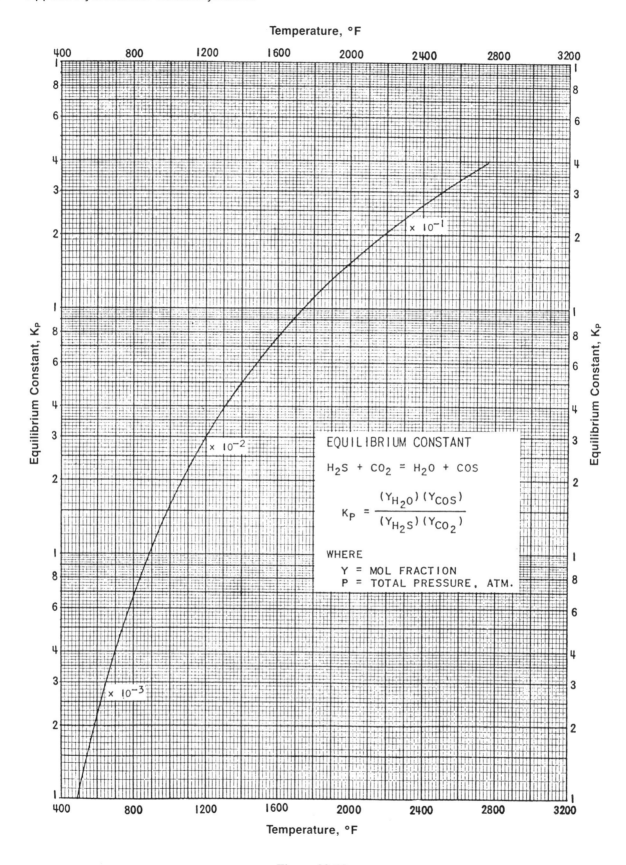

Figure 22.10

Temperature, °F

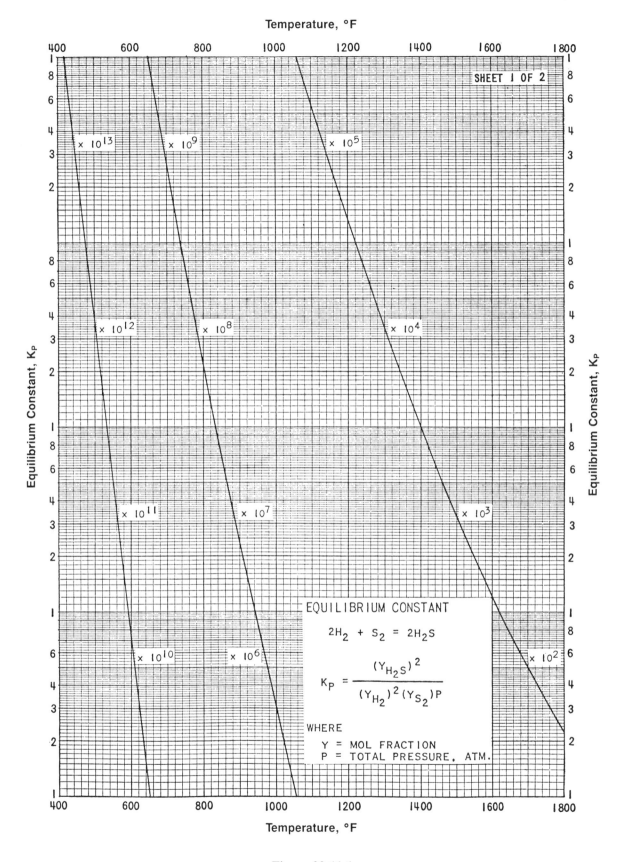

EQUILIBRIUM CONSTANT

$$2H_2 + S_2 = 2H_2S$$

$$K_P = \frac{(Y_{H_2S})^2}{(Y_{H_2})^2(Y_{S_2})P}$$

WHERE

Y = MOL FRACTION
P = TOTAL PRESSURE, ATM.

SHEET 1 OF 2

Temperature, °F

Figure 22.11-1

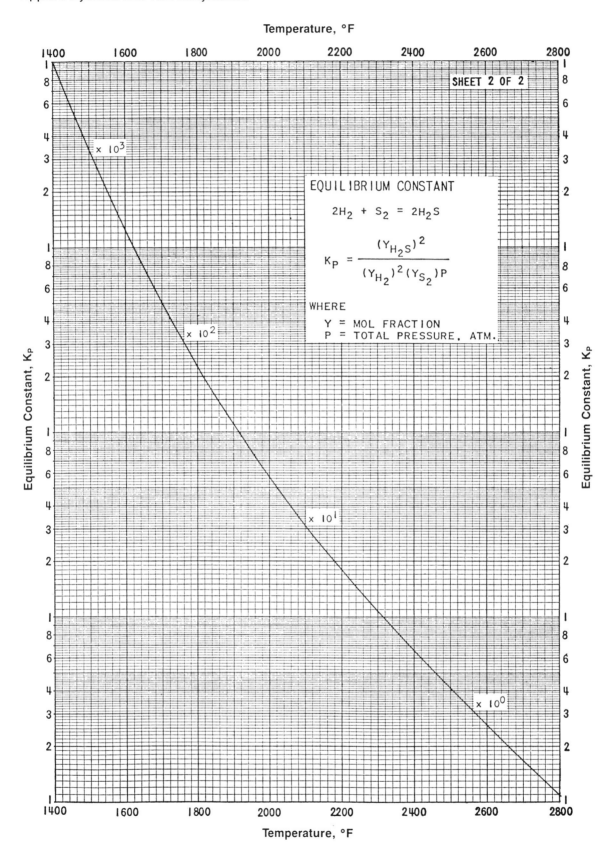

Figure 22.11-2

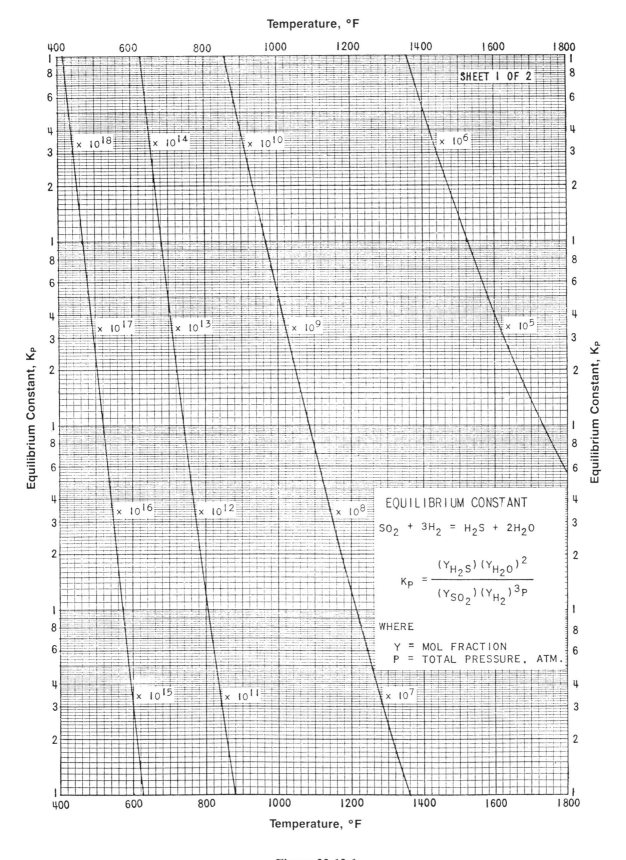

Figure 22.12-1

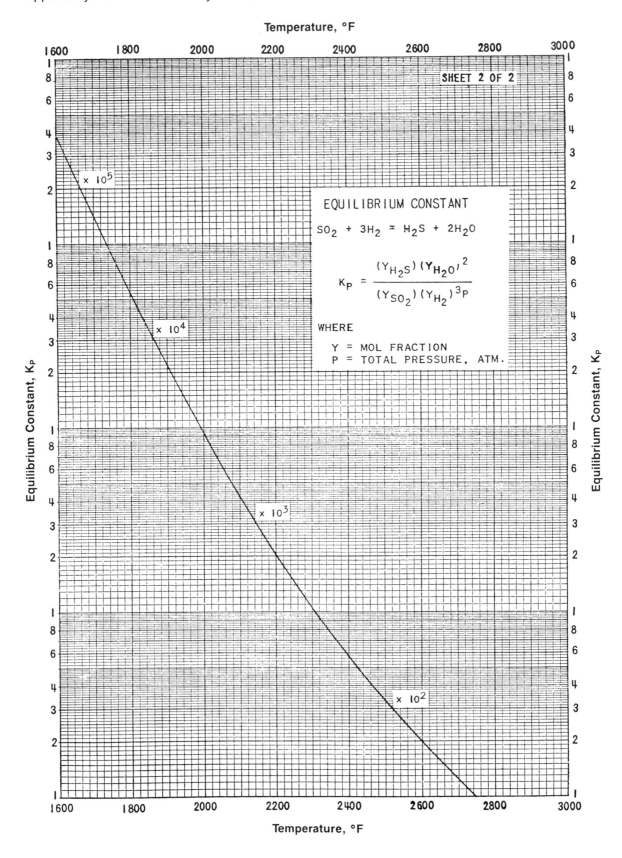

Figure 22.12-2

Gas	cal/gmol @ 298°K		$C_p^o = \alpha + \beta T$	
	ΔH_f^o	ΔG_f^o	α	β
H_2	0	0	6.65	0.00070
CO	$-26{,}416$	$-32{,}808$	6.89	0.00038
CH_3OH	$-48{,}080$	$-38{,}620$	2.00	0.030

(text continued from page 157)

Examples

Three examples will illustrate the evaluation and application or reaction equilibrium constants. These are for the following reactions: (a) methanol synthesis from carbon monoxide and hydrogen; (b) propane from polymerization of ethylene and methane; and (c) hydrogen production from partial oxidation of methane in presence of steam.

Each of these three examples illustrates different points. In the methanol example, a comparison is made between experimentally and thermodynamically derived *K* values. Effects of temperature, pressure, and conversion level, defined by C_2H_4 disappearance, are shown in the second example. The combination of stoichiometric and equilibrium expressions to determine the reactor effluent composition and heat balance are shown in the third example.

Methanol Example. Methanol is synthesized by passing carbon monoxide and hydrogen over a catalyst. The process is carried out so that the reaction $CO_{(g)} + 2H_{2(g)} = CH_3OH_{(g)}$ is equilibrium controlled. The heat of reaction and the equilibrium constant can be found experimentally or by thermodynamic calculations. Von Wettberg and Dodge (5) studied this reaction in the laboratory and obtained the following expression for the equilibrium constant:

$$\log_{10} K_a = \frac{4{,}570}{T} - 9.010 \log_{10} T$$
$$+ 0.00312\, T + 11.90 \qquad (22.27)$$

where T = temperature in °K

Equation 22.27 is an empirical expression fitted to values of K_a that had been derived from experimental measurements of the reactor effluent compositions at different reactor temperatures.

Values of K_a for this reaction are found via the following form of Equation 22.13

$$K_a = \left(\frac{Y_{CH_3OH}}{Y_{CO} Y^2_{H_2} P^2}\right)\left(\frac{\phi_{CH_3OH}}{\phi_{CO} \phi^2_{H_2}}\right) \qquad (22.28)$$

where P and the Y_i values are from measurements, ϕ_i values are calculated from $\overline{f}_i/PY_i$ relationships, applied at system temperature, pressure and composition.

In many cases it is possible to assume that $\phi = f/P$, which is independent of composition. For comparison, the value of K_a will be estimated by thermodynamic calculations, using the following data:

The solution is made by the following steps:

Step 1. For the reaction at 298°K

$$\Delta H^o = -48{,}080 + 26{,}416$$
$$= -21{,}664 \text{ cal/g mol}$$

Step 2. Find $\Delta\alpha$ and $\Delta\beta$ for the reaction

$$\Delta\alpha = [2 - 2(6.65) - 6.89] = -18.19$$

$$\Delta\beta = [0.03 - 2(0.0007) - 0.00038] = +0.0282$$

Step 3. Find integration constant ΔH^o_o by applying Equation 22.19 at 298°K.

$$\Delta H^o_o = -21{,}644 + (18.19)(298) - (0.0141)(298)^2$$
$$= -17{,}494 \text{ cal/g mol}$$

Step 4. Find integration constant *I* by applying Equation 22.20

$$\Delta G^o_{298} = -38{,}620 + 32{,}808$$
$$= -5{,}812 \text{ cal/g mol}$$

$$I = -\frac{5{,}812}{298} + \frac{17{,}494}{298} - \frac{18.12}{298} 298 \ln 298$$
$$+ \frac{0.0141}{298}(298)^2 = I = -63$$

Step 5. Find ΔG^o and K_a at the desired temperature of 350°C by using the above values of $\Delta\alpha$, $\Delta\beta$, ΔH^o_o, and *I* in Equation 22.20;

$$\Delta G^o = -17{,}494 + 18.19(623)\ln 623 - 0.0141(623)^2$$
$$- 63(623) = 10{,}726 \text{ cal/g mol}$$

$$\ln K_a = -\frac{\Delta G^o}{RT} = -\frac{10{,}726}{1.987(623)} = -8.68$$

$$K_a = 1.66 \times 10^{-4}$$

The experimentally based empirical equation of Von Wettberg and Dodge, i.e., Equation 22.27, gave

$K_a = 0.61 \times 10^{-4}$

The difference between these two values, i.e., 1.66×10^{-4} vs 0.61×10^{-4}, is equivalent to a 10% difference in ΔG^o values and could be due to small errors in the basic thermodynamic data used in the ΔG^o calculations. Assuming that Equation 22.27 is correct, it follows that 22.20 gives a value of ΔG^o that is about 10% low. The other possibility is that Equation 22.27 is not exactly correct.

Propane Example. In this example, the expression for G^o is found from heats of combustion and entropy of formation values. The K_a values, thus obtained, are corrected to high pressure, via fugacity coefficients, and then reactor effluent compositions are computed. The following data are to be used for the reaction $CH_4 + C_2H_4 = C_3H_8$.

Component	Heat of combustion cal/gmol @ 298°K	$Cp^o = \alpha + \beta T$		Entropy of formation e.u., cal/gmol °K @ 298°
		α	β	
Hydrogen ..	68,318	6.50	0.0009	...
Carbon	94,030	2.673	0.00262	...
Methane ...	212,790	4.313	0.01444	43.9
Ethylene ...	337,250	4.033	0.02167	57.4
Propane	530,570	9.00	0.0300	65.8

Solution is by the following steps:

Step 1. The heat of reaction at 298°K is found as follows:

$$\Delta H^o_{298} = 530,570 - 212,790 - 337,250$$
$$= -19,470 \text{ cal/g mol}$$

Step 2. Values of $\Delta\alpha$ and $\Delta\beta$ are found for the reaction,

$$\Delta\alpha = [9.0 - 4.313 - 4.033] = 0.654$$
$$\Delta\beta = [0.030 - 0.01444 - 0.02167] = -0.0064$$

Step 3. The value of ΔH^o for the reaction is found next,

$$\Delta H^o_o = -19,470 - 0.654(298) + \frac{0.00611}{2}(298)^2$$
$$= -19,390 \text{ cal/g mol}$$

Step 4. The value of ΔS^o_{298} is found for the reaction,

$$\Delta S^o_{298} = 65.8 - 57.4 - 43.9$$
$$= -35.5 \text{ e.u. cal/g mol °K}$$

Step 5. Find $\Delta G^o_{298} = \Delta H^o_{298} - T\,\Delta S^o_{298}$

$$\Delta G^o_{298} = -19,390 - 298(-35.5)$$
$$= -8,810 \text{ cal/g mol}$$

Step 6. The integration constant I is found next,

$$I = -\frac{810}{298} + \frac{19,390}{298} + \frac{0.654}{298}\,298 \ln 298$$

$$-\frac{0.00611}{2(298)}(298)^2$$

$$I = 38.165$$

Step 7. Using Equations 22.16 and 22.20, with the above values of the integration constants, values of ΔG^o and K_a are computed for several temperatures. The results are given in Table 22.1.

Step 8. Values of K_P were found at two pressures (1800 and 5000) for each of several temperatures by the following relationships:

where $K_p = K_a/K\phi$

$$K\phi = \frac{\phi_{C_3H_8}}{\phi_{CH_4}\,\phi_{C_2H_4}}$$

$\phi_i = f_i/P$ (by simplifying assumption)

Values of f/P were found from generalized correlations of f/P as a function of T_r and P_r. These calculations are straightforward and are not included here. Results are given in Table 22.1.

Step 9. Stoichiometric calculations are next made at both pressures and at two levels of ethylene content in products gases. These calculations were made using the values of K_P and the following relationships:

At 5 mol% C_2H_4 in products,

$$K_p = \frac{X}{(95-X)\,5\,P}$$

At 10 mol% C_2H_4 in products,

$$K_p = \frac{X}{(90-X)10\,P}$$

where X = mol percent propane in products.

Table 22.1
Thermodynamic Calculation for Propane Example

T°K	$\Delta G°$	K_a	P = 1,800 Lb/in.² Abs		P = 5,000 Lb/in.² Abs	
			K_ϕ	K_P	K_ϕ	K_P
298	−8,856.9	3.12×10^6	0.404	7.7×10^6	0.293	10.7×10^6
400	−5,264.5	0.75×10^3	0.531	1.41×10^3	0.474	1.58×10^3
500	−1,576.4	0.486	0.766	6.37	0.622	7.848
600	2,099.2	0.172	0.857	0.201	0.739	0.233
700	5,829.6	0.0152	0.892	0.017	0.771	0.0197
783*	8,955.9	0.002575	0.898	0.00298	0.786	0.0034

* 950°F

Results of these stoichiometric calculations are given in Table 22.2.

Table 22.2
Stoichiometric Calculations for Propane Example

	Mol % Propane in Product			
	5 mol % C_2H_4 in Product		10 mol % C_2H_4 in Product	
T°K	1,800 Lb/in.² Abs	5,000 Lb/in.² Abs	1,800 Lb/in.² Abs	5,000 Lb/in.² Abs
298	94.99	95.00	90.00	90.00
400	94.99	95.00	89.94	90.00
500	94.98	95.00	89.88	90.00
600	94.22	94.76	89.80	89.88
700	86.66	92.28	85.84	88.67
783*	61.45	81.45	70.54	82.84

* 950°F

Hydrogen Example. This illustration is concerned with the production of hydrogen by the partial oxidation of methane in the presence of steam. The reactor in this example is charged 100 lb of CH_4 at 800°F, 100 lb of O_2 at 800°F, and 100 lb of H_2O at 1,800°F. Product gases are assumed to leave the reactor at 1,700°F in chemical equilibrium. The superheated steam is added during and after, rather than before, the partial combustion reaction.

From the previous information, it is possible to calculate the composition of the product gases, which will contain only CO, CO_2, H_2, and H_2O. All of the oxygen and methane will be consumed. Four equations are required for calculating the amounts of each of the four gases in the reactor effluent. These four expressions are obtained from: a carbon balance, an oxygen balance, a hydrogen balance, and the equilibrium relationship, as follows:

Let M = moles of carbon supplied
S = moles of steam supplied

X = moles of oxygen supplied
CO_2 = moles of carbon dioxide in product
CO = moles of carbon monoxide in product
H_2 = moles of hydrogen in product
H_2O = moles of steam in product

In terms of these symbols, a carbon balance is

$$CO_2 + CO = M \qquad (22.29)$$

A hydrogen balance is

$$H_2 + H_2O = 2M + S \qquad (22.30)$$

And an oxygen balance is

$$\tfrac{1}{2}CO + CO_2 + \tfrac{1}{2}H_2O = X + \tfrac{1}{2}S \qquad (22.31)$$

These relationships are combined for numerical solution. A combination of Equations 22.29 and 22.31 gives

$$CO_2 + H_2O = 2X + \tfrac{1}{2}S \qquad (22.32)$$

Combining Equations 22.29, 22.30, and 22.31 gives

$$CO + H_2 = 4M − 2X \qquad (22.33)$$

Equations 22.29, 22.32, and 22.33 are three of the four expressions that will be used in making the stoichiometric calculations for this process. The fourth expression is the equilibrium relationship, given in Figure 22.5.

At the reactor temperature of 1,700°F the value of $K_p = 0.7$ is read from Figure 22.5. This gives

$$[(CO_2)(H_2)]/[(CO)(H_2)] = 0.7 \qquad (22.34)$$

The numerical solution of the stoichiometric relationships for this problem is given in Table 22.3. Using these compositions, a heat balance is made and presented in Table 22.4. This heat balance shows that the input and output heat quantities are nearly equal so only a very small amount of heat must be added to the reactor. This could be done by increasing the preheat of the methane and oxygen feeds and/or by increasing the superheat of the steam.

An alternate to adding heat via the charge streams is to allow the reactor to operate at a temperature lower than

Table 22.3
Hydrogen by Partial Oxidation of Methane—Material Balance

Charge: 100 lb Methane, 100 lb Oxygen, 100 lb Steam
Product gas at 1,700°F:

$$CO_2 + CO = \frac{74.9}{12} = 6.24 \text{ moles}$$

$$CO_2 + H_2O = \frac{100}{16} + \frac{100}{18.016} - \frac{74.9}{12} = 5.56 \text{ moles}$$

$$CO + H_2 = \frac{74.9}{12} - \frac{100}{16} + \frac{25.1}{2.016} = 18.69 \text{ moles}$$

$$\frac{CO_2 \times H_2}{CO \times H_2O} = 0.7$$

Solving: CO = 5.07 moles
CO$_2$ = 1.17 moles
H$_2$ = 13.62 moles
H$_2$O = 4.39 moles

1,700°F. In this connection, it is of interest to point out that the product gas would change with a change in the outlet temperature.

The effluent from the 1,700°F reactor are next cooled and sent to a shift converter. This part of the process is not included here.

Computer Applications

The equilibrium constant, K_a, for a chemical reaction is found from Equation 22.16, after having found the free energy change, (G^*_o), for the reaction. The free energy change for the reaction is found by combining the free energies of formation for the components involved in the reaction, as shown by Equation 22.7. These free energies of formation can be evaluated from the physical and thermal properties information given in BLOCK DATA.

There are two options for making these free energy of formation evaluations: a) by coefficient combinations as in Equations 21.4–21.7 and 22.19–22.20, and b) by solution of Equation 7.33. Consistent and correct calculations by either procedure should give the same results. The choice of method might be based on the available starting information and the type of system into which the free energies are being integrated.

Coefficient Combinations

Equation 22.20, an example of this procedure, is based on the heat capacity relationship of Equation 21.4, and standard enthalpy and free energy of formation, the latter being derived from Third Law research as previously described.

Table 22.4
Hydrogen by Partial Oxidation of Methane—Heat Balance

Gas	Lb	Moles	Temp °F	$H° - H°_o$	$\Delta H°_f$	Btu
				Btu/lb/mole		
Input						
CH$_4$	100	6.25	800	12,354	− 28,777	−102,600
O$_2$	100	31.125	800	9,091	0	28,400
H$_2$O	100	5.55	1800	20,128	−102,793	−459,000
Total	300	14.925				−533,200
Output						
H$_2$	27.3	13.62	1700	15,157	0	+206,500
H$_2$O	79.1	4.39	1700	19,080	−102,793	−368,000
CO	142.1	5.07	1700	15,948	− 48,963	−167,500
CO$_2$	51.5	1.17	1700	23,157	−169,143	−170,800
Total	300.0	24.32				−499,800
Net endothermic heat to be added to reactor						33,400

An alternative to the Equation 22.20 procedure would be based upon Equations 7.29 through 7.31, and the Passut-Danner (5) ideal gas state correlations for hydrocarbons. In these correlations, the coefficients are defined by the enthalpy expression, i.e., Equation 7.29, as compared with the heat capacity definition of Equation 21.4. The heat capacity relationship of Passut-Danner, i.e., Equation 7.30, should be noted and compared with Equation 21.4 to appreciate the differences in these two sets of coefficients.

A similar expression to Equation 22.20 is derived from the Passut-Danner expressions as follows: Combine Equations 7.29 and 7.31,

$$[(H^* - H^*_o) - TS^*] = A + B(T - T\ln T) - CT^2 - (1/2)DT^3 - (1/3)ET^4 - (1/4)FT^5 - GT \quad (22.35)$$

$$\Delta G^*_{f,T} = \Delta H^*_{f,o} + \Delta[(H^* - H^*_o) - TS^*] \quad (22.36)$$

Equations 22.35 and 22.36 are applied to chemical reactions by combining the values of the coefficients A through G for the reactants and products, in their stoichiometric amounts, in a similar manner to the combination in Equation 22.20. Values of these A through G are given in BLOCK DATA as DS(25,I) through DS(31,I) for the discrete components presented there. These get read as HCF(1) through HCF(7) for use in program EQUIL.

This coefficient method can be applied to reactions for which consistent sets of coefficients are available for the reactants, products, and their elements. This is not the case, presently, for the pseudo components of petroleum, for which the Kesler-Lee (2) heat capacity equation is used in EQUIL. This expression, given as Equation

7.34, uses coefficients that are defined in a similar way to those in Equation 21.4. Also, there is no G coefficient for conventional entropy calculations by the Kesler-Lee correlation for petroleum.

Equation 7.33

In view of the previous considerations, the direct free energy computation alternative seems best for the current programming. As given in Chapter 7 of Volume 1, the last term of Equation 7.33 is not correct for this application. A correction gives the following:

$$(\Delta G^*)_{f,T} = (\Delta G^*)_{f,536.7} + \left[\Delta H^* - \sum_1^N n_i\,(\Delta H^*_i)_{536.7}\right]$$

$$- \left\{TS^* - 536.7(S^*)_{536.7} - \sum_1^N n_i[TS^*_i\right.$$

$$\left. - 536.7(S^*_i)_{536.7}\right\} \quad (22.37)$$

where subscript i denotes terms for the elements.

A computer program, named EQRK3, has been prepared to make reaction equilibrium constant calculations based upon Equation 7.33, as modified by Equation 22.37. This program, which is given in Table 22.5, calculates the enthalpies and free energies of formation for the reactants and products and also for the reaction. The application of this program is demonstrated for the propane cracking and polymerization of ethylene and methane, and the water gas shift reactions. The equilibrium constants for these reactions are reported in Tables 22.6 and 22.7.

(text continued on page 186)

Table 22.5
Chemical Reaction Equilibrium Constants by
FORTRAN Program EQRK3, Using PROPVT and DATBK

```
      COMMON /PRINT/ TITLE(20),CNAME(7,30),FD(30),WM(30),BP(30),
     1       SG(30),ICOD,KHF,KYF,KTL,IKV,IHS,IHD,IOUT,NC,
     2       TI,PI,EF,TO,PO,TTOL,PTOL,XTOL,SUMF,HF,SF
      COMMON /CRIT/ ID(30),TC(30),PC(30),WC(30)
      COMMON /HSTR/ HCF(9,30),HSI(30),HOF(30)
      COMMON /HFOR/ FOF(30),CHONS(30)
      COMMON /DLIQ/ RKW(30),VST(30),HOV(30)
      COMMON /DSDATA/DS(33,216)
      DIMENSION KE(30,5),XJE(30,5),HONS(30),ONS(30),NS(30),S(30),HCR(30)
     * ,HCT(30),HER(30,5),HET(30,5),SER(30,5),SET(30,5),SCR(30),SCT(30),
     * DLH(30,5),DLS(30,5),SDLH(30),SDLS(30),DELH(30),DELS(30)
     * HOFT(30),FOFT(30),TF(7),EQK(30)
      WRITE(*,102)
  102 FORMAT(////,' A Chem Reaction input data file must be named.'/
     *' A name of 3 to 6 characters + extension of .DAT is suggested.'/)
```

(table continued on next page)

Table 22.5
Continued

```
      OPEN(5,FILE = ' ',STATUS = 'NEW')
      WRITE(*,12)
 12   FORMAT(/////)
      WRITE(*,103)
103   FORMAT(///,' Name a results file; same root name as input file'/
     *' with a different extension, such as .RST is suggested. '//)
      OPEN(6,FILE = ' ',STATUS = 'NEW')
C PROMPT FOR TITLE
      WRITE(*,104)
104   FORMAT(//,' Enter problem title; maximum of 80 characters.'/)
      READ(*,105) (TITLE(I),I = 1,20)
105   FORMAT(20A4)
      WRITE(5,105) (TITLE(I),I = 1,20)
C PROMPT FOR NUMBER OF REACTANT AND PRODUCT COMPOUNDS
      WRITE(*,106)
106   FORMAT(//,' Enter total number of components in reaction.'/)
      READ(*,*) NC
      WRITE(*,108)
108   FORMAT(//,' Enter number (1 to 7) of temperatures for calcs.'/)
      READ(*,*) KJ
      WRITE(*,109)
109   FORMAT(//,' Enter temperature values, degrees F.'/)
      READ(*,*) (TF(K),K = 1,KJ)
      WRITE(5,110) NC,KJ,(TF(K),K = 1,KJ)
110   FORMAT(2I3,1X,7F10.2)
C PROMPT FOR ID NUMBERS AND NAMES OF REACTANT & PRODUCT COMPONENTS
      DO 117 I = 1,NC
111   WRITE(*,112) I
112   FORMAT(//,' Enter all component',I3,' ID number and moles.'/
     *' If you need to see the component ID and name list enter'/
     *' enter  - 1 as the ID number and the component quantity;'/
     *' The ID & component name will be displayed. Select and enter'/
     *' the correct ID and amount for that component and continue.',/
     *' Quantities should be negative for reactant components and'/
     *' positive for product components.')
      READ(*,*) ID(I),FD(I)
      IF(ID(I).NE. - 1) GO TO 115
      KI = 0
      DO 113 K = 1,214,2
      KI = KI + 1
      WRITE(*,114) (DS(J,K),J = 1,8), (DS(J,K + 1),J = 1,8)
114   FORMAT(1X,2(3X,F4.0,1X,7A4))
      IF(KI.LT.20) GO TO 113
      KI = 0
      PAUSE
113   CONTINUE
      GO TO 111
115   WRITE(5,116) ID(I),FD(I)
116   FORMAT(I3,7X,F10.3)
117   CONTINUE
      REWIND 5
      READ(5,105) (TITLE(I),I = 1,20)
      WRITE(6,2) (TITLE(I),I = 1,20)
 2    FORMAT(//,3X,20A4,/3X,14('-----'))
      READ(5,3) NC,KJ,(TF(K),K = 1,KJ)
```

(table continued on next page)

Table 22.5
Continued

```
   3 FORMAT(2I3,1X,7F10.2)
     DO 15 I = 1,NC
  15 READ(5,6) ID(I),FD(I)
   6 FORMAT(I3,7X,F10.3)
     DO 250 K = 1,KJ
     T = TF(K) + 459.7
C HEADING FOR COMPONENT RESULTS
     WRITE(6,4) TF(K), T
   4 FORMAT(/,3X,'FORMATION AND REACTION EQUILIBRIUM CONSTANTS',/
    * 5X,'AT TEMPERATURE OF ',F8.2,' F = ',F8.2,' R')
     WRITE(6,5)
   5 FORMAT(/5X,'COMPONENT RESULTS',13X,'FORMATION BTU/LBMOLE',5X,'LOGE
    * OF KF', 2X,'MOLES',/,33X,' ENTHALPY ',2X,'FREE ENERGY')
     DO 200 I = 1,NC
     CALL PHPROP(ID(I),CNAME(1,I),WM(I),BP(I),SG(I),TC(I),PC(I),
    * WC(I),RKW(I),VST(I),HOV(I),HOF(I),FOF(I),CHONS(I),HCF(1,I))
C FIND NUMBERS OF ELEMENTS (C,H,O,N,S) FOR EACH COMPOUND
C Item DS(24,I) in BLOCK DATA gives atoms of carbon, hydrogen, oxygen,
C nitrogen and sulfur. Following reads and interprets these values.
     KE(I,1) = 610
     KE(I,2) = 602
     KE(I,3) = 603
     KE(I,4) = 604
     KE(I,5) = 611
C AMOUNTS OF ELEMENTS IN COMPONENTS
     JJ = INT((0.5 + CHONS(I))/1.E5)
     XJE(I,1) = FLOAT(JJ)
     IHONS = INT(0.5 + CHONS(I) − XJE(I,1)*1.E5
     JJ = IHONS/1000
     XJE(I,2) = FLOAT(JJ)
     IONS = IHONS(I) − 1000*JJ
     JJ = IONS/100
     XJE(I,3) = FLOAT(JJ)
     INS = IONS − 100*JJ
     JJ = INS/10
     XJE(I,4) = FLOAT(JJ)
     IS = INS − JJ*10
     XJE(I,5) = FLOAT(IS)
C CORRECT TO ATOMIC WEIGHT BASIS FOR GASES (H2,N2,O2)
     XJE(I,2) = XJE(I,2)/2.
     XJE(I,3) = XJE(I,3)/2.
     XJE(I,4) = XJE(I,4)/2.
C GET ELEMENT H AND S CONSTANTS AND MAKE CALCS FOR ELEMENTS
     DO 201 J = 1,5
     CALL PHPROP(KE(I,J),CNAME(1,I),WM(I),BP(I),SG(I),TC(I),PC(I),
    * WC(I),RKW(I),VST(I),HOV(I),HOF(I),FOF(I),CHONS(I),HCF(1,J))
 201 CONTINUE
C CALCULATE IDEAL H & S VALUES FOR ELEMENTS:
C HER(I,J) = H VALUE AT 536.7; HET(I,J) = H VALUE AT T;
C SER(I,J) = T*S VALUE AT 536.7; SET(I,J) = T*S VALUE AT T.
C   ENTHALPY CALCS FOR ELEMENTS
     CALL IDEAHS(1,1,5,536.7,2)
     DO 202 J = 1,5
 202 HER(I,J) = XJE(I,J)*HSI(J)
     CALL IDEAHS(1,1,5,T,2)
```

(table continued on next page)

Table 22.5
Continued

```
      DO 203 J = 1,5
  203 HET(I,J) = XJE(I,J)*HSI(J)
      DO 204 J = 1,5
  204 DLH(I,J) = HET(I,J) − HER(I,J)
      SDLH(I) = 0.0
      DO 205 J = 1,5
  205 SDLH(I) = SDLH(I) + DLH(I,J)
C ENTROPY CALCS FOR ELEMENTS
      CALL IDEAHS(1,1,5,536.7,3)
      DO 206 J = 1,5
  206 SER(I,J) = 536.7*XJE(I,J)*HSI(J)
      CALL IDEAHS(1,1,5,T,3)
      DO 207 J = 1,5
  207 SET(I,J) = T*XJE(I,J)*HSI(J)
      DO 208 J = 1,5
  208 DLS(I,J) = SET(I,J) − SER(I,J)
      SDLS(I) = 0.0
      DO 209 J = 1,5
  209 SDLS(I) = SDLS(I) + DLS(I,J)
  200 CONTINUE
C SETUP FOR COMPONENT CALCULATIONS
      DO 301 I = 1,NC
      CALL PHPROP(ID(I),CNAME(1,I),WM(I),BP(I),SG(I),TC(I),PC(I),
     * WC(I),RKW,(I),VST(I),HOV(I),HOF(I),FOF(I),CHONS(I),HCF(1,I))
  301 CONTINUE
C ENTHALPY CALCS FOR COMPONENTS
      CALL IDEAHS(1,1,NC,536.7,2)
      DO 210 I = 1, NC
  210 HCR(I) = HSI(I)
      CALL IDEAHS(1,1,NC,T,2)
      DO 211 I = 1, NC
  211 HCT(I) = HSI(I)
      DELHF = 0.0
      DO 212 I = 1, NC
      DELH(I) = HCT(I) − HCR(I)
C Solution of Equation 7.32
      HOFT(I) = HOF(I) + (DELH(I) − SDLH(I))
      DELHF = DELHF + FD(I)*HOFT(I)
  212 CONTINUE
C ENTROPY CALCS FOR COMPONENTS
      CALL IDEAHS(1,1,NC,536.7,3)
      DO 213 I = 1, NC
  213 SCR(I) = 536.7*HSI(I)
      CALL IDEAHS(1,1,NC,T,3)
      DO 214 I = 1, NC
  214 SCT(I) = T*HSI(I)
      DELFE = 0.0
      DO 215 I = 1, NC
      DELS(I) = SCT(I) − SCR(I)
C Solution of Equation 7.33
      FOFT(I) = FOF(I) + (DELH(I) − SDLH(I)) − (DELS(I) − SDLS(I))
      EQK(I) = − FOFT(I)/1.98719/T
      WRITE(6,7) (CNAME(J,I),J = 1,7),HOFT(I),FOFT(I),EQK(I),FD(I)
```

(table continued on next page)

Table 22.5
Continued

```
    7 FORMAT(5X,7A4,3(F10.2,2X),F8.4)
      DELFE = DELFE + FD(I)*FOFT(I)
  215 CONTINUE
      WRITE(6,8) DELHF,DELFE
    8 FORMAT(/,5X,'NET CHANGES FOR PROCESS',5X,2(F10.3,2X))
      ALNK = − DELFE/1.98719/T
      BLOGK = ALNK/2.3025851
      ACKJ = EXP(ALNK)
      WRITE(6,20) ALNK, BLOGK, ACKJ
   20 FORMAT(/,5X,'VALUE OF LOGE OF K FOR REACTION = ',F10.4,/,5X,'VALUE
     * OF LOG10 K FOR REACTION = ',F10.4,/,5X,'VALUE OF K FOR THE
     * REACTION = ',E13.5,/,5X,14('-----'))
  250 CONTINUE
      WRITE(*,125)
  125 FORMAT(//,' Reaction Equilibrium calculations are completed,'/
     1' Results can be viewed or printed.')
      REWIND 5
      REWIND 6
      STOP
      END
```

Table 22.6
Results of Calculations by Program EQRK3 for Propane Formation and Cracking Reactions at
$1,000\,°F = 1,459.7\,°R$

Exothermic Polymerization Reaction (Methane + Ethene = Propane)

Component Results	Formation Btu/lbmole		Log_E of K_f	Moles
	Enthalpy	Free Energy		
Methane	− 37,509.41	− 483.08	.17	− 1.000
Ethene	17,313.48	44,613.80	− 15.38	− 1.0000
Propane	− 54,358.33	56,335.79	− 19.42	1.0000
Net changes for process	− 34,162.400	12,205.070		

Value of Log_E of K for reaction	=	− 4.2076
Value of Log_{10} K for reaction	=	− 1.8273
Value of K for the reaction	=	0.014882

Endothermic Cracking Reaction (Propane = Methane + Ethene)

Component Results	Formation Btu/lbmole		Log_E of K_f	Moles
	Enthalpy	Free Energy		
Propane	− 54,358.33	56,335.79	− 19.42	− 1.0000
Methane	− 37,509.41	− 483.08	.17	1.0000
Ethene	17,313.48	44,613.80	− 15.38	1.000
Net changes for process	34,162.400	− 12,205.070		

Value of Log_E of K for reaction	=	4.2076
Value of Log_{10} K for reaction	=	1.8273
Value of K for the reaction	=	67.197

Table 22.7
Results of Calculations by Program EQRK3 for Water Gas Shift Reaction at 1,000°F = 1,459.7°R

Exothermic Reaction (Water + Carbon Monoxide = Hydrogen + Carbon Dioxide)

| Component Results | Formation Btu/lbmole | | Log$_E$ of K$_f$ | Moles |
	Enthalpy	Free Energy		
Water	− 106,026.20	− 87,289.75	30.09	− 1.0000
Carbon Monoxide	− 47,761.17	− 79,077.16	27.26	− 1.0000
Hydrogen	.00	.00	.00	1.0000
Carbon Dioxide	− 169,601.00	− 170,192.31	58.67	1.000
Net changes for process	− 15,813.640	− 3,825.406		

Value of Log$_E$ of K for reaction	=	1.3188
Value of Log$_{10}$ K for reaction	=	0.5727
Value of K for the reaction	=	3.7309

The water gas shift K values of 3.7309 at 1,000°F obtained by the calculations in Table 22.7 agree with the values read from Figure 22.4.

Subroutine REACK

After completing and testing program EQRK3.FOR for reaction equilibrium constants, a new subroutine was prepared for EQUIL, combining both reaction heat and equilibrium functions. This new subprogram is called REACK and is called with an ICOD = 14 specification. Operating with EQUIL is more efficient because of convenient access to BLOCK DATA and other EQUIL subprograms.

Free Energy of Petroleum Components

A method for calculating the standard free energy of formation for pseudo components of a petroleum frac-

tion was recently presented (3). This method is discussed in Chapter 24.

References

1. API Research Project 44, Texas A & M, "Selected Values of Thermodynamic Properties of Hydrocarbons and Other Gases."
2. Kesler, M. G. and B. I. Lee, *Hydrocarbon Process.* 55(3) 153 (1976).
3. Montgomery, R. L., "Estimation of Thermochemical Properties of Undefined Hydrocarbon Mixtures: Heats of Combustion of Petroleum Fractions," presented at the American Chemical Society National Meeting, Kansas City, Missouri, September 15, 1982.
4. Layng, E. T., private commmunication, December 27, 1971 and June 14, 1972.
5. Passut, C. A. and R. P. Danner, *I & EC Process Des. & Dev.*, 11, 543 (1972).
6. U.S. Bureau of Mines Bulletin 406.
7. U.S. Bureau of Mines Bulletin 1694.
8. Von Wettberg and Dodge, *Ind. Engr. Chem.*, 22, 1040 (1930).

Chapter 23

Enthalpy-Composition Diagrams

Enthalpy composition diagrams for two component mixtures have been useful thermodynamic tools for the past five decades, since the usefulness of such a chart was demonstrated by McCabe (11) in an application to the aqueous solutions of sodium hydroxide. After this start on the NaOH-H$_2$O binary system, enthalpy-concentration charts have been developed for aqueous solutions of H$_2$SO$_4$, HCl, CaCl$_2$, NH$_3$, and C$_2$H$_5$OH. In addition, H-X diagrams have also been prepared for other than aqueous solutions, examples being oxygen-nitrogen and isopropanol-toluene systems.

Weight and molar bases have both been used in the construction of the H-X charts. Coordinates for the weight basis charts are enthalpy in Btu/lb or cal/gm vs. weight fraction of the solute in the solution. Aqueous system H-X charts are usually prepared in this way.

The coordinates for the molar based charts are enthalpy in Btu/lb-mole or cal/gm-mole vs. mole fraction of the solute in the solution. Either set of coordinates is correct, the basis for choosing one or the other being application convenience.

Weight-basis charts are more convenient for most calculations on evaporators or concentrators processing aqueous solutions. For distillation calculations, in which vapor and liquid mixtures are involved and where molar composition ratios, i.e., vapor-liquid $K = y/x$ ratios, are used in the phase equilibria predictions, molar coordinates are more convenient for the enthalpy–composition charts.

Heat and Mass Balances by H-X Diagrams

H-X diagrams are constructed on rectangular coordinates with enthalpy plotted against composition in weight or molar units for lines of constant temperature, all at the same pressure. When the effect of pressure is significant, a separate H-X chart might be needed for each pressure of interest. Pressure can affect the H-X diagram in two ways, i.e., via the T-X or phase diagram, and via the enthalpy values at different pressures.

A feature of the H-X diagram that makes it an outstanding calculation tool is the "heat balance line," which is illustrated in Figure 23.1 and the following discussion. Figure 23.1 represents a section of an H-X diagram for a binary mixture at constant pressure. Short segments of three isotherms, T_1, T_2, and T_3, are shown. Enthalpy values along these isotherms include the heats of mixing and may be represented by either of the following expressions:

$$\underline{H} = \overline{H}_i x_i + \overline{H}_j x_j \tag{23.1}$$

$$\underline{H} = \underline{H}_i x_i + \underline{H}_j x_j + H_{ij}^E \tag{23.2}$$

where $\underline{H}$ = mixture enthalpy per weight or mole unit,
$\underline{H}_i$ = enthalpy of pure component i
$\overline{H}_i$ = partial enthalpy of component of i in j
H_{ij}^E = excess enthalpy or heat of mixing components i and j.

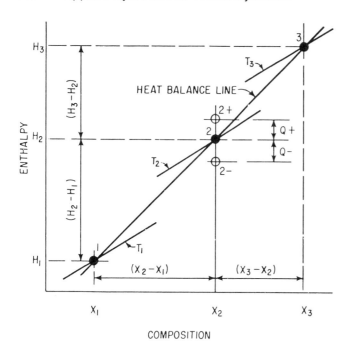

Figure 23.1. *HX* diagram for binary showing heat balance line.

If a final temperature is T_{2+}, or T_{2-} is desired for the mixture, the amount of heat that must be added or removed during the operation may be read from the chart, as illustrated by the quantities $Q+$ and $Q-$ on Figure 23.1. The straight-line heat balance feature of the enthalpy-composition diagram makes such charts very useful tools for a wide variety of problems on binary systems.

H-X Diagram Preparation Methods

The evaluation of enthalpies for use in preparing enthalpy-composition diagrams for binary systems requires the following information: (a) phase $(T-X)$ conditions; (b) zero enthalpy datum states; (c) effects of temperature and pressure on the enthalpies of both components and their mixtures in each phase (usually liquid and vapor); (d) heats of mixing at several conditions of temperature and composition, and (e) heats of vaporization or condensation at several conditions of temperature and composition.

The zero H datum state depends upon the system and the datum selected influences the procedures that are used in constructing the $H-X$ diagram. Two examples of frequently used datum states are pure solid or liquid at ambient temperature and pressure, and ideal gas at zero pressure and zero absolute temperature.

For mixtures of two substances that are normally liquid in their pure states at atmospheric pressure and ambient temperature, the zero enthalpy datum points are the pure liquids at a convenient temperature, such as 0 or 25°C. In this case the $H-X$ diagram is prepared by starting with the liquid phase and using heat capacities plus heats of vaporization for the mixture to locate the saturated liquid line and the saturated vapor line from the base line. A similar procedure is used for aqueous solutions of salt, acid, or caustic soda.

As an illustration, Figure 23.2 is a combination two-in-one diagram, in which the temperatures and the enthalpies of the coexisting equilibrium vapor and liquid phases are both plotted against composition for the acetone-water system at atmospheric pressure. Temperature tie-lines are shown as dashed lines in both the *T-X* and the *H-X* portions of the diagram.

The enthalpy datum point is $H=0$ for pure liquid water and pure liquid acetone at 100°F. Molar enthalpies and mole fraction compositions are the coordinates of the *H-X* part of the chart. Subcooled liquid and superheated vapor isotherms could have been drawn on this diagram, had these been needed. These subcooled and superheated isotherms are not required in most applications of enthalpy-composition diagrams, however.

Equations 23.1 and 23.2 are next applied at points on each isotherm on Figure 23.1 and an expression for the "heat balance line" is formulated, as follows: Assume that N_1 moles of the mixture at x_1 mole fraction composition and T_1 temperature are blended with N_3 moles of the x_3 mole fraction and T_3 mixture. If no heat is added or removed, the property values of the mixture, i.e., H_2, x_2 and T_2 will fall on a straight line through the point of the two blended solutions, i.e., N_1 and N_2, as shown in Figure 23.1. This is written mathematically as follows:

Heat and material balances for the mixing operation are

$$N_1 \underline{H}_1 + N_3 \underline{H}_3 = (N_1 + N_3)\, \underline{H}_2 \tag{23.3}$$

$$N_1 x_1 + N_3 x_3 = (N_1 + N_3)\, x_2 \tag{23.4}$$

Combining the heat and material balances gives

$$\frac{N_1}{N_3} = \frac{\underline{H}_3 - \underline{H}_2}{\underline{H}_2 - \underline{H}_1} = \frac{x_3 - x_2}{x_2 - x_1} \tag{23.5}$$

Equation 23.5 is an expression for a straight line on the H vs. X coordinates, as can be seen by examination of Figure 23.1, proving that straight lines are heat balance lines on an enthalpy-composition diagram with rectangular coordinates.

For mixtures of a liquid solvent and a gas solute, the pure liquid reference state is not practical and it is much more convenient to choose the ideal gas states for both components as the zero enthalpy starting points in the enthalpy evaluations for the construction of an *H-X* diagram.

Calculation and construction procedures differ for the preparation of the liquid-based and the vapor-based *H-X* diagrams. For the liquid-based chart, the enthalpies of the subcooled and saturated liquids, of different compositions, are calculated and plotted first. Then the heats of vaporization are found for different mixture compositions and the saturated vapor enthalpies are found by addition.

For the vapor-based *H-X* chart, the enthalpy values of several ideal gas mixtures are first calculated at the dew point temperatures of the mixtures at the system pressure. Each of these ideal gas mixture enthalpies is combined with the isothermal enthalpy pressure correction to obtain the enthalpy values at the saturated vapor state for each mixture composition. Then the heats of condensation are obtained for different mixture compositions and the saturated liquid enthalpy values are calculated by subtracting the heats of condensations.

Enthalpy-composition (*H-X*) diagrams are often constructed by using a combination of experimental data and thermodynamic calculations, used in different portions of the chart depending upon the system and the conditions involved.

In many cases, *H-X* diagrams can be prepared entirely from thermodynamic calculations, made with physical constants and properties of the pure components, generalized correlations and/or equations of state, but with no experimental data on mixtures of the two pure components. The other extreme is to obtain complete experimental data, such as: the compositions of the coexisting equilibrium vapor and liquid phases, and heats of vaporization or condensation and mixing, so that essentially no thermodynamic calculations are required.

Shah and Donnelly (17) prepared an experimentally derived *H-X* diagram for isopropanol-toluene at one atmosphere. An intermediate case, in which experimental vapor-liquid equilibria data are available for the binary mixture, and nothing more, is the more usual case in *H-X* diagram preparation.

The calculation of the enthalpies for the coexisting equilibrium vapor and liquid, using experimental or derived pressure-temperature-composition data, is the main objective of this chapter. In addition to the P-V-T-X-Y data, the preparation calculations use an equation of state or generalized correlations in computing the required properties. When possible, these calculations should be made by digital computers, using equations. When this is not possible, manual shortcut methods can be used. These are illustrated by some examples.

VLE and H Approximations

Calculating temperature-composition and enthalpy-composition points for non-ideal binary solutions, for which *H-X* diagrams might be needed, is not like making similar calculations for regular solutions, such as the multi-component mixtures for which the equations and computer programs presented in previous chapters of this book are intended. For present purposes, we need some other tools, such as activity coefficients and enthalpy approximation methods. Acccordingly, such

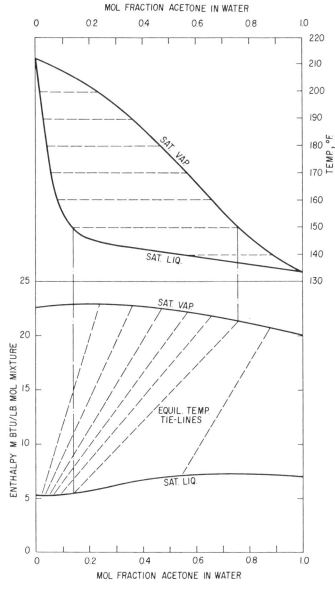

Figure 23.2. *TX* and *HX* diagrams for acetone-water at 1 atm.

items will be included here and then applied to examples later.

Van Laar Equation for van der Waals Fluid

The well known van Laar Equation was originally derived 60 years ago from a consideration of thermodynamic changes occurring on the mixing of pure liquids, using the van der Waals Equation ($P=RT/(V-B)-a/V^2$), assuming no volume change on mixing, change in entropy equal to that of an ideal solution, i.e., maximum randomness.

In terms of the van der Waals constants, the van Laar Equation is developed on pages 56–58 of Robinson and Gilliland[14]

$$\ln \gamma_1 = \frac{B'/T}{\left(1+A'\dfrac{x_1}{x_2}\right)^2} \tag{23.6}$$

$$\ln \gamma_2 = \frac{A'B'/T}{\left(A'+\dfrac{x_2}{x_1}\right)^2} \tag{23.7}$$

where $A' = \dfrac{b_1}{b_2}$ (23.8)

$$B' = \frac{b_1}{R}\left(\frac{\sqrt{a_1}}{b_1} - \frac{\sqrt{a_2}}{b_2}\right)^2 \tag{23.9}$$

a and b are the van der Waals constants. These equations can be put into a more symmetrical form, which involves differently defined constants. By algebraic manipulations

$$\ln \gamma_1 = \frac{A}{\left[1+\dfrac{A}{B}\dfrac{x_1}{x_2}\right]^2} \tag{23.10}$$

$$\ln \gamma_2 = \frac{B}{\left[1+\dfrac{Bx_2}{Ax_1}\right]^2} \tag{23.11}$$

where $A = b_1/RT\,(\sqrt{a_1}/b_1 - \sqrt{a_2}/b_2)^2$ (23.12)

$$B = b_2/RT\,(\sqrt{a_1}/b_1 - \sqrt{a_2}/b_2)^2 \tag{23.13}$$

It will be recalled that the van der Waals constants are found from the critical constants, i.e.,

$$a = \frac{27}{64}\frac{R^2 T_c^2}{P_c} \quad \text{and} \quad b = RT_c/8P_c$$

The van der Waals method for evaluating the van Laar constants is only an approximation, to be used when no other information is available.

Regular Solutions and Solubility Parameter

Scatchard (16) and Hildebrand (7) obtained a similar expression for the liquid activity coefficient without assuming the van der Waals fluid. This development is described in Hildebrand and Scott (7) where the entropy of mixing of two components is expressed in "free volumes" form, as follows:

Case I—Free volumes of the two components are proportional to their molar volumes, i.e.,

$$\frac{(V_f)_1}{(V_f)_2} = \frac{V_1}{V_2}$$

For this case the partial entropies of mixing were shown to be:

$$\overline{\Delta S_1} = -R\left[\ln X_{v_1} + X_{v_2}\left(1 - \frac{V_1}{V_2}\right)\right] \tag{23.14}$$

$$\overline{\Delta S_2} = -R\left[\ln X_{v_2} + X_{v_1}\left(1 - \frac{V_2}{V_1}\right)\right] \tag{23.15}$$

where V_1 and V_2 = molal volumes of components 1 and 2

$\overline{\Delta S_1}$ and $\overline{\Delta S_2}$ = partial entropy of mixing for components 1 and 2

$$X_{v_1} = \frac{x_1 V_1}{x_1 V_1 + x_2 V_2}$$

= liquid volume fraction of component 1

$$X_{v_2} = \frac{x_2 V_2}{x_1 V_1 + x_2 V_2}$$

= liquid volume fraction of component 2

Case II—Free volumes and also molar volumes are equal, giving the following for the partial entropy of mixing:

$$\overline{\Delta S_1} = -R \ln x_1 \tag{23.16}$$

$$\overline{\Delta S_2} = -R \ln x_2 \tag{23.17}$$

Case II is the entropy of mixing for an ideal solution. For Case I the activity coefficient relationship is

$$\ln \gamma_1 = \ln \frac{V_1}{V_{av}} + X_{v_2} \left(1 - \frac{V_1}{V_2}\right)$$
$$+ \frac{V_1 X^2_{v_2}}{RT} (\delta_1 - \delta_2)^2 \tag{23.18}$$

When the ratio (V_1/V_2) is close to unity the first two right hand members drop out giving

$$\ln \gamma^1 = \frac{V_1 X^2_{v_2}}{RT} (\delta_1 - \delta_2)^2 \tag{23.19}$$

The development of this equation follows:

For an ideal solution where $\bar{P}_i = x_i P_i^o$ (Raoult's Law) the heat of mixing is zero and the partial free energy of mixing is

$$\overline{\Delta G_i} = RT \ln \frac{\bar{P}_i}{P_i^o} \; RT \ln x_i \tag{23.20}$$

where $\Delta \bar{G}_i =$ partial free energy of mixing for component i
$\bar{P}_i =$ partial pressure of "i" in solution
$P_i^o =$ vapor pressure of "i"
$x_i =$ mol fraction

Very few substances form ideal solutions. Many more form what is known as regular solutions, which is non-ideal to the extent that there is a heat of mixing but the entropy of mixing is the same as for ideal solutions, i.e., Case II above and

$$\overline{\Delta S_i} = -R \ln x_i \tag{23.21}$$

Equation 23.21 represents the maximum randomness during mixing.

For a regular solution the partial free energy of mixing is obtained by combining Equation 23.21 with the following thermodynamic equation

$$\overline{\Delta G_i} = \overline{\Delta H_i} - T\overline{\Delta S_i} \tag{23.22}$$

giving

$$\overline{\Delta G_i} = \overline{\Delta H_i} + RT \ln x_i \tag{23.23}$$

where $\overline{\Delta H_i} =$ partial enthalpy of mixing for component i

An expression for $\overline{\Delta H_i}$ must be obtained and combined with Equation 23.23 to give an expression for $\overline{\Delta G_i}$.

For regular solutions it has been shown in Chapter VII of Hildebrand and Scott (7) that the heat of mixing term can be expressed as a function of the cohesive-energy densities of the components, as in the following equation

$$\overline{\Delta H_1} = V_1 X_{v_2}^2 \left[\left(\frac{\Delta E_1}{V_1}\right)^{1/2} - \left(\frac{\Delta E_2}{V_2}\right)^{1/2}\right]^2 \tag{23.24}$$

where ΔE_1 and $\Delta E_2 =$ molal energy to vaporize the liquid to infinite volume.

Combining Equations 23.23 and 23.24 and introducing a solubility parameter, δ, which is defined below, gives the following expression for the partial free energy of mixing

$$\overline{\Delta G_1} = RT \ln x_1 + V_1 (\delta_1 - \delta_2)^2 X_{v_2}^2 \tag{23.25}$$

$$\overline{\Delta G_2} = RT \ln x_2 + V_2 (\delta_1 - \delta_2)^2 X_{v_1}^2 \tag{23.26}$$

where $\delta = \left(\frac{\Delta E}{V}\right)^{1/2} =$ solubility parameter $\tag{23.27}$

$\Delta E =$ energy of vaporization of liquid to infinite attenuation.

Equations 23.25 and 23.26 are for regular solutions. $\overline{\Delta G}$ from these equations will always be positive. The sign of $(\delta_1 - \delta_2)$ does not matter because squaring will always give a positive term anyway.

By the definition of fugacity and the choice of the reference state, the following equation is obtained for the activity coefficient:

$$\ln \gamma_1 = \ln \frac{\bar{f}_1}{f_1 x_1} = \frac{V_1}{RT} (\delta_1 - \delta_2)^2 X_{v_2}^2$$
$$= \frac{V_1}{RT} (\delta_m - \delta_1)^2 \tag{23.28}$$

and

$$\ln \gamma_2 = \ln \frac{\bar{f}_2}{f_2 x_2} = \frac{V_2}{RT} (\delta_1 - \delta_2)^2 X_{v_1}^2$$
$$= \frac{V_2}{RT} (\delta_m - \delta_2)^2 \tag{23.29}$$

Equations 23.28 and 23.29 differ by the liquid volume and liquid volume fraction. It will be noted that there are two alternate forms of each activity coefficient equation. This results from the fact that

$$X^2_{v_2} (\delta_1 - \delta_2)^2 = (\delta_m - \delta_1)^2$$

and

$$X^2_{v_1} (\delta_1 - \delta_2)^2 = (\delta_m - \delta_2)^2$$

where $\delta_m = \delta_{\text{mixture}} = \sum X_{v_i} \delta_i = \dfrac{\sum x_i V_i \delta_i}{\sum x_i V_i}$

Van Laar Equation and Solubility Parameter

Equations 23.28 and 23.29 are frequently referred to as the Scatchard-Hildebrand relationships. They may be used to estimate the liquid phase activity coefficients, but it is more convenient to put them in another form. These equations reduce to the familiar van Laar equation by defining the two constants A and B as follows.

$$A = \frac{V_1}{RT} (\delta_1 - \delta_2)^2 \qquad (23.30)$$

$$B = \frac{V_2}{RT} (\delta_1 - \delta_2)^2 \qquad (23.31)$$

Combining Equations 23.28 with 23.30 and 23.29 with 23.31 replacing X_{v_2} and X_{v_1} by their equivalents $(x_2 V_2)/(x_1 V_1 + x_2 V_2)$ and $(x_1 V_1)/(x_1 V_1 + x_2 V_2)$, and rearranging gives

$$\ln \gamma_1 = \frac{A}{\left[\dfrac{A}{B}\dfrac{x_1}{x_2} + 1\right]^2} \qquad (23.32)$$

$$\ln \gamma_2 = \frac{B}{\left[\dfrac{B}{A}\dfrac{x_2}{x_1} + 1\right]^2} \qquad (23.33)$$

This is the well known van Laar Equation, with the constants A and B given by Equations 23.30 and 23.31 as functions of the component physical properties and temperature. If the subscripts 1 and 2 denote the solute and the solvent, respectively, the above terms have the following meanings:

V_1 and V_2 = liquid molar volume of solute and solvent
δ_1 and δ_2 = solubility parameter of solute and solvent

x_1 and x_2 = liquid mole fraction of solute and solvent liquid phase
γ_1 and γ_2 = activity coefficient of solute and solvent

For multicomponent mixtures the solvent is the total of all the other components except the solute. The liquid volume and the solubility parameter for the multicomponent solvent are the average V and δ values for the components of the mixture; i.e.,

$$(V_2)_{\text{mix}} = \sum x_i V_i \qquad (23.34)$$

$$(\delta_2)_{\text{mix}} = \sum X_{v_i} \delta_i \qquad (23.35)$$

It will be noted that V_2 is a molar average while δ_2 is a liquid volume average.

Advantages of this method for estimating activity coefficients are: 1) A and B can be estimated from physical properties, 2) the effect of concentration on activity is included, 3) the effect of temperature on A and B is included.

A disadvantage is that the values of V and δ cannot be calculated directly for all substances and conditions. The lighter components of a mixture are frequently in solution at a temperature and pressure where these components could not exist as a liquid in the pure state. This might be regarded as a hypothetical liquid state for these gases. The values of V and δ for this hypothetical state must be obtained by extrapolation or by back-calculations from experimental vapor liquid equilibria data.

Liquid Molar Volume

Molar volumes of the liquid mixture components are required at the system conditions for evaluating A and B, the van Laar Coefficient. The choice of V values is important because V_1 and V_2 have a direct effect on A and B.

In this connection it is of interest to point out that Equation 23.28 and 23.29 were derived with no volume change on mixing so it is not theoretically rigorous, especially for solutions containing one component above its critical.

For the present purposes, Equations 23.30, 23.31, 23.32, and 23.33 should be regarded as semi-empirical relationships with adjustable parameters that can be made to apply to all hydrocarbon vapor-liquid hydrocarbon mixtures. The liquid molar volume was estimated by the use of Watson's expansion factor (21) in previous work (13)

$$V = (V_1 \omega_1) (5.7 + 3.0 \ T_r) \qquad (23.36)$$

where V_1 = liquid molar volume at a low reference temperature T_1

ω_1 = expansion factor at T_1

T_r = reduced temperature

The product $(V_1\omega_1)$ is a constant for each component. Values of this liquid volume characteristic constant were evaluated and are given in Table 23.1. Equation 23.36 when applied with the characteristic constants in Table 23.1, will give liquid volumes of an incompressible liquid, i.e., a liquid at pressures 5 to 10 times the critical pressure.

Table 23.1
Constants for Activity Coefficient Calculations

	Hildebrand Solubility Parameter δ	Watson Expansion Factor $V_1\omega_1$
H_2	3.33	1.05
CH_4	5.45	5.00
C_2H_4	5.8	6.63
C_2H_6	5.88	7.77
C_3H_6	6.2	8.70
C_3H_8	6.00	9.70
iC_4H_8	6.70	10.97
$1\text{-}C_4H_8$	6.70	10.91
iC_4H_{10}	6.25	11.69
nC_4H_{10}	6.70	11.62
iC_5H_{12}	6.75	14.08
nC_5H_{12}	7.05	14.07
nC_6H_{14}	7.30	16.52
nC_7H_{16}	7.45	18.96
nC_8H_{18}	7.55	21.39
nC_9H_{20}	7.65	23.83
$nC_{10}H_{22}$	7.75	26.28
Cyclopentane	8.10	11.22
Cyclohexane	8.2	14.07
Methyl-cyclohexane	7.85	16.51
Benzene	9.15	11.64
Toluene	8.90	14.15
Ethylbenzene	8.80	16.59

In this pressure region liquid volume is a straight line function of temperature and substantially below the saturated liquid volume. Equation 23.36 is a special and simplified case of the complete Watson liquid density correlation in which the reduced pressure is also a variable.

Comparisons with observed saturated liquid volumes for hydrocarbons show that Equation 23.36 is low, as illustrated below for the ethane-benzene data of Kay and Nevins (8).

Conditions: Saturated liquid at 800 psia
Volume of Mixture cc/gm mol

Mole Fraction Ethane in Benzene	Observed	Equation 23.35	Complete Watson Correlation
0.15	134	95	143
0.38	94	84	101
0.50	89	80	92
0.72	87	75	85
0.85	91	72	89

These mixture volumes were calculated as the molar averages, i.e., $V_{mix} = \Sigma x_i V_i$.

The large differences between the observed and the Equation 23.35 mixture volumes indicate the need for more work to improve this method for evaluating the van Laar A and B coefficients. A volume correlation that will give correct and consistent activity coefficients is required.

Heat of Vaporization from *K*-Values

Two enthalpy of vaporization approximation methods are described next. One method is for a differential heat of vaporization and the other is for an integral isobaric heat of vaporization. Both are for ideal solutions.

Real and Hypothetical States

Two phenomena occur in some mixtures, i.e., "gas-in-liquid solution" and "liquid-in-vapor-mixture." For these component-types, enthalpy-temperature charts have been modified to include "gas-in-liquid" and "liquid-in-vapor" parameters, as shown in Figure 23.3,

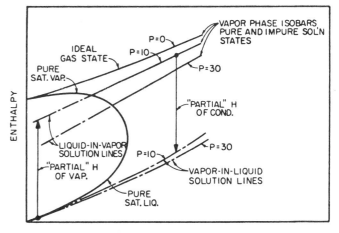

Figure 23.3. Typical *H-T* plot for hydrocarbon showing "gas-in-liquid" and "liquid-in-vapor" lines.

which illustrates the way Peters (12) modified several hydrocarbon H-T charts in 1949. In Figure 23.3, typical "partial" heats of vaporization and condensation are shown by lines and arrows.

For ideal mixture components, the partial heats of vaporization can be calculated from the vapor-liquid K-ratios by an expression that is derived by starting with the following previously presented equation:

$$\left(\frac{\partial \ln \bar{f}_i}{\partial T}\right)_{P,N_i} = \frac{H_i^o - \overline{H}_i}{RT^2} \tag{23.37}$$

Equation 23.37 applies to either vapor or liquid phases. In terms of the fugacities, the vapor-liquid K values is

$$K_i = \frac{\bar{f}_i^L/x_i}{\bar{f}_i^V/y_i} \tag{23.38}$$

For ideal mixtures, the Lewis and Randall Fugacity Rule states that

$$\bar{f}_i^L = x_i f_i^L \text{ and } \bar{f}_i^V = y_i f_i^V$$

Combining gives

$$K_I = f_i^L/f_i^V \tag{23.39}$$

Equation 23.39 gives the ideal solution K values in terms of the pure component fugacities in the vapor and liquid phases at the system conditions.

These fugacities are functions of temperature and pressure for each component of the ideal mixtures. This means that one of these phases could be real and the other hypothetical for some components of the mixture. For the lighter components of a mixture, the f_i^V value will be for a real gas state whereas the f_i^L value will be for a "gas-in-liquid-solution" hypothetical state. For the heavier components of a mixture, the f_i^L value will be for a real liquid state whereas the f_i^V value will be for a "liquid-in-vapor-mixture" hypothetical state.

Evaluating these hypothetical pure component state values of the fugacity is not necessary for present purposes. Differentiating the logrithmic form of Equation 23.39 gives:

$$\left(\frac{\partial \ln K_I}{\partial T}\right)_P = \left(\frac{\partial \ln f_i^L}{\partial T}\right)_P - \left(\frac{\partial \ln f_i^V}{\partial T}\right)_P \tag{23.40}$$

For a pure component or an ideal solution component

$$\left(\frac{\partial \ln f_i^L}{\partial T}\right)_P = \frac{H_i^o - H_i^L}{RT^2} \tag{23.41}$$

and

$$\left(\frac{\partial \ln f_i^V}{\partial T}\right)_P = \frac{H_i^o - H_i^V}{RT^2} \tag{23.42}$$

Combining Equations 23.40, 23.41, and 23.42 gives

$$H_i^V - H_i^L = RT^2 \left(\frac{\partial \ln K_I}{\partial T}\right)_P \tag{23.43}$$

Equation 23.43 gives the partial heat of vaporization for an ideal mixture component in terms of the isobaric temperature derivative of $\ln K_I$, it being assumed that the ideal K value is independent of composition. Equation 23.43 is the relationship that Peters (12) used to locate points on the "gas-in-liquid" and the "liquid-in-vapor" lines.

Integral Heat of Vaporization

Computing the isothermal enthalpy difference for saturated liquid mixtures via the application of the Gibbs-Duhem Equation to isobaric vapor-liquid equilibrium data has been suggested (19) and studied for binaries (9). The method was reviewed and applied later (6). Isobaric versions of the Gibbs-Duhem Equation may also be applied to saturated vapor and liquid. Such an application leads to a convenient approximation for the isobaric integral heat of vaporization.

Differential forms of the isobaric Gibbs-Duhem for the coexisting equilibrium vapor and liquid are as follows:

$$\sum y_i \left(\frac{\partial \ln \bar{f}_i^V}{\partial T}\right)_P = -\frac{H^V - H^o}{RT^2} = -\frac{\Delta H^V}{RT^2} \tag{23.44}$$

and

$$\sum x_i \left(\frac{\partial \ln \bar{f}_i^L}{\partial T}\right)_P = -\frac{H^L - H^o}{RT^2} = -\frac{\Delta H^L}{RT^2} \tag{23.45}$$

The temperature derivatives of the logarithm of fugacity are taken along the isobaric dew point and the isobaric bubble point curves.

The temperature derivatives in Equation 23.44 and 23.45 differ from the temperature derivatives in the partial enthalpy expression, used by Edmister and Ruby (3, 3a), in that the former are restricted to constant pressure while the latter are restricted to constant pressure and composition. Although $(\partial \ln \bar{f}_i^V/\partial T)_P \neq (\partial \ln \bar{f}_i^V/\partial T)_{Py}$ the summations are equal, i.e.,

$$\sum y_i \left(\frac{\partial \ln \bar{f}_i^V}{\partial T}\right)_P = \sum y_i \left(\frac{\partial \ln \bar{f}_i^V}{\partial T}\right)_{P,y} \qquad (23.46)$$

The equality of Equation 23.46 must be true to give the same value of the enthalpy difference for the mixture via both routes, i.e., partial enthalpy relationship and the isobaric Gibbs-Duhem Equation.

In terms of the fugacity coefficient $\phi_i = \bar{f}_i^V/Py_i$ and the equilibrium constant $K_i = y_i/x_i$, Equations 23.44 and 23.45 are transformed to a more practical form in the following.

From $\bar{f}_i^V = \phi_i Py_i$

$$\left(\frac{\partial \ln \bar{f}_i^V}{\partial T}\right)_P = \left(\frac{\partial \ln \phi_i}{\partial T}\right)_P + \left(\frac{\partial \ln y_i}{\partial T}\right)_P \qquad (23.47)$$

Noting that $\sum y_i \left(\dfrac{\partial \ln y_i}{\partial T}\right)_P = 0$ gives

$$\sum y_i \left(\frac{\partial \ln \bar{f}_i^V}{\partial T}\right)_P = \sum y_i \left(\frac{\partial \ln \phi_i^V}{\partial T}\right)_P \qquad (23.48)$$

From the definitions of ϕ_i and K_i and the equal-fugacity criteria of equilibria, $\bar{f}^L = \phi_i K_i P x_i$, we find

$$\left(\frac{\partial \ln \bar{f}_i^L}{\partial T}\right)_P = \left(\frac{\partial \ln \phi_i}{\partial T}\right)_P + \left(\frac{\partial \ln K_i}{\partial T}\right)_P + \left(\frac{\partial \ln x_i}{\partial T}\right)_P \qquad (23.49)$$

Noting that $\sum x_i \left(\dfrac{\partial \ln x_i}{\partial T}\right)_P = 0$ gives

$$\sum x_i \left(\frac{\partial \ln \bar{f}_i^L}{\partial T}\right)_P = \sum x_i \left(\frac{\partial \ln \phi_i K_i}{\partial T}\right)_P \qquad (23.50)$$

Combining Equations 23.44 and 23.48 gives an expression for the vapor enthalpy difference. A similar expression is found for the liquid phase by combining Equations 23.45 and 23.50. These two expressions follow:

$$\sum y_i \left(\frac{\partial \ln \phi_i}{\partial T}\right)_P = -\frac{H^V - H^o}{RT^2} = -\frac{\Delta H^V}{RT^2} \qquad (23.51)$$

and

$$\sum x_i \left(\frac{\partial \ln \phi_i K_i}{\partial T}\right)_P = -\frac{H^L - H^o}{RT^2} = -\frac{\Delta H^L}{RT^2} \qquad (23.52)$$

In both expressions ϕ_i is the fugacity coefficient of the components of the dew point vapor. Inherent in the deri-

vation of Equation 23.52 for the liquid mixture is the requirement that $\bar{f}_i^V = \bar{f}_i^L$ for coexisting equilibrium vapor. Equations 23.44 and 23.45 were combined with this equality [i.e., $(\partial \ln \bar{f}^V/\partial T)_P = (\partial \ln \bar{f}^L/\partial T)_P$] to give an expression for the enthalpy difference for equilibrium vaporization, i.e.,

$$\sum (x_i - y_i) \left(\frac{\partial \ln \bar{f}_i^V}{\partial T}\right)_P = \frac{\Delta \underline{H} - \Delta \underline{H}^o}{RT^2} \qquad (23.53)$$

where $\quad \Delta \underline{H} = (\underline{H}_{DP}^V - \underline{H}_{BP}^L)$

$$\Delta \underline{H}^o = (\underline{H}_{DP}^o - \underline{H}_{BP}^o)$$

The two ΔH terms are at the same temperature, i.e., the equilibrium temperature for coexisting vapor and liquid phases. H^V and H^L are at system pressure while the H^o terms are at zero pressure. H^o_{DP} and H^o_{BP} differ because of composition.

Evaluation of the left hand side of Equation 23.53 may be done via Equation 23.47 and the derivative of $\ln y_i$ in terms of K_i and x_i, i.e.,

$$\left(\frac{\partial \ln y_i}{\partial T}\right)_P = \left(\frac{\partial \ln K_i}{\partial T}\right)_P + \left(\frac{\partial \ln x_i}{\partial T}\right)_P \qquad (23.54)$$

Combining Equations 23.53 and 23.47 and 23.54 and recognizing that $\sum y_i (\partial \ln v_i/\partial T_i)_P = 0$ and $\sum x_i (\partial \ln x_i/\partial T_i)_P = 0$ gives

$$\sum (x_i - v_i) \left(\frac{\partial \ln \phi_i}{\partial T}\right)_P + \sum x_i \left(\frac{\partial \ln K_i}{\partial T}\right)_P$$

$$= \frac{\Delta \underline{H} - \Delta \underline{H}^o}{RT^2} \qquad (23.55)$$

Equation 23.55 is identical to the difference between Equations 23.51 and 23.52. This combination is not as convenient in applications as the separate forms, i.e., Equations 23.51 and 23.52.

The first left hand member of Equation 23.55 is usually very small compared to the second left hand member. When this first term can be neglected, an approximation equation for the isobaric integral heat of vaporization was suggested previously (2,6) in the form:

$$\Delta \underline{H}_{Vap} = \frac{RT_1 T_2}{T_2 - T_1} \sum x_i \ln \frac{K_2}{K_1} + \Delta H^o \qquad (23.56)$$

where K_1 and K_2 are the K-ratios at the bubble and dew point temperatures, T_1 and T_2

In this approximation, the point derivative of Equation 23.55 was replaced by a difference ratio, i.e. ($\ln K_2 - \ln$

$K_1)/(T_2 - T_1)$. Also it was assumed that $T^2 = T_1 T_2$. The $\Delta \underline{H}_{Vap}$ approximated by Equation 23.56 is an average value over the temperature T_1 to T_2. When T_1 and T_2 are close together, $\Delta \underline{H}_{Vap}$ will be a better approximation.

Equations 23.43 and 23.56, derived and discussed above, are used in two of the examples that follow.

Examples

The preparation of enthalpy-composition diagrams from a minimum of experimental data is illustrated for four binary systems: methanol-benzene, ethanol-benzene, ethanol-toluene, and hydrogen chloride-normal hexane. Each of these examples illustrates a different feature of such calculations. All binaries are non-ideal solutions, three of them forming constant boiling mixtures or azeotropes of the minimum boiling point type.

For all four of the binary systems, the enthalpies are referred to the ideal gas state zero enthalpy datum. This determines the order in which the enthalpy calculations are made. Experimentally determined temperature-composition and heat of vaporization data were used in the preparation of the methanol-benzene diagram.

For the two ethanol binaries, which form azeotropes, the van Laar (18) activity coefficient equation is used in constructing the temperature-composition diagram. The Scatchard-Hildebrand (7,16) activity coefficient equation is used for the fourth mixture, as was a method of estimating heats of vaporization from ideal K-values (2,6).

Methanol-Benzene. For this binary, temperature-composition data and isobaric integral heats of vaporization were available at a pressure of 735 mm Hg from experimental measurements (18). These data were used in preparing an H-X diagram for the vapor-liquid two phase region, for which the values are given in Table 23.2. As can be seen, the T-X-Y values identify a minimum constant boiling mixture.

The first step in the preparation of the values for Table 23.2 was to plot the temperature-composition data and read values of x and y (mole fraction of methanol in the equilibrium vapor and liquid) at even intervals of temperature. These values are in the first three columns of Table 23.2. Next, enthalpies of the saturated vapor at each composition and temperature were calculated. These values, which are listed in the last column of Table 23.2, are ideal gas state enthalpies.

At this low pressure the effect of pressure on the vapor enthalpy is nil, so it is assumed that the saturated vapor enthalpy at the dew point temperature and pressure is the same as the ideal gas of the same composition. For example, the enthalpy at the 140°F dew point of the 51% methanol saturated vapor mixture has an enthalpy of

**Table 23.2
Temperature-Composition and
Enthalpy-Composition Values for Methanol-Benzene
at 735 mm Hg**

M.F. Methanol			Enthalpy, Btu/lb mole	
liq.	vap.	T° F	Sat. liq.	Sat. vap.
0	0	174.4	−5,080	8,200
0.01	0.09	170.0	−5,190	7,900
0.023	0.183	165.0	−5,280	7,600
0.035	0.262	160.0	−5,360	7,320
0.047	0.330	155.0	−5,470	7,100
0.064	0.390	150.0	−5,600	6,890
0.095	0.450	145.0	−5,810	6,690
0.150	0.510	140.0	−6,190	6,500
0.221	0.556	137.0	−6,580	6,370
0.400	0.580	136.0	−7,340	6,300
0.606	0.606	135.5	−8,200	6,240
0.724	0.655	136.0	−8,700	6,150
0.835	0.700	137.0	−9,150	6,050
0.871	0.732	138.0	−9,270	5,980
0.919	0.794	140.0	−9,400	5,900
0.954	0.856	142.0	−9,500	5,780
0.990	0.957	145.0	−9,580	5,670
1.00	1.00	147.0	−9,600	5,630

$$H^o = 0.51 \ (5,600) + 0.49 \ (7,400) = 6,500 \text{ Btu/lb mole}$$

where $5,600$ = enthalpy of methanol as vapor at 140°F, Btu/lb mole

$7,400$ = enthalpy of benzene as vapor at 140°F, Btu/lb mole

Theoretically, this enthalpy value for the 51% mixture should be corrected for pressure but this effect (between 0 and 735 mm Hg) is nil, as previously stated. Also the heat of mixing is neglected, giving a point on the saturated vapor curve. Other points on the vapor curve are found in a similar way. Note that a 140°F dew point also exists for a 79.4% methanol solution.

The isobaric integral heat of vaporization is the heat required to vaporize a mixture from its bubble point to its dew point, both points being at the same pressure but at different temperatures, for a constant composition mixture. For the 51% methanol mixture this isobaric integral heat of vaporization was obtained from the experimental data and is 14,300 Btu/lb mole. The corresponding bubble point is 135.7°F. The enthalpy of the saturated liquid of 51% methanol composition and at 135.7°F is

$$H^L = 6,500 - 14,300 = 17,800 \text{ Btu/lb mole}$$

All of the enthalpy values in Table 23.2 were obtained this way.

Ethanol-Benzene. For this binary, the temperature-composition curves at 760 mm Hg pressure were obtained via calculations based upon vapor pressure and the azeotropic conditions, i.e., composition and temperature. Ethanol-benzene forms a minimum constant boiling mixture at 44.8 mole % ethanol and 68.24°C. In these calculations, the assumptions are made that the van Laar Equations 23.32 and 23.33 represent the departure from ideal vapor-liquid equilibria behavior and that heats of vaporization can be calculated from the K-ratios and temperatures by Equation 23.56.

As can be seen by an examination of Equations 23.32 and 23.33, values of the activity coefficients at one composition can be used to determine the values of the constants A and B. Then the van Laar Equation can be applied to other compositions with these values of A and B.

Expressions for A and B are found by combining Equations 23.32 and 23.33 and then rearranging to obtain the following functions of x_1, x_2, γ_1 and γ_2;

$$A = \ln \gamma_1 \left(1 + \frac{x_2 \ln \gamma_2}{x_1 \ln \gamma_1}\right)^2 \qquad (23.57)$$

$$B = \ln \gamma_2 \left(1 + \frac{x_1 \ln \gamma_1}{x_2 \ln \gamma_2}\right)^2 \qquad (23.58)$$

The details of these derivations are not included. This is an interesting exercise for the reader. The key step is finding the relationship

$$\frac{A}{B} = \left(\frac{x_2}{x_1}\right)^2 \left(\frac{\ln \gamma_2}{\ln \gamma_1}\right)$$

Equations 23.57 and 23.58 are used in the calculations for the ethanol-benzene system as follows:

Step 1. Using the azeotrope point (44.8 mole percent ethanol at 68.24°C and 760 mm Hg) and the vapor pressures at 68.24°C (495 mm Hg for ethanol and 510 mm Hg for benzene), the activity coefficients at this azeotrope point are found via the expression

$$Py_i = \gamma_i \, p_i^o \, x_i \qquad (23.59)$$

Equation 23.59 defines an activity coefficient that is based upon Raoult's and Dalton's Laws and gives values of gamma that are empirical and limited to the conditions of the data used. At the azeotrope point $x_i = y_i$, giving for ethanol (in benzene)

$$\gamma_1 = 760/495 = 1.537$$
$$\ln \gamma_i = 0.430$$

and for benzene (in ethanol)

$$\gamma_2 = 760/510 = 1.490$$
$$\ln \gamma_2 = 0.399$$

Substituting into Equations 23.57 and 23.58 gives

$$A = 0.430 \left(1 + \frac{(0.448)(0.430)}{(0.552)(0.399)}\right)^2 = 1.978$$

$$B = 0.399 \left(1 + \frac{(0.552)(0.399)}{(0.448)(0.430)}\right)^2 = 1.401$$

These A and B constants are for use in Equations 23.32 and 23.33.

Step 2. Values of γ_i and γ_2 are next calculated via Equations 23.32 and 23.33 at several compositions of the liquid phase. Sometimes it is convenient to plot these γ_i values against liquid mole fraction, x_i. A few typical values are given below for this system.

x_1 mole fract. ethanol	γ_1 ethanol	γ_2 benzene
0.1	4.26	1.023
0.3	2.15	1.219
0.5	1.505	1.614
0.7	1.110	2.275
0.9	1.012	3.33

Step 3. Using the above activity coefficients, and vapor pressures, equilibrium values of y_i were calculated for each value of x_i. This is a successive approximation calculation in which the equilibrium temperature is assumed, vapor pressures found, activity coefficients obtained from Step 2, K-values estimated ($K_i = p^o_i \, \gamma_i / 760$), y_i values calculated ($y_i = K_i \, x_i$) and tested for convergence, the test being $\Sigma y_i = 1$. A few typical results are:

x_1	T°C	p_1^o	p_2^o	K_1	K_2	y_1	y_2
0.1	71.2	562	565	3.15	0.761	0.315	0.865
0.3	68.4	500	513	1.414	0.8225	0.424	0.576
0.5	68.3	499	511	0.923	1.084	0.4615	0.542
0.7	69.2	518	528	0.7565	1.577	0.530	0.470
0.9	73.4	613	609	0.816	2.67	0.733	0.267

From these results, temperature-composition curves can be plotted for the coexisting equilibrium vapor and liquid phases. We are now ready to calculate the enthalpies.

Step 4. Enthalpies of the saturated vapor mixtures were found via ideal gas state heat capacities and the assumption that the isothermal enthalpy differences between zero and 760 mm Hg pressure are negligible. Constants for the following molar heat capacity expression, with temperature in degrees Kelvin, were estimated by a group contribution method.

$$C^o{}_p = a + b\,T + cT^2 \tag{23.60}$$

	a	b	c
ethanol ...	4.55	26.99×10^{-3}	-8.24×10^{-6}
benzene ..	0.23	77.83×10^{-3}	-27.16×10^{-6}

A temperature integration of Equation 23.60 gives the following expression for the enthalpy above a reference temperature of 0°K, at which $H^o = 0$:

$$H^o - H^o_O = a\,T + \tfrac{1}{2}\,b\,T^2 + \tfrac{1}{3}\,c\,T^3 \tag{23.61}$$

Using Equation 23.61, with the above values of the constants and the vapor temperature-composition values from Step 3, and the ideal gas state values of H^o_{Oi}, the enthalpies were calculated for the saturated vapor mixtures at different points of temperature and composition. These are the H values given in the fifth column of Table 23.3 for saturated vapor.

Step 5. Saturated liquid enthalpies for this binary mixture were formulated by combining the ideal solution liquid enthalpies and the excess (heat of mixing) enthalpy. The ideal-solution heat of vaporization for each component is found from

$$\underline{H}^V - \underline{H}^L = R\,T^2 \left(\frac{d \ln K_l}{d\,T}\right)_P \tag{23.43}$$

In this case $K_l = p^o_i/P$ and $\ln p^o_i = C_i + D_i/T$, so that Equation 23.43 reduces to

$$(H^V - H^L)\ \text{ideal} = -D_i\,R \ \text{in cal./g mole}$$

where $D_i = -5280$ for ethanol
$= -4200$ for benzene
$R = 1.9875$

Step 6. Excess enthalpy is formulated from the van Laar Equation, as follows: The partial excess enthalpy for a component of a mixture is

$$\underline{H}^E_i = -RT^2 \left(\frac{\partial \ln \gamma_i}{\partial\,T}\right)_P \tag{23.62}$$

The isobaric temperature derivative of $\ln \gamma_i$ can be estimated via the Scatchard-Hildebrand form of the van Laar Equation, in which the constants A and B are functions of the molar volume and the solubility parameter of each component, as shown in Equations 23.30 and 23.31.

Such a combination leads to the following expression for the activity coefficient

$$\ln \gamma_i = \frac{V^L_i}{RT} (\delta_i - \delta_m)^2 \tag{23.63}$$

where $\delta_m = \dfrac{\sum x_i\,V^L_i\,\delta_i}{\sum x_i\,V^L_i}$

= liquid volume averaged solubility parameter for mixture

Assuming that the molal volume, V^L_i, and the solubility parameter, δ_i, are independent of temperature, as has been done in previous work on the applications of this expression to vapor-liquid equilibrium predictions, leads to the following derivative

$$\left(\frac{\partial \ln \gamma_i}{\partial T}\right)_P = -\frac{V^L_i}{RT^2}(\delta_i - \delta_m)^2 \tag{23.64}$$

Combining Equations 23.62, 23.63, and 23.64 gives

$$\overline{H}^E_i = RT \ln \gamma_i \tag{23.65}$$

The excess enthalpy for the liquid mixture is

$$\underline{H}^E = \sum x_i\,\overline{H}^L - \sum x_i\,\underline{H}^L = RT \sum x_i \ln \gamma_i \tag{23.66}$$

Combining Equations 23.61 and 23.66 gives the following expression for the enthalpy of the saturated liquid mixture in terms of the vapor pressure curve slope, the activity coefficient and the enthalpy of the saturated vapor enthalpy:

$$\underline{H}^L = \underline{H}^V + R \sum x_i\,B_i + RT \sum x_i \ln \gamma_i \tag{23.67}$$

Typical values of the last two terms in Equation 23.67 for the ethanol-benzene system are given in the following tabulation:

	cal/g mole	
x_1	$R \sum x_iB_i$	$RT \sum x_i \ln \gamma_i$
0.1	-8560	118
0.3	-9000	250
0.5	-9420	278
0.7	-9850	213
0.9	-10280	89

Step 7. Saturated liquid mixture enthalpies are next calculated by the combination of Equation 23.67, using the results of Steps 4, 5 and 6. This calculation is made with the vapor enthalpy $\underline{H}^V$ values for the components involved at the temperatures and compositions of the liquids, i.e., the same x_i values used in Step 6. It is not correct to use equilibrium vapor compositions in this calculation.

Saturated liquid enthalpies, found in this way, are given in the fourth column of Table 23.3.

Table 23.3
Temperature-Composition and Enthalpy-Composition Values for Ethanol-Benzene at 760 mm Hg

M.F. Ethanol			Enthalpy, g cal/g mole	
liq.	vap.	T° C	Sat. liq.	Sat. vap.
0	0	79.7	−3,790	4,560
0.01	0.075	78.0	−3,800	4,390
0.023	0.150	76.0	−3,850	4,200
0.045	0.217	74.0	−3,920	4,100
0.095	0.305	71.0	−4,100	3,950
0.200	0.395	69.0	−4,450	3,800
0.300	0.425	68.4	−4,790	3,760
0.447	0.447	68.2	−5,290	3,720
0.600	0.490	68.4	−5,800	3,680
0.700	0.540	69.0	−6,110	3,630
0.770	0.586	70.0	−6,360	3,600
0.858	0.673	72.0	−6,720	3,500
0.918	0.765	74.0	−7,000	3,410
0.963	0.877	76.0	−7,180	3,300
1.0	1.0	78.1	−7,320	3,150

Ethanol-Toluene

The calculations for this binary were made by the same method used for the ethanol-benzene system, with minor modifications. The van Laar constants were found from the 76.65° C and 81 mole percent ethanol azeotrope to be $A = 1.52$ and $B = 1.73$. Vapor pressures were fitted to the equation:

$$\ln p_i^o = C_i + \frac{D_i}{T} + E_i \ln T \qquad (23.68)$$

where p_i^o = vapor pressure in mm Hg
T = temperature, °K

The constants for Equation 23.68 were found by fitting to experimental vapor pressure data, using a digital computer, obtain the following:

	C_i	D_i	E_i
ethanol ...	59.960	−6922.85	−5.7392
toluene ...	48.657	−5893.57	−4.4820

Ideal gas state enthalpies were computed via Equation 23.64 for ethanol, using the same constants used in the previous example. For toluene, Equation 23.61 was also used, with the following constants:

	a	b	c
toluene ..	0.59	95.48×10^{-3}	$−33.04 \times 10^{-6}$

Expressions for the ideal heats of vaporization were obtained via Equation 23.43, as in the previous example, giving

$$(H_i^V - H_i^L)_{Ideal} = R \, (- D_i + E_i \, T)$$

The values of the constants D_i and E_i are those given under Equation 23.68.

Proceeding in the same way as in the ethanol-benzene calculations, the temperature-composition-enthalpy values were computed for the ethanol-toluene system at 760 mm Hg. These results are given in Table 23.4.

Table 23.4
Temperature-Composition and Enthalpy-Composition Values for Ethanol-Toluene at 760 mm Hg

M.F. Ethanol			Enthalpy, g cal/g mole	
liq.	vap.	T° C	Sat. liq.	Sat. vap.
0	0	110.6	−1,650	6,650
0.025	0.253	101.55	−2,050	5,580
0.050	0.387	95.80	−2,370	5,080
0.100	0.529	88.85	−2,820	4,520
0.150	0.604	84.84	−3,115	4,260
0.20	0.649	82.29	−3,254	4,094
0.25	0.679	80.64	−3,562	3,992
0.30	0.700	79.48	−3,765	3,912
0.35	0.715	78.73	−4,032	3,863
0.40	0.726	78.18	−4,149	3,824
0.45	0.735	77.78	−4,337	3,795
0.50	0.743	77.52	−4,532	3,769
0.55	0.749	77.27	−4,722	3,750
0.60	0.757	77.12	−4,910	3,731
0.65	0.766	76.96	−5,120	3,708
0.70	0.775	76.81	−5,315	3,677
0.75	0.788	76.71	−5,529	3,645
0.81	0.81	76.65	−5,839	3,601
0.85	0.831	76.70	−5,957	3,542
0.90	0.867	76.90	−6,176	3,459
0.95	0.920	77.4	−6,404	3,331
1.0	1.0	78.8	−6,598	3,152

Hydrogen Chloride-Normal Hexane

An enthalpy-composition diagram was also prepared for the hydrogen chloride-normal hexane system at 100 psia pressure. Such a diagram was constructed for use in making separation calculations for the stripping, via multistage distillation, of HCl from a mixture of HCl and n-hexane, the latter coming from an isomerization reactor. The thermodynamic calculations for this *H-X* chart construction were made by using various approximation methods. The work was done in 1961 by two graduate students (15) and is presented here as an example of this application of thermodynamics.

The temperature-composition diagram for the HCl-nC_6H_{14} binary was based on *K*-ratios predicted by using generalized fugacities obtained from Figures 6.5 and 6.6 and on activity coefficients from the van Laar Equation. Constants for the van Laar Equation were found via the solubility parameters and molar volumes, using Equations 23.30 and 23.31. Values of V_i^L and δ_i were read from tables for *n*-hexane. For HCl these two parameters were estimated from experimental vapor-liquid equilibrium data of Ashley and Brown (1) on the Ethane-HCl binary.

Activity coefficients were found for ethane and HCl from the vapor and total pressures, the compositions of the equilibrium vapor and liquid phases and the assumption that

$$\gamma_i = y_i P / x_i p_i^o$$

These "vapor pressure" activity coefficients were calculated for ethane and HCl along the four isotherms (-30, -40, -50 and $-60°C$) covered in the experimental data. The γ_i values were then plotted on the Lu-Lavergne (10) coordinates. Figure 23.4 shows the $-40°C$ isotherm. The other isotherms are parallel and displaced at very small intervals from the one shown.

The coordinates of Figure 23.4 are obtained from an expression obtained by combining Equations 23.32 and 23.33 to eliminate the liquid mole fraction ratios, x_1/x_2 and x_2/x_1, which manipulation gives

$$(\ln \gamma_1)^{1/2} = B^{1/2} - (B/A)^{1/2} (\ln \gamma_1)^{1/2} \qquad (23.69)$$

$$(\ln \gamma_1)^{1/2} = A^{1/2} - (A/B)^{1/2} (\ln \gamma_2)^{1/2} \qquad (23.70)$$

Lu and Lavergne (10) proposed plotting the activity coefficients of binary components on the coordinates $(\ln \gamma_1)^{1/2}$ vs. $(\ln \gamma_2)^{1/2}$. When a straight line is drawn through the points of such a plot, the intercepts are the values of $A^{1/2}$ and $B^{1/2}$. This is a way of finding the average or overall values of the constants of the van Laar Equation by graphical means.

Figure 23.4 is this type of plot for the ethane-hydrogen chloride system. Values of the *A* and *B* constants may be

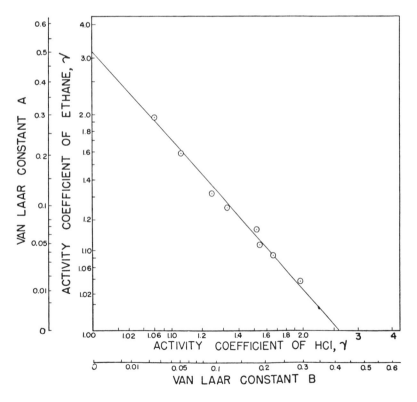

Figure 23.4. Lu-Lavergne (10) plot for $-40°C$ isotherm of the Ashley-Brown (1) data for C_2H_6-HCl.

read directly from the auxiliary scales of Figure 23.4. The numerical values of these scales correspond to logarithm of the activity coefficient to the base 10 in Equations 23.32 and 23.33. The A and B scale values in Figure 23.4 should be multiplied by 2.303 to obtain appropriate value of A and B for the natural logarithm form of the van Laar Equation.

Points for the $-40°F$ isotherm are shown on Figure 23.4. A straight line is drawn through these points to obtain the intercepts. Similar lines were obtained for other isotherms. Values of the van Laar constants were found to be

T°C	A	B
−30	0.480	0.385
−40	0.503	0.415
−50	0.525	0.437
−60	0.545	0.457

These A and B values are based on $\log_{10} \gamma_i$ values and should be multiplied by 2.303 to be consistent with $\ln \gamma_i$ values.

Using these values of the van Laar constants, Equations 23.30, 23.31 and 23.36, and the values of ethanes constants from Table 23.1, values of the unknown constants in the following expressions were found for HCl:

$$A = \frac{V_1\omega_1(5.7 + 3\ T_r)}{RT}(\delta_1 - \delta_2)^2 \qquad (23.71)$$

$$B = \frac{V_2\omega_2(5.7 + 3\ T_r)}{RT}(\delta_1 - \delta_2)^2 \qquad (23.72)$$

For ethane, $V_2\omega_2 = 7.77$ cc/g mole and $\delta_2 = 5.88$ (cal./cc)$^{1/2}$. Using the above values of A and B for the HCl—C_2H_6 binary, Equations 23.71 and 23.72 were applied at each temperature to obtain the values of $V_1\omega_i$ and δ_1 for HCl, i.e., the two unknowns in these two equations. These were found to be the following:

T°C	$V_1\omega_1$, cc/g mole	δ_1 (cal./cc)$^{1/2}$
−30	6.30	2.95 or 8.85
−40	6.44	2.93 or 8.83
−50	6.56	2.92 or 8.84
−60	6.66	2.92 or 8.84

There are two roots for δ_1, the one of interest being selected from the results of another calculation made for ethane at $-30°C$, as follows:

$$\delta_1 = \left(\frac{\Delta E}{V}\right)^{1/2} = \left(\frac{34,200}{467}\right)^{1/2} = 8.56(\text{cal/cc})^{1/2}$$

The value of $\delta_1 = 8.56$ is approximate but good enough to indicate that the higher of the back-calculated values of δ_1 should be used. Therefore the parameters chosen for HCl are:

$$V_1\omega_1 = 6.49 \text{ cc/g mole}$$
$$\delta_1 = 8.83 \text{ (cal/cc)}^{1/2}$$

which values are averages.

Using these constants for HCl plus constants from Table 23.1 for nC_6H_{14}, values of the van Laar constants at $100°F$ were found by substituting into Equations 23.71 and 23.72 to obtain

$$A = \frac{6.49\ [5.7 + 3(0.958)]}{(1.987)\ (311)}(8.83 - 7.30)^2 = 0.209$$

$$B = \frac{16.52\ [5.7 + 3(0.611)]}{(1.987)\ (311)}(8.83 - 7.30)^2 = 0.473$$

Assuming that $x_1 = 0.13$ for HCl, the activity coefficients are found by substituting into Equations 23.32 and 23.33 and giving $\gamma_1 = 1.193$ and $\gamma_2 = 1.003$.

Ideal solution K-ratios for HCl and nC_6H_{14} at $100°F$ and 100 psia were estimated to be 5.98 and 0.0726 by the use of Figures 6.5 and 6.6. Combining these ideal K's and the above activity coefficients gives actual K ratios of 7.20 and 0.0727, respectively. A check of the assumed liquid composition with these K-values completes the calculation, as follows:

$$(0.13)\ (7.20) + (0.87)\ (0.0727) = 1.00$$

Similar equilibrium vapor and liquid composition calculations were made at several temperatures between -40 and $290°F$. From these results, a temperature-composition plot was made. This is given as Figure 23.5.

Enthalpies of saturated and superheated vapors were computed by combining the ideal gas state enthalpies and the isothermal pressure corrections. With these enthalpies and the vapor dew points from Figure 23.5, the vapor portion of the enthalpy-composition diagram, i.e., the upper part of Figure 23.6, was plotted.

The next steps are the calculations of the isobaric integral heat of vaporization and the enthalpies of saturated liquids, both calculations being made at several compositions. Equation 23.56 is used in making the isobaric heat of vaporization calculations. This expression is not rigorous, as was shown in its derivation, but it should be a fairly good approximation for "regular" solutions, such

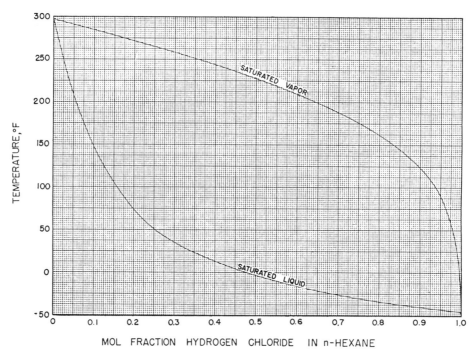

Figure 23.5. Temperature-composition diagram for n-C_6H_{14}-HCl at 100 psia.

as HCl-nC$_6$H$_{14}$. Equation 23.56 was not used in the previous examples as it is unpredictable and supersensitive for an azeotrope forming binary mixture. A sample calculation of the isobaric integral heat of vaporization of a 20 mole percent HCl, 80 mole percent n-hexane binary follows:

	Bubble point		Dew point	
	HCl	nC$_6$H$_{14}$	HCl	nC$_6$H$_{14}$
$T°$F	73	73	271	271
$K = y/x$	4.87	0.044	18.2	0.81
$H°$, Btu/lb mole	3600	11,700	4990	19,440

These temperatures and the K_i and H_i^o values are next used in the solution of Equation 23.56, as follows:

$$\frac{RT_1T_2}{T_2 - T_1} = 3900 \text{ Btu/lb mole}$$

	HCl	nC$_6$H$_{14}$
K_2/K_1	3.74	18.4
$\ln (K_2/K_1)$	1.32	2.91
$\dfrac{RT_1T_2}{T_2 - T_1} \ln \dfrac{K_2}{K_1}$	5,150	11,360
$\Delta H°$, Btu/lb mole	1,390	7,740
	6,540	19,100

$$\Delta H_{vap} = (0.20)(6540) + (0.80)(19,100)$$
$$= 16,610 \text{ Btu/lb mole}$$

Similar calculations were made at other compositions, taken at even intervals of x, from which saturated liquid enthalpy points were located on Figure 23.6 by measuring down from the saturated vapor points ΔH values equal to these isobaric integral heats of vaporization.

Isotherms in the subcooled liquid region were next constructed. As experimental data on heats of mixing and mixture heat capacities were not available for this system, it was necessary to make an approximation, i.e., to assume additivity of component liquid heat capacities. This is equivalent to straight line isotherms in the subcooled liquid region. The pure hexane ends of these isotherms were determined easily via heat capacities of liquid nC$_6$H$_{14}$.

Two points on each liquid isotherm were then known, i.e., the pure nC$_6$H$_{14}$ and the saturated liquid mixture ends. These should be fairly reliable. The straight line assumption is questionable, however. Nevertheless, these lines are drawn on Figure 23.6 to complete the enthalpy-composition diagram.

Another feature of the enthalpy-composition diagram is illustrated on Figure 23.6 and is the "Auxiliary Line" for locating the terminals of other equilibrium temperature tie-lines. This "Auxiliary Line" is the locus of the compositions of the coexisting equilibrium vapor and liquid at the same temperature and is used in interpolating to locate additional isotherm tie-lines when these are needed.

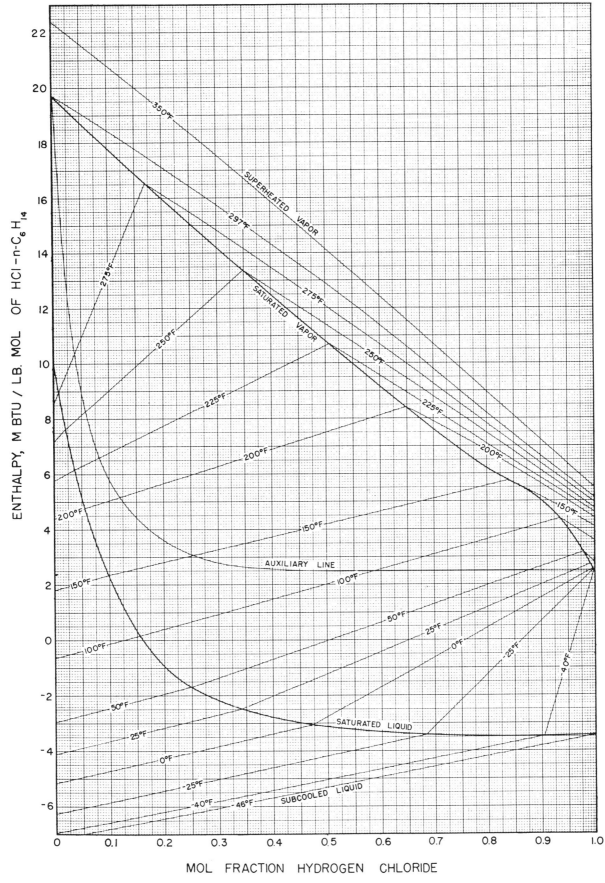

Figure 23.6. Enthalpy-composition diagram for n-C_6H_{14}-HCl at 100 psia.

Multicomponent Enthalpy-Composition Diagrams

The *H-X* diagrams have proven to be so useful for binary mixtures that one wonders if there might be a single composition variable that would permit applying this technique to complex mixtures of more components. Pseudo and hypothetical hydrocarbon components, i.e., narrow boiling cuts of petroleum, are characterized by

the normal atmospheric boiling point and the specific gravity for the calculation of thermodynamic properties, as has been explained in Chapter 15. If narrow cuts require two variables for composition characterization, it seems too much to expect that a single variable would be acceptable for a complex mixture.

For some purposes, compositions of multicomponent hydrocarbon mixtures have been characterized by the

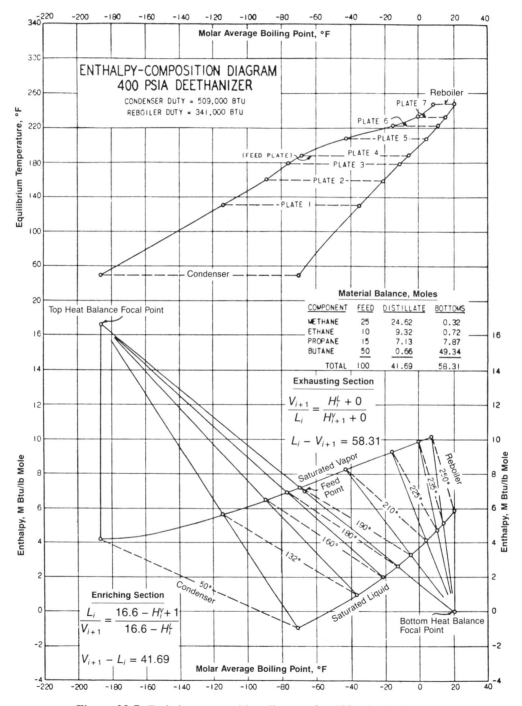

Figure 23.7. Enthalpy-composition diagram for 400-psia deethanizer.

molal average of the atmospheric boiling points. The Kellogg *P-T-C K* Charts and the Edmister-Ruby generalized fugacity correlations, described in Chapter 10 (Volume 1), both used the molal average boiling point, MABP, as the characterizing variable for aliphatic hydrocarbon mixtures.

With this background, MABP was tried as a composition variable in the preparation of *H-X* type diagrams for

two systems: a 400 psia deethanizer and an atmospheric pressure debutanizer. For each of these systems, there are two diagrams, i.e., temperature vs. MABP and enthalpy vs. MABP. These diagrams appear in Figure 23.7 for the deethanizer and Figure 23.8 for the debutanizer. Temperature and heat balance tie-lines are shown, as are the component material balances and the duties of the reboilers and condensers.

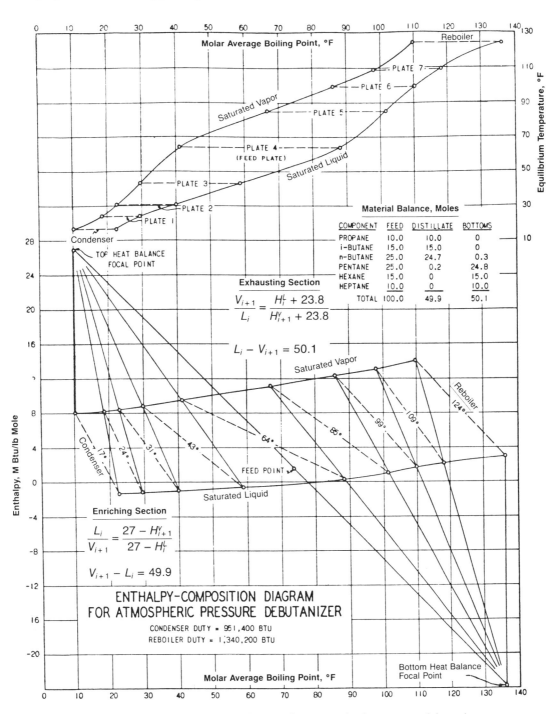

Figure 23.8. Enthalpy-composition diagram for atmospheric pressure debutanizer.

These diagrams were prepared from the results of manual plate-to-plate calculations and heat balances, prepared as part of course work in thermodynamics and distillation. Composition dependent K and H values were used on the deethanizer. Most of the original worksheets have been lost but the K and H values used in the final converged solution of the 400-psia deethanizer are given in Table 23.5.

Table 23.5
K and H Values Used in 400-psia Deethanizer Calculations

T°F	Methane	Ethane	Propane	Butane	Liquid	Vapor
		Vapor-Liquid K-Values			H-Values, M Btu/lb mole	
50	4.95	1.15	0.36	0.113	−0.95	4.20
132	5.90	1.85	0.78	0.347	+0.95	5.65
160	6.00	2.05	0.97	0.460	2.00	6.45
180	5.90	2.14	1.12	0.545	2.80	7.16
190	5.80	2.18	1.19	0.590	3.20	7.56
210	5.60	2.25	1.34	0.690	4.05	8.45
235	5.20	2.30	1.47	0.82	5.20	9.55
250	5.00	2.32	1.54	0.90	5.85	10.20

If you are tempted to use these diagrams to find the effect of changing reflux ratio on the separation, or some other change, you should realize that any operating condition change will probably result in different T-X and H-X curves for the vapor and liquid. In view of this, these diagrams do not seem to be very useful.

When computer progams, using the K and H prediction capabilities described in Chapter 17, become available for distillation, such diagrams as Figures 23.7 and 23.8 can be prepared more easily.

References

1. Ashley, J. H. and G. M. Brown, *Chem. Eng. Prog. Sym. Ser.* Vol. 50, No. 10 p. 129 (1954).
2. Edmister, W. C. *AIChE J.*, 1, 1, 38 (1955).
3. Edmister, W. C. and C. L. Ruby, *Chem. Eng. Prog.* 51, p. 95F (1955).
3a. Edmister, W. C. "Applied Hydrocarbon Thermodynamics; Part 19, Generalized Correlation of Fugacity Coefficients for Hydrocarbon Mixture Components," *Petroleum Refiner* 39, 7, 133 (1960).
4. Edmister, W. C. "Applied Hydrocarbon Thermodynamics; Part 24, Liquid Phase Fugacity and Activity Coefficients," Petroleum Refiner, Vol. 39, No. 12, pp. 159–168 (December 1960).
5. Edmister, W. C. "Applied Hydrocarbon Thermodynamics; Part 40, Partial Enthalpies of Ideal Mixture Components," *Hydrocarbon Processing*, March 1971, pp. 88–92.
6. Edmister, W. C. "Applied Hydrocarbon Thermodynamics; Part 48, Isothermal Enthalpy Differences from Isobaric Vapor-Liquid Equilibria Data," *Hydrocarbon Processing*, April 1973, pp. 175–181.
7. Hildebrand and Scott, *Solubility of Non-Electrolytes,* Reinhold (1950).
8. Kay, W. B. and T. D. Nevins, *Chem. Eng. Prog. Sym. Ser.* 48, No. 3 (1952).
9. Lee, B. I. and W. C. Edmister, *AIChE J.*, 15 615 (1969).
10. Lu, B. C. Y. and B. A. Lavergne, *Chem. Engr.* 64, 132 (August 1958).
11. McCabe, W. L., *Trans. AIChE* 31, 129 (1935).
12. Peters, H. F., *Petroleum Refiner* 28, No. 5, p. 109, (1949).
13. Prausnitz, J. M., W. C. Edmister and K. C. Chao, *AIChE J.*, 6, 214 (1960).
14. Robinson and Gilliland, *Elements of Fractional Distillation,* 4th Edition, McGraw-Hill (1950).
15. Robinson, R. L., Jr. and L. Yarborough, private communication, homework project, Oklahoma State University, January 1961.
16. Scatchard, Transactions Faraday Society 33, 160 (1937).
17. Shah, V. D. and H. G. Donnelly, *Chem. Eng. Prog. Sym. Ser.* Vol. 63, No. 81 (1967).
18. Tallmadge, J. A., D. W. Schroeder, L. N. Canjar and W. C. Edmister, *Chem. Eng. Prog. Sym. Ser.* Vol. 51, No. 2 (1955).
19. Tao, L. C., *AIChE J.*, 15, 362 (1969).
20. Van Laar, J. J., *Z. Physik Chem.* Vol. 72, 723 (1910); Vol. 83, 599 (1913).
21. Watson, K. M. "Thermodynamics of the Liquid State," *Ind. Eng. Chem,* 35; 398 (1943).

24

Thermochemical Properties of Petroleum Fractions*

by R. L. Montgomery

There are many regularities in the properties of petroleum fractions. Some of these regularities are shown by correlations used in previous chapters of this book. Thermochemical properties of petroleum fractions are likewise expected to follow regular patterns. This chapter describes an approach toward discovering those regularities for the enthalpies of formation and free energies of formation.

In this chapter the term "pseudo component" is used to describe a mixture of discrete components that are chemically similar but may have significantly different boiling points. In all other chapters of this book the definition of pseudo component is as given in Chapter 14.

Use of Estimated Properties

The heats (enthalpies) and free energies of formation estimated as described in this chapter can be used with the procedures and programs described elsewhere in this book. However, the techniques in this chapter were treated separately in preparing their FORTRAN implementation. For each petroleum fraction, only one value of enthalpy of formation and one of free energy of formation can be transferred to another procedure without the danger of causing internal inconsistencies, possibly leading to incorrect results.

These possible inconsistencies arise because the variations with temperature of the enthalpy and free energy of formation depend upon the enthalpies and entropies of the petroleum fraction and of the elements contained in it. The values assumed (explicitly or implicitly) for these enthalpies and entropies are not necessarily the same in this chapter as in the other chapters.

Chapter 25 describes specifically how enthalpy of formation estimation methods from this chapter can be applied to determining the absolute enthalpies of petroleum fractions, including those divided into narrow boiling cuts as in Chapter 15.

Limitation

The general method described here is applicable in principle to any hydrocarbon mixture, provided that sufficient chemical analysis data are available. However, the specific technique described in the section "Detailed Procedure" is applicable to straight-run fractions only. That is, fractions that are prepared by distillation of petroleum without any additional chemical reactions such

---

* Development of the correlations in this chapter and determination of the experimental heats of combustion were supported by the M. W. Kellogg Company. The correlations were released to the public by the Company at the September 1982 American Chemical Society National Meeting in Kansas City, Missouri. The idea for these correlations was originally suggested by Stanley B. Adler.

as cracking. The chemical compositions of straight-run fractions are particularly simple, because they are made up almost entirely of paraffins, naphthenes, and aromatics.

See the section on "Other Development Possibilities for Method" later in this chapter.

Verification of Estimates

So far the only experimental test of the techniques in this chapter has been for the enthalpies of formation at 77°F. They are directly related to the heats of combustion at that temperature, and they were tested by comparing calculated and experimental heats of combustion. One problem with such a test is to identify which measurements are accurate. Carrying out accurate heat of combustion measurements is difficult and requires special skills and equipment. For example, heat of combustion data used in a study of correlations for synfuels (2) suffered from unresolved discrepancies of about 1%, although the experimental method has a nominal precision of 0.1% (3). The data used in testing the present techniques were obtained with personnel and equipment, which periodically demonstrated high precision and accuracy in measurements on pure compounds of known properties. Further information about the comparison of heats of combustion is given in a later section called "Results for 19 Petroleum Fractions."

Once the enthalpies of formation at 77°F have been verified, the enthalpies of formation at other temperatures depend only on the enthalpy versus temperature values for the petroleum fractions and of the elements contained in them. A reasonable selection of such values can be used to test the estimated enthalpy of formation values at other temperatures.

Likewise, the variation with temperature of the free energy of formation of a petroleum fraction depends only upon these enthalpy versus temperature data (and the free energy of formation at one temperature). This leaves only the absolute values of the free energy of formation without a test method. It might be possible to test them if the method was extended to petroleum fractions processed by chemical reactions such as cracking. The free energy differences between reactants and products could then be calculated and checked for consistency with the observed directions of reaction under various reaction conditions.

Basis of Estimation Technique

Huang (4) has summarized published methods for characterizing undefined hydrocarbon mixtures and estimating their thermodynamic properties. The new

method described in this chapter resembles most the pseudo component approach of Huang and Daubert (5), which was derived for calculating enthalpies. The present method extends the pseudo component technique to thermochemical property estimation. New pseudo components are defined that do not represent individual compounds but are hypothetical entities. Their properties are averages of the properties of actual compounds.

Each pseudo component of the mixture is fully characterized by its molecular weight, sulfur content, number of rings per molecule, and hydrocarbon type (paraffin, naphthene, or aromatic). The enthalpy of formation and free energy of formation of each pseudo component is a weighted average of experimental data on pure compounds having the same molecular weight, number of rings per molecule, and hydrocarbon type as the pseudo component. The experimental data are extrapolated as necessary, and a correction is applied for the sulfur content.

The properties of a straight-run petroleum fraction lead to a unique description of an equivalent mixture of pseudo components. The enthalpy of formation and free energy of formation of the petroleum fraction are then taken as an average over the pseudo components.

Data Required for Estimation Method

The information required for estimating thermochemical properties of a straight-run petroleum fraction includes the API gravity, ASTM distillation data, sulfur content, molecular weight, and hydrocarbon type analysis. An estimate of the heat of combustion can be obtained by a simpler related technique, using only the API gravity, ASTM distillation data, and sulfur content. This simpler technique is described in the section, "Revision of the Hougen-Watson-Ragatz Correlation."

Summary of Estimation Procedure

A complete description of the calculation method is given later in a section entitled "Detailed Procedure." Briefly, the heat and free energy of vaporization are estimated by Trouton's rule and the integration of Watson's correlation (6) for heat of vaporization versus temperature. Once the heat (enthalpy) and free energy of the mixture have been estimated for the ideal gas state, the vaporization information can be used to convert them to the corresponding quantities for the liquid state.

Pseudo components are selected to represent groups of discrete compounds that might exist in the complex mixture of the petroleum fraction. Simple assumptions about the average chemical composition of these groups lead to the number of carbon atoms per molecule consistent with the average molecular weight.

Since the average number of carbon atoms per molecule and the average number of rings per molecule are usually not integers, the pseudo components are chosen to include those with the next smaller and next larger integral numbers of carbon atoms and rings per molecule. Thus, there are ten possible pseudo components: two paraffins with different numbers of carbon atoms per molecule, four naphthenes (two different numbers of carbon atoms and two different numbers of rings per molecule), and four aromatics (same as for naphthenes).

For instance, if a mixture has an average of 13.84 carbon atoms per molecule, its naphthenes have an average of 1.955 rings per molecule, and its aromatics have an average of 1.57 rings per molecule, then the mixture is assumed to contain C_{13} and C_{14} paraffins, C_{13} and C_{14} naphthenes, each with one or two rings per molecule, and C_{13} and C_{14} aromatics, each with one or two rings per molecule.

At present, olefin pseudo components are not required, since the olefin content of straight-run petroleum fractions is small. In future extensions of this work, pseudo components representing mono-olefins, di-olefins, etc., may be required to describe other mixtures, such as cracked stocks in petroleum processing. For instance, one of these stocks might require six more pseudo components representing mono-, di-, and cyclo-olefins (two of each with different numbers of carbon atoms per molecule).

The method of calculating pseudo component thermochemical properties as averages over pure compounds of the same general description (hydrocarbon type, number of carbon atoms, and rings per molecule) is explained in the section entitled "Averaging Pure Compounds." Each atom of sulfur is assumed to replace a -CH2- group, producing a heat effect that is a function of temperature only. This approximate treatment is satisfactory, since the sulfur content of straight run petroleum fractions is a few wt% at most.

The standard enthalpy and free energy of formation of the petroleum fraction as an ideal gas are calculated by assuming it to be an ideal mixture of ideal gas pseudo components. The previously estimated enthalpy and free energy of vaporization are then used to calculate the enthalpy and free energy of formation of the liquid petroleum fraction.

The amounts of carbon, hydrogen, and sulfur in the hypothetical mixture are calculated from the chemical formulas and mole fractions of the pseudo components. The heat of combustion of the petroleum fraction is then evaluated as the heat of combustion that these elements would have if they were separated, less the heat (enthalpy) of formation of the petroleum fraction.

Revision of the Hougen-Watson-Ragatz Correlation

In 1947, Hougen, Watson, and Ragatz (7) published a correlation of the heats of combustion of petroleum fractions with their API gravities and Watson characterization factors. This is equivalent to correlating them with API gravity and normal boiling point. Results from the present method of estimation suggested that sulfur content also has a large effect on the heat of combustion. Therefore, sulfur content was tested as another parameter in the correlation.

The estimation method described in this chapter was used to compute the gross heats of combustion of a large number of hypothetical petroleum fractions ranging from light naphthas to heavy gas oils and containing no sulfur. The values used for molecular weight, content of each hydrocarbon type, and average rings per molecule were estimates typical of analytical results on petroleum fractions. Changing the estimates (other than mean average boiling point and API gravity) within reasonable limits had fairly small effects on the calculated heats of combustion.

The results of the computations were plotted as in the original Hougen-Watson-Ragatz correlation. The new curves required considerable smoothing, as might have been expected, but were definitely of the same shapes as in the original correlation. The modified correlation is shown in Figure 24.1. The Watson characterization factor used in this revision of the correlation is based on the mean average boiling point (Chapter 7, Volume 1).

The effect of sulfur was determined empirically as 120 Btu/lb per wt% sulfur in the fraction. This quantity was obtained from the three petroleum fractions in Table 24.1, which contained over 1% sulfur, by determining their deviations from the modified correlation. The empirical numerical value for the effect of sulfur is consistent with its effect as calculated by the more complex estimation method.

The last column (column 10) of Table 24.1 lists the heats of combustion of 19 petroleum fractions as derived from Figure 24.1, the modified Hougen-Watson-Ragatz correlation. The experimental heats of combustion are fitted as well by this correlation as by the more complex estimation method. Results from that method are listed in column 9 of Table 24.1. (The actual experimental heats of combustion have not yet been released by the M. W. Kellogg Company.)

The correlation in Figure 24.1 has been implemented as a FORTRAN computer program, which includes points from the curves plotted in the figure plus a routine to interpolate and extrapolate from them. This correlation is not expected to be accurate for other than straight-run petroleum fractions.

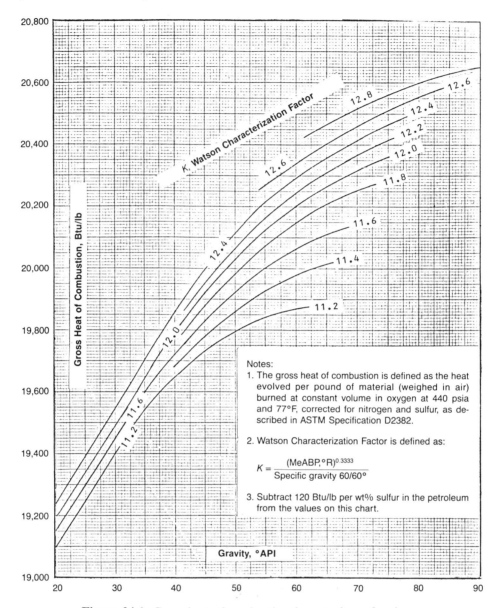

Figure 24.1. Gross heat of combustion for petroleum fractions.

The Hougen-Watson-Ragatz correlation can be used with other estimation methods for carbon/hydrogen ratio and molecular weight to estimate heats of formation from gravity, Watson characterization factor and sulfur content. Chapter 25 describes how this and other information can be applied to determining absolute enthalpies of petroleum fractions, including those separated into narrow boiling cuts as described in Chapter 15.

Results for 19 Petroleum Fractions

The gross heats of combustion used to test the new calculation method were determined by the ASTM high precision method (3), which has a nominal precision of

0.1% or about 20 Btu/lb. The actual experimental technique differed only in insignificant ways from the ASTM procedure.

Computations for comparison with experimental heats of combustion were carried out as described under "Detailed Procedure." The data used in these computations are listed in the first eight columns of Table 24.1. As is shown by the modified Hougen-Watson-Ragatz correlation previously described, the main determinants of the heat of combustion are the mean average boiling point, API gravity, and sulfur content. These quantities were determined by experimental measurements in every case. The remaining quantities include results of direct measurements by various techniques as well as indirect

Thermochemical Properties of Petroleum Fractions

estimates obtained from published and proprietary correlations. The values used for these remaining quantities are not necessarily accurate. They give satisfactory results in the computations, and they were not derived from heat data.

The relatively small effect of the assumed hydrocarbon type distribution and rings per molecule can be explained as follows: Suppose that a pseudo component is selected erroneously and contains a larger percentage of hydrogen and a smaller percentage of carbon than the pseudo component that would best represent the petroleum fraction. The separated elements in the erroneous pseudo component will have a greater (more exothermic) heat of combustion than those in the best pseudo component, on a weight basis. Also, the erroneous pseudo component will probably have a more exothermic enthalpy of formation on a weight basis, likewise because of its greater hydrogen content. The two heat effects tend to cancel each other, since the enthalpy of formation is subtracted from the heat of combustion of the separated elements in calculating the heat of combustion of the petroleum fraction.

The petroleum fractions ranged from light naphtha to heavy gas oil and their origins included Texas, Kuwait, Libya, and Nigeria. Results from two of the 24 fractions measured could not be used because the necessary API gravity and distillation data had not been obtained. Re-

sults from two more were rejected because duplicate API gravity measurements had disagreed greatly. Results from a fifth fraction were rejected because it was a synthetic mixture, rather than a petroleum fraction. Gross heats of combustion of the 19 remaining fractions, estimated as described under "Detailed Procedure," are given in column 9 of Table 24.1. Even though no heat of combustion data for petroleum fractions were used in developing the procedure (only for testing it), its deviation from experimental data is satisfactorily small. (The experimental heats of combustion and their average deviation from calculated values have not yet been released by the M. W. Kellogg Company.)

Detailed Procedure

This section describes in detail the thermochemical property estimation procedure and illustrates the steps of the procedure with calculations for an example petroleum fraction.

The example petroleum fraction has a gravity of 39.8°API. Its ASTM D-158 distillation assay indicated a volumetric average boiling point (VABP) of 488.2°F and a 10%–90% distillation slope of 2.675°F/% off. (Various boiling points are described in Chapter 7.) The distillation slope is equal to the temperature rise between the 10 volume % off point and the 90% off point, divided by

Table 24.1
Properties, Including Heats of Combustion, for 19 Petroleum Fractions

				Naphthenes		Aromatics		Heats of Combustion, Btu/lb		
MeABP °F	Gravity °API	Wt% Sulfur	Molecular Weight	Mole % Naphthenes	Rings per Molecule	Mole % Aromatics	Rings per Molecule	Calculated Values	Figure 24.1 Values	Chart-Calc Difference
130.1	80.0	0.01	81	9	1.000	6	1.000	20525.	20537.	12.0
176.4	74.9	0.00	92	11	1.000	2	1.000	20534.	20496.	− 38.0
232.0	64.4	0.04	107	18	1.000	9	1.000	20261.	20308.	47.0
271.2	55.7	0.04	118	25	1.000	13	1.000	20108.	20109.	1.0
228.0	57.8	0.00	104	28	1.000	11	1.000	20170.	20118.	− 52.0
182.9	70.8	0.04	93	21	1.000	5	1.000	20389.	20402.	13.0
265.9	56.6	0.10	116	41	1.000	7	1.000	20123.	20123.	0.0
233.2	58.1	0.01	106	50	1.000	14	1.000	19967.	20134.	167.0
474.4	40.3	0.45	195	38	1.806	15	1.365	19715.	19724.	9.0
501.8	39.7	0.22	208	30	2.380	22	1.000	19717.	19749.	32.0
632.2	34.5	0.46	277	36	2.762	24	1.300	19568.	19584.	16.0
398.3	41.2	0.12	160	34	2.466	20	1.000	19686.	19734.	48.0
390.7	40.7	0.06	157	36	2.474	18	1.000	19694.	19716.	22.0
551.1	36.5	1.33	231	27	2.402	26	1.068	19518.	19520.	2.0
573.1	35.3	0.11	242	38	2.580	15	2.090	19628.	19629.	1.0
699.6	26.8	2.52	308	26	3.297	46	1.424	19045.	19066.	21.0
800.1	24.0	2.89	375	43	4.278	25	2.099	18961.	18949.	− 12.0
805.8	22.9	0.41	376	37	4.053	37	2.685	19189.	19206.	17.0
580.5	33.9	0.13	245	28	3.755	25	1.417	19572.	19577.	5.0

eighty. The sulfur content of the fraction is 0.45% by weight. Its molecular weight is 193.13 and its hydrocarbon type distribution is 49.3 mole % paraffins, 37.5 mole % naphthenes with an average of 1.955 rings per molecule, and 13.2 mole % aromatics with an average of 1.57 rings per molecule.

Several of the twenty-three calculation steps that follow are from the American Petroleum Institute *Technical Data Book* (8), which will be abbreviated as the API-TDB. The intermediate and final results of these sample calculations are found in Tables 24.2–24.6. The steps are:

Step 1. From the *VABP* and the distillation slope, use the API-TDB methods to determine the molar average boiling point (*MABP*) as 463.25°F and the mean average boiling point (*MeABP*) as 472.13°F.

The procedure for determining the *MABP* and *MeABP* is graphical, employing Figure 2B1.1 of the API-TDB. If the API-TDB is not available, the *MABP* could be calculated from the true boiling point curve prepared as described in Chapter 14, Example 1, Step 1. The definition of the *MABP* given by Equation 7.4 could then be applied as a summation over the mole fractions and true boiling points of the cuts comprising the petroleum fraction.

Figure 14.1 gives the API gravities as well as the moles per unit volume of each cut. Equation 7.8 could be used to convert API gravity to specific gravity, which could then be used to calculate the weight fraction in each cut. Equation 7.6 could be applied as a summation over the weight fractions and true boiling points of the cuts to calculate the cubic average boiling point. Equation 7.7 could then be used to obtain the *MeABP* as the average of the *MABP* and the cubic average boiling point.

For several cases examined, the use of the *MABP* in place of the *MeABP* made insignificant differences in the calculated heats of combustion.

Step 2. Obtain the pseudocritical temperature as 807.3°F by the API-TDB method. The pseudocritical temperature is used as a correlation parameter in place of the critical temperature (which would be used for a pure compound). The equation is

$$T_c = 24.2787 * (MeABP ** 0.58848)$$
$$* (SG ** 0.3596) \qquad (24.1)$$

where *MeABP* = mean average boiling point, °R
SG = specific gravity (60/60°F)

The relationship between API gravity and specific gravity is given in Chapter 7.

Step 3. Find the latent heat of vaporization (*HOV*) at the *MABP* from Trouton's rule using the constant 21.5 cal$_{IT}$ per °K. Trouton's rule states that the molar heat of vaporization equals a constant times the absolute boiling point temperature. One International Table calorie (abbreviated cal$_{IT}$) per gram-mole per °K is equal to one Btu per lb-mole per °F. The absolute temperature scale corresponding to 0°F is 0°R, which is 0°F plus 459.67. The latent heat at the *MABP* equals (21.5)(463.25 + 459.67) = 19,843 Btu/lb-mole.

Step 4. Calculate the latent heat of vaporization as a function of temperature using Watson's correlation (6,8):

$$HOVT = HOV * ((T_c - T)/(T_c - T_b)) ** 0.38 \qquad (24.2)$$

where *HOVT* = heat of vaporization at T
HOV = heat of vaporization at boiling point
T_c = critical temperature, °R
T = temperature, °R
T_b = boiling point temperature, °R

For a mixture of pseudo components, T_c would be the pseudo critical temperature and T_b the molar average boiling temperature. For the above example, the latent heat of vaporization at 77°F is 26,413 Btu/lb-mole.

Step 5. At the *MABP*, the vapor pressure is about one atmosphere. Assuming ideal gas behavior, then, the standard free energy of vaporization (i.e., to a hypothetical ideal gas at one atmosphere pressure) is zero at the *MABP*. Calculate the standard free energy of vaporization by integrating the heat of vaporization as a function of temperature:

$$FOVT = -T \int_{T_b}^{T} (HOVT/TT) \, dT \qquad (24.3)$$

where *FOVT* = free energy of vaporization at T
$TT = T * T$

Combining Equations 24.2 and 24.3 and performing the indicated integration leads to:

$$FOVT = \{(T/T_c) * HOV/[(1.0 - (T_b/T_c)) ** 0.38]\} \sum_{i=0}^{\infty}$$

$$\{(1.0/(i + .38)) * [[1.0 - (T/T_c)]$$
$$** (i + 0.38) - [1.0 - (T_b/T_c)]$$
$$** (i + 0.38)]\} \qquad (24.4)$$

For this example, the free energy of vaporization at 77°F is 10,026 Btu/lb-mole.

Table 24.2
Pseudo Component Types and Mole Fractions for Numerical Example Calculations

Hydrocarbon Component Type	Chemical Formula	Mole Fractions			
		Of the Hydro-Carbon Type	Of the *n* Value	Of the Aromatic + Naphthene Rings per Molecule	Of the Specific Pseudo Component
Paraffin	$C_{13}H_{28}$	0.493	0.16	—	0.0789
Paraffin	$C_{14}H_{30}$	0.493	0.84	—	0.4141
Naphthene (monocyclic)	$C_{13}H_{26}$	0.375	0.16	0.045	0.0027
Naphthene (bicyclic)	$C_{13}H_{24}$	0.375	0.16	0.955	0.0573
Naphthene (monocyclic)	$C_{14}H_{28}$	0.375	0.84	0.045	0.0142
Naphthene (bicyclic)	$C_{14}H_{26}$	0.375	0.84	0.955	0.3008
Aromatic (monocyclic)	$C_{13}H_{20}$	0.132	0.16	0.43	0.0091
Aromatic (bicyclic)	$C_{13}H_{14}$	0.132	0.16	0.57	0.0120
Aromatic (monocyclic)	$C_{14}H_{22}$	0.132	0.84	0.43	0.0477
Aromatic (bicyclic)	$C_{14}H_{16}$	0.132	0.84	0.57	0.0632
				Total	1.0000

Table 24.3
Ideal Gas State Enthalpy and Free Energy Calculations for the Example Mixture at 77°F Values in kcal/g-mole

Chemical Formula	Enthalpy of Formation	Mole Fraction Times Enthalpy of Formation	Free Energy of Formation	Mole Fraction Times Free Energy of Formation
$C_{13}H_{28}$	− 79.99	− 6.311	9.21	0.727
$C_{14}H_{30}$	− 85.22	− 35.290	10.89	4.510
$C_{13}H_{26}$	− 86.97	− 0.235	4.41	0.012
$C_{13}H_{24}$	− 75.85	− 4.346	9.35	0.536
$C_{14}H_{28}$	− 95.30	− 1.353	4.02	0.057
$C_{14}H_{26}$	− 84.18	− 25.321	8.96	2.695
$C_{13}H_{20}$	− 32.58	− 0.296	27.17	0.247
$C_{13}H_{14}$	+ 8.18	+ 0.098	46.17	0.554
$C_{14}H_{22}$	− 39.84	− 1.900	27.05	1.290
$C_{14}H_{16}$	+ 0.92	+ 0.058	46.05	2.910
Sub totals		− 74.896		13.538
Mixing effect		0.000		− 1.011
Sulfur correction		0.371		0.122
Mixture totals		− 74.525		12.649

Table 24.4
Sample Calculation of Standard Free Energy of Formation for the Pseudo Component Represented by Hexanes

Isomer Name	ΔF_i k-cal/ g-mole	Exp $(-\Delta F_i/RT)$	y_i	Contributions to ΔF	
				$y_i(\Delta F_i)$	$RT\, y_i \ln y_i$
n-Hexane	− 0.07	1.1254	0.01587	− 0.00111	− 0.03896
2-Methylpentane	− 1.20	7.5782	0.10689	− 0.12827	− 0.14160
3-Methylpentane	− 0.51	2.3650	0.03335	− 0.01701	− 0.06719
2,2-Dimethylbutane	− 2.37	54.6075	0.77015	− 1.82526	− 0.11917
2,3-Dimethylbutane	− 0.98	5.2283	0.07373	− 0.07226	− 0.11390
Totals		70.9054	0.99999	− 2.04391	− 0.48082

$\Delta F = -RT \ln 70.9054 = -2.5247$. This result may also be obtained by:
$\quad = -2.04391 - 0.48082 = -2.5247$

Table 24.5
Total Standard Ideal Gas Free Energies of Formation of Pseudo Components Calculated from Data on Individual Isomers—kcal/g-mole

Hydrocarbon Type	n Value	Temperature, °K						
		298.15	300	400	500	600	800	1000
Paraffins	5	− 4.02	− 3.81	7.87	19.87	32.13	57.08	82.37
	6	− 2.52	− 2.27	11.84	26.32	41.06	70.99	101.24
	7	− 0.66	− 0.37	15.84	32.59	49.72	84.55	119.61
	8	1.64	1.97	20.48	39.63	59.19	98.89	138.99
	9	2.76	3.14	24.22	45.84	67.84	112.58	157.80
	10	4.18	4.61	28.20	52.24	76.68	126.34	176.51
Monocyclic naphthenes	6	7.48	7.70	20.07	32.71	45.59	71.81	98.36
	7	6.50	6.77	21.67	36.96	52.43	83.54	114.94
	8	6.75	7.06	24.51	42.43	60.36	96.85	133.26
Mono olefins	5	14.18	14.34	22.69	31.40	40.33	58.53	76.97
	6	15.67	15.86	26.38	37.36	48.58	71.46	94.58
Monocyclic aromatics	6	30.99	31.06	35.01	39.24	43.66	52.84	62.27
	7	29.23	29.34	35.39	41.81	48.48	62.24	76.32
	8	28.10	28.24	36.55	45.27	54.28	72.76	91.59
	9	27.57	27.76	38.55	49.72	61.18	84.55	108.22
	10	27.80	28.04	41.27	54.79	68.41	96.90	125.36

Figure 24-2. Assumed arrangement of rings in compounds with several rings per molecule.

Step 6. Assume that all rings are six-membered and that compounds with more than one ring per molecule have the second ring sharing two carbon atoms with the

first ring, the third ring sharing two carbon atoms with the second ring, the fourth ring sharing two carbon atoms with the third, etc. Also assume that whenever a ring shares carbon atoms with two others, the pairs of carbon atoms shared are directly opposite each other. The assumed arrangement is illustrated in Figure 24.2.

Each type of hydrocarbon present as a pseudocomponent will then have the general chemical formula:

$$C_n H_{2n+m}$$

where $m = 2$ for paraffins
$\quad\quad\quad m = (2 - 2Q_N)$ for naphthenes

Table 24.6
Data for Calculating Standard Ideal Gas Enthalpies and Free Energies of Formation for Pseudo Components in kcal/g-mole

Hydrocarbon Type	Mole Group	298.15°K		600°K		800°K	1,000°K	
		Free Energy	Enthalpy	Free Energy	Enthalpy	Enthalpy	Free Energy	Enthalpy
Paraffins	Base	− 12.55	− 12.04	− 12.60	− 13.08	− 13.67	− 11.94	− 14.03
	Per C atom	+ 1.674	− 5.227	+ 8.909	− 5.800	− 6.022	+ 18.845	− 6.099
Naphthenes	Base	+ 9.49	+ 21.33	+ 0.52	+ 14.41	+ 10.51	− 6.47	+ 6.60
	Per C atom	− 0.3905	− 8.331	+ 7.447	− 7.868	− 7.480	+ 17.431	− 6.916
	Per extra ring	+ 4.94	+ 11.12	+ 0.39	+ 7.20	+ 7.39	− 4.26	+ 8.19
Mono olefins	Base	+ 6.79	+ 14.64	− 1.04	+ 14.40	+ 13.72	− 10.95	+ 13.61
	Per C atom	+ 1.447	− 5.022	+ 8.242	− 5.585	− 5.801	+ 17.586	− 5.872
Aromatics	Base	+ 28.79	+ 61.76	− 3.30	+ 58.81	+ 56.08	− 43.14	+ 53.67
	Per C atom	− 0.1246	− 7.257	+ 7.157	− 7.397	− 7.364	+ 16.842	− 7.173
	Per extra ring	+ 19.00	+ 40.76	− 3.89	+ 42.72	+ 43.63	− 35.50	+ 44.16
Substitution of S for CH$_2$		+ 4.50	+ 13.69	− 5.11	+ 11.39	+ 4.87	− 15.49	− 1.65

Note: Use experimental data from Tables 24.5 and 24.7 for aromatics with 6 or 7 carbon atoms.

$m = -6Q_A$ for aromatics

Q_N = the number of naphthenic rings per naphthene molecule

Q_A = the number of aromatic rings per aromatic molecule

Calculate the average value of m' for the mixture, from the individual m values and mole fractions of each hydrocarbon type. The general result is

$$m' = 2*X_P + (2 - 2Q_N')*X_N - 6Q_A'*X_A \qquad (24.5)$$

where Q_N' and Q_A' represent averages for the mixture and X_P, X_N, and X_A represent the mole fractions of paraffins, naphthenes, and aromatics, respectively.

The presence of sulfur is temporarily neglected. For the present numerical example, $m' = -0.974$.

Step 7. From the general chemical formula, the molecular weight, the atomic weights of carbon and hydrogen, and the value of m', calculate n', the average value of n. The general result is

$$n' = (MW - 1.0079 \, m')/14.0268 \qquad (24.6)$$

$n' = 13.84$ for the present example

where MW = molecular weight. For the present numerical example, $n' = 13.84$.

Step 8. Assume that the values of n in the chemical formula for each of the pseudo components have the integer values next larger and next smaller than n'. For the average value of n to be n', the mole fraction with the larger value of n must be equal to n' minus the smaller value of

n. In the present numerical example the mole fraction having $n = 14$ is $(13.84 - 13)$, or 0.84. The mole fraction having $n = 13$ is, of course, $(1.0 - 0.84)$, or 0.16.

Step 9. Assume that the naphthene pseudo components have the integer values of Q_N, next larger and next smaller than Q_N', and that the aromatic pseudo components have the integer values of Q_A, next larger and next smaller than Q_A'. The mole fractions of naphthene pseudo components, having each value of Q_N can be calculated just as for n in step 8 above. In the present example, the mole fractions are 0.43 for $Q_A = 1$, 0.57 for $Q_A = 2$, 0.045 for $Q_N = 1$, and 0.955 for $Q_N = 2$.

Step 10. Assume that the mole fractions having each value of n are the same for all types of pseudo components. List all the pseudo components with their chemical formulas and calculate the mole fraction of each by multiplying the mole fraction of the hydrocarbon type by the mole fraction having the corresponding value of n, and (in the cases of aromatics and naphthenes) multiplying the result by the mole fraction having the corresponding Q_A or Q_N. For the present numerical example, the results are shown in Table 24.2.

Step 11. Let f represent the weight fraction of sulfur in the mixture. Assume that each atom of sulfur replaces a -CH2- group in the overall formula C_nH_{2n+m}. Assume that the molecular weight is constant and that m' is constant; that is, assume that

$$n_H = 2n_C + m'$$

in the new average chemical formula including sulfur,

$$C_{n_C}H_{n_H}S_{n_S}$$

and the results are:

$$n_S = (MW)(f)/32.064$$

$$n_C = [(MW)(1.0 - f) - 1.0079\, m']/14.0268$$

$$n_H = 2n_C + m'$$

The result for the present numerical example is

$$C_{13.777}H_{26.580}S_{0.0271}$$

Step 12. Calculate the enthalpies and free energies of each of the pseudo components from the data in the following section on "Averaging Pure Compounds." The calculation of pseudo component properties is summarized in the following three paragraphs.

Table 24.6 can be used to calculate the standard ideal gas free energies or enthalpies of formation of any pseudo component involved in the present method. For example, consider a bicyclic naphthene with 13 carbon atoms per molecule, containing 0.45% sulfur by weight, at 900°K.

For pseudo components without sulfur, add the base to 13 times the increment per carbon atom and add the total to the increment for one extra ring. Interpolate between 800°K and 1,000°K for enthalpy and between 600°K and 1,000°K for free energy in Table 24.6. The interpolated values in kcal/g-mole are enthalpy, base + 8.555, increment per carbon atom − 7.198, increment per extra ring + 7.79; free energy, base − 4.7225, increment per carbon atom + 14.935, increment per extra ring − 3.0975.

The overall free energy of formation = − 4.7225 + (13)(14.935) − 3.0975 = + 186.335 kcal/g-mole; and the overall enthalpy of formation = 8.555 + (13)(− 7.198) + 7.79 = − 77.229 kcal/g-mole. The interpolated increments per -CH2- group replaced by sulfur are + 1.61 for enthalpy and − 12.895 for free energy. This type of information is used in Step 16.

The results at 77°F for the present example are shown in Table 24.3.

Step 13. Multiply the mole fraction of each of the pseudo components by its enthalpy and free energy of formation to determine the contributions of that component to the overall enthalpy and free energy of formation, respectively, of the mixture.

The results at 77°F for the present numerical example are shown in Table 24.3.

Step 14. Add together the contributions of all of the pseudo components to obtain their total contributions to the enthalpy and free energy of formation of the mixture.

For the present numerical example, the results at 77°F are shown as "sub totals" in Table 24.3.

Step 15. Add to the free energy of formation the additional contributions due to the ideal entropy of mixing. Let $X_1, X_2, \ldots, X_9, X_{10}$ represent the mole fractions of the ten pseudo components numbered in any order. Assuming a random distribution of sulfur atoms in the pseudo components, the additional contribution to the free energy of formation of the mixture is

$$RT\left[n_S \ln n_S + (1 - n_S) \ln (1 - n_S) + \sum_{i=1}^{10} X_i \ln X_i\right]$$

where R = gas constant

For the present numerical example at 77°F, the entropy of mixing contribution is − 1.011 kcal/g-mole.

Step 16. Add to the enthalpy and free energy of formation the contributions due to the effect of sulfur on the pseudo components, calculated as described in the following section on "Averaging Pure Compounds." (In brief, use the increments per -CH2- group replaced by sulfur, determined as in Step 12, and multiply them by n_S.) At 298.15°K, the increments are + 13.69 for the heat of formation and + 4.50 for the free energy of formation, per sulfur atom. The total enthalpy and free energy of formation are the final calculated values for the ideal gas state.

The results for the present numerical example at 77°F are shown in Table 24.3.

Step 17. Convert the enthalpy and free energy of formation of the ideal gas mixture to Btu/lb-mole by multiplying them by 1,798.80. (The conversion factor is not exactly 1800, because the thermochemical values are expressed in thermochemical calories instead of International Table calories.) Subtract the heat and free energy of vaporization at 77°F to obtain the enthalpy and free energy of the liquid mixture at 77°F.

For the present numerical example, the enthalpy of formation of the liquid at 77°F is − 160,469 Btu/lb-mole and its free energy of formation is 12,727 Btu/lb-mole.

Step 18. Calculate the standard gross heat of combustion at 77°F of the separated elements contained in one lb-mole of mixture. ("Standard" refers to the case in which all reactants and products are in their standard states. In this case these states are graphite, orthorhombic sulfur, liquid water, and hypothetical ideal gas hydrogen, carbon dioxide, and sulfur dioxide, all at one atmosphere pressure.) The general result in Btu/lb-mole is given by (8)

$$- 169,179\, n_C - 61,442.5\, n_H - 127,614\, n_S$$

For the present numerical example, the result is $-3,967,379$ Btu/lb-mole.

Step 19. Calculate the standard gross heat of combustion of the liquid at 77°F by subtracting the enthalpy of formation from the heat of combustion of the separated elements.

For the numerical example the standard gross heat of combustion at 77°F is $-3,806,910$ Btu/lb-mole.

Step 20. In the experimental method described by ASTM procedure D2382-76, the mixture is burned at constant volume. Convert the standard (constant pressure) gross heat of combustion to the constant volume gross heat of combustion by adding the work done on the contracting amount of gas in the former process. That work is equal to $P\Delta V$ where ΔV is the volume change during the reaction. The volumes of the liquid mixture and of liquid water produced in combustion are small compared to the volume of gases involved in the combustion, so liquid volume changes can be neglected. Since gases are ideal in the standard heat of combustion, then,

$$P\Delta V = RT\Delta n$$

where Δn is the change in the number of moles of gas in the combustion of one mole of liquid. The stoichiometry of the combustion reaction requires Δn to be $-n_H/4$. Therefore the quantity to be added is $RTn_H/4$.

For the present numerical example, $RTn_H/4 = 7,082$ Btu/lb-mole and the constant volume gross heat of combustion is $-3,799,828$ Btu/lb-mole.

Step 21. In the experimental method ASTM D2382-76, the mixture is weighed in air and no correction for buoyancy is applied. The correction factor for air buoyancy is given with sufficient accuracy by

$$1 + 0.001184\,[(1/d) - (1/7.76)]$$

where d = the density of the petroleum fraction

The specific gravity is an adequate approximation to the density. (See Chapter 7 for the relationship between specific gravity and API gravity.) If the weights used in determining the amount of the mixture are known to have a density other than 7.76 g/cm³, their actual density may be substituted for that value in the previous expression. Multiply the constant volume gross heat of combustion by the correction factor to determine the constant volume gross heat of combustion for the mixture weighed in air.

For the present numerical example, the correction factor is 1.00128, and the constant volume gross heat of combustion for the mixture weighed in air is $-3,804,692$ Btu/lb-mole.

Step 22. The heat of combustion calculated in Step 21 still involves the usual standard states, including ideal gases. The heat correction for differences between standard states and the conditions in a combustion experiment such as one using ASTM procedure D2382-76 is called the Washburn correction. It is described fully by Hubbard, Scott, and Waddington (10). The Washburn correction is usually applied in combustion calorimetry on pure compounds to calculate standard heats of combustion.

For completeness, approximate Washburn corrections were applied in obtaining the calculated values for Table 24.1. However, calculation of the corrections is complex, and the total correction is usually less than 0.05% of the heat of combustion. For the present numerical example it is 0.03% of the heat of combustion and can be neglected. The final FORTRAN implementation of this calculation method does not include a Washburn correction.

Step 23. Divide the constant volume gross heat of combustion for the mixture weighed in air by the molecular weight to obtain the calculated heat of combustion comparable to the results of measurements by ASTM procedure D2382-76. Delete the negative sign to agree with the sign convention used in the ASTM procedure.

For the present numerical example, the calculated heat of combustion is 19,700 Btu/lb. For comparison, the experimental value was $19,729 \pm 20$ Btu/lb.

Averaging Pure Compounds

Each pseudo component is described by its hydrocarbon type (paraffin, naphthene, olefin, or aromatic), number of carbon atoms per molecule, and number of rings per molecule. Each pseudo component represents a mixture of all isomers fitting its description.

The relative proportions of the different isomers that would best represent the pseudo component are unknown. A reasonable estimate is that the proportions are those determined by chemical equilibrium. Pseudo components resembling these have previously been defined by Smith (12).

For paraffins, mono-olefins, monocyclic naphthenes, and monocyclic aromatics, ideal gas free energies of formation were obtained from API Project 44 (9) and Stull et al. (11). Estimates were made for the free energies of formation of a few isomers not provided by those references. A few of the least stable isomers were assumed not to contribute significantly to the descriptions of the pseudo components. The free energy data for the remaining isomers were treated as follows: The conditions for chemical equilibrium in the ideal gas state were expressed as

$$y_j/y_i = \exp\left[(\Delta F_i - \Delta F_j)/RT\right]$$

where the isomers are numbered sequentially 1, 2, 3, ... , $N - 1$, N in arbitrary order, y_i is the mole fraction of isomer number i, etc., ΔF_j is the free energy of formation of isomer number j, etc., R is the gas constant, and T is the absolute temperature.

Applying these conditions along with the fact that the mole fractions must total unity, one obtains

$$y_i = \exp\left(-\Delta F_i/RT\right) \Big/ \sum_{k=1}^{\infty} \exp\left(-\Delta F_k/RT\right)$$

The total standard free energy of formation of the ideal gas pseudo component is the sum of the contributions from each isomer and the ideal free energy of mixing according to the following equation:

$$\Delta F = \sum_{i=1}^{N} y_i\,\Delta F_i + RT \sum_{i=1}^{N} y_i \ln y_i$$

where ΔF = total standard free energy of formation of the ideal gas pseudo component.

When the previous expression for y_i is substituted into the equation for ΔF and the result is simplified, one obtains the equation used to calculate ΔF:

$$\Delta F = -RT \ln\left[\sum_{k=1}^{N} \exp\left(-\Delta F_k/RT\right)\right]$$

A sample calculation for hexane at 298.15°K is given in Table 24.4 along with the corresponding values of y_i and the contributions of each isomer to ΔF. Overall standard free energies of formation were calculated for a number of pseudo components at 298.15°K, at 100°K intervals from 300°K to 600°K, and at 200°K intervals from 600°K to 1,000°K. The results are given in Table 24.5.

The six values between 300°K and 1,000°K for each pseudo component were fitted by least squares to two linear equations joined at 600°K, where there was a change of slope. The four coefficients of each pair of these equations were then fitted to linear equations as a function of n, the number of carbon atoms per molecule. Each hydrocarbon type required a different set of equations in n. Subjective judgment was applied in deciding upon the constants to be used in the equations. The re-

sults of this curve fitting are shown in Table 24.6, which is described in the following:

Standard ideal gas heats (enthalpies) of formation, designated by ΔH, were obtained using

$$\Delta H = \sum_{i=1}^{N} y_i\,\Delta H_i$$

where ΔH_i = standard ideal gas enthalpy of formation of isomer number i

These ideal gas heats of formation were used only as guides in establishing a method for calculating the ΔH values from the ΔF values. With such a method established, extrapolation of ΔF allows concurrent extrapolation of ΔH values consistent with them.

The values of ΔH for each pseudo component were first assumed to be linear functions of temperature with a change of slope at 600°K. The constants in the linear equations for ΔH were adjusted so that the ΔF values calculated by the following expression were consistent with the previous equations for ΔF vs. T.

$$\Delta F = -T \int (\Delta H/T^2)\,dT$$

This procedure resulted in discrepancies with the data at 1,000°K. Therefore, the values of ΔH at 1,000°K were adjusted by averaging them with ΔH values produced by a different equation (a linear equation adjusted to be consistent with ΔF values at 300°K, 600°K, and 1,000°K). This results in three final linear equations for ΔH vs. T for each pseudo component, with changes in slope at 600°K and 800°K.

It is assumed that all rings are six-membered, and that in multi-ring isomers they are attached to each other as shown in Figure 24.2 and described in Step 6 of the section on "Detailed Procedure." It is further assumed that there is a constant increment in ΔH or ΔF at any constant temperature for each additional ring. The values for this increment shown in Table 24.6 are based on experimental data for fused-ring compounds; subjective judgment was exercised in selecting the numerical values of the increment.

It is assumed that each atom of sulfur replaces a -CH2 group, and that there is a constant increment in ΔH or ΔF at any constant temperature for each additional sulfur atom. The values for this increment shown in Table 24.6 are based on experimental data for organic sulfur compounds; subjective judgment was exercised in choosing the numerical values of the increment.

Table 24.6 can be used to calculate the standard ideal gas free energies or enthalpies of formation of any pseudo component involved in the present method. For example, consider a bicyclic naphthene with 13 carbon atoms per molecule, containing 0.45% sulfur by weight, at 900°K. For the pseudo component without sulfur, add the base to 13 times the increment per carbon atom and add the total to the increment for one extra ring. Interpolate between 800°K and 1,000°K for enthalpy and between 600°K and 1,000°K for free energy in Table 24-6. The interpolated values in kcal/g-mole are: for enthalpy, base $+ 8.555$, increment per carbon atom $- 7.198$, increment per extra ring $+ 7.79$; for free energy, base $- 4.7225$, increment per carbon atom $+ 14.935$, increment per extra ring $- 3.0975$. The overall free energy of formation is $-4.7225 + (13)(14.935) - 3.0975 = +186.335$ kcal/g-mole; the overall enthalpy of formation is $8.555 + (13)(- 7.198) + 7.79 = - 77.229$ kcal/g-mole.

The interpolated increments per -CH2- group replaced by sulfur are $+ 1.61$ for enthalpy and $- 12.895$ for free energy. The formula of a bicyclic naphthene with 13 carbon atoms per molecule, using our assumption regarding the attachment of the rings to each other, is $C_{13}H_{24}$ (see the section on "Detailed Procedure" for the general formula). The presence of 0.45 wt% sulfur replacing -CH2- groups would change the formula to

$$C_{12.943}H_{23.88}S_{0.0253}$$

The effects of sulfur content on the heat and free energy at 900°K would then be $(0.0253)(1.61) = 0.041$ on enthalpy, and $(0.0253)(- 12.895) = - 0.326$ on free energy of formation.

There is an additional effect of sulfur on the free energy due to the ideal free energy of mixing sulfur-containing molecules with hydrocarbon molecules. In the present example, the correction for this effect is approximately:

$$RT [0.0253 \ln 0.0253 + (1 - 0.0253) \ln (1 - 0.0253)] = - 0.211 \text{ kcal/g-mole.}$$

The total ideal gas heat and free energy of formation at 900°K of the bicyclic naphthene pseudo component with 13 carbon atoms per molecule and 0.45% sulfur by weight are $- 77.229 + 0.041 = - 77.188$ kcal/g-mole and $186.335 - 0.326 - 0.211 = 185.798$ kcal/g-mole, respectively.

It is easier to apply the correction for sulfur content to the mixture as a whole (as is described in the "Detailed Procedure") than to apply it to each pseudo component.

From Table 24.6, the corrections for replacement of a -CH2- group by sulfur at 298.15°K are $+ 13.69$ to the heat and 4.50 to the free energy. For the numerical example in the "Detailed Procedure," there is an average of 0.0271 atoms of sulfur per molecule. The corrections to the heat and free energy due to sulfur content are therefore $(0.0271)(13.69) = 0.371$ and $(0.0271)(4.5) = 0.122$ kcal/g-mole, respectively, as shown under "Detailed Procedure." The correction for the entropy of mixing of the sulfur was calculated separately in the example there.

The average and maximum deviations of enthalpies and free energies in Tables 24.5 and 24.7 (except for aromatics with 6 or 7 carbon atoms per molecule) from values calculated using the data in Table 24.6 are shown in Table 24.8.

FORTRAN Program for Applying "Detailed Procedure"

Table 24.9 is a program listing for applying the "Detailed Procedure" to the calculations of thermochemical properties for petroleum fractions. The program is in FORTRAN and has been run on IBM compatible and Macintosh machines. As written, data entries and results are shown on the screen and also printed as the calculations proceed. In addition, this information is stored in a disk file for the record.

Other Applications

As previously mentioned, the method might be extended to include petroleum fractions that have been processed beyond a mere distillation from the petroleum. This would require use of olefin pseudo components plus either analysis for the olefin content or a method for estimating the amounts and types of olefins. Another possibility would be extension of the method to chemical kinetic calculations, if the reactions could be expressed in a generalized or averaged way involving pseudo components.

Adler and Hall (1) have applied the modified Hougen-Watson-Ragatz correlation as an aid in estimating other properties of petroleum fractions. They have also presented a preliminary extension of this correlation to heats of combustion of pyrolyzed and cracked petroleum stocks (Figure 3 of Reference 1). For these processed stocks, the heats of combustion are less exothermic than predicted by the modified Hougen-Watson-Ragatz correlation. The difference is 750 Btu/lb at a Watson Characterization Factor of 10.7 and 1,000 Btu/lb at a Watson Factor of 10.2.

Table 24.7
Total Standard Ideal Gas Enthalpies of Formation of Pseudo Components Calculated from Data on Individual Isomers—kcal/g-mole

Hydrocarbon Type	n Value	Temperature, °K						
		298.15	300	400	500	600	800	1000
Paraffins	5	− 38.30	− 38.33	− 39.54	− 40.83	− 41.99	− 43.59	− 44.31
	6	− 43.74	− 43.78	− 45.40	− 46.77	− 47.96	− 49.66	− 50.43
	7	− 48.13	− 48.18	− 50.38	− 52.29	− 53.85	− 55.94	− 56.87
	8	− 52.61	− 52.66	− 55.21	− 57.40	− 59.17	− 61.59	− 62.64
	9	− 59.09	− 59.14	− 61.24	− 63.35	− 65.17	− 67.61	− 68.60
	10	− 65.60	− 65.62	− 66.99	− 69.09	− 71.02	− 73.62	− 74.64
Monocyclic naphthenes	6	− 28.78	− 28.82	− 29.97	− 31.11	− 32.27	− 33.98	− 34.71
	7	− 36.85	− 36.90	− 38.91	− 39.94	− 40.66	− 41.67	− 42.01
	8	− 43.81	− 43.87	− 46.50	− 47.94	− 48.00	− 48.83	− 48.80
Mono olefins	5	− 9.95	− 9.98	− 11.54	− 12.80	− 13.76	− 15.02	− 15.56
	6	− 13.91	− 13.95	− 15.91	− 17.49	− 18.72	− 20.32	− 20.91
Monocyclic aromatics	6	+ 19.82	+ 19.79	+ 18.56	+ 17.54	+ 16.71	+ 15.51	+ 14.82
	7	+ 11.95	+ 11.92	+ 10.34	+ 9.05	+ 8.02	+ 6.65	+ 6.01
	8	+ 4.24	+ 4.20	+ 2.37	+ 0.84	− 0.39	− 2.06	− 2.86
	9	− 3.39	− 3.44	− 5.38	− 6.96	− 8.18	− 9.70	− 10.28
	10	− 10.06	− 10.11	− 12.30	− 13.88	− 15.09	− 16.96	− 17.63

Table 24.8
Average and Maximum Deviations of Enthalpies and Free Energies in Tables 24.5 and 24.7 from Values Calculated Using Data in Table 24.6*
Deviations in kcal/g-mole

Hydrocarbon Type	Deviations of Enthalpy		Deviations of Free Energy	
	average	maximum	average	maximum
Paraffins	0.27	1.29	0.20	0.80
Naphthenes	0.36	1.51	0.30	0.60
Alkenes	0.51	1.58	0.12	0.21
Aromatics	0.39	0.85	0.15	0.40

* Aromatics with 6 or 7 carbon atoms are not included.

Table 24.9
Program for Applying "Detailed Procedure"

```
      PROGRAM DETPRC
C THIS IS THE EXECUTIVE ROUTINE
      OPEN(2)
      CALL READIN
      CALL COMPUT
      END
      SUBROUTINE READIN
C ROUTINE TO ACCEPT INPUT
      COMMON /READS/ DEGAPI, BPMAV, BPMEA, SULFUR, WTMOL,
     1                PCTMP, PCTMN, PCTMA, RINGSN, RINGSA
      SAVE /READS/
      WRITE(*,101)
      WRITE(2,101)
      WRITE(6,101)
      READ(*,*) DEGAPI
      WRITE(*,102)
      WRITE(2,102)
      WRITE(6,102)
      READ(*,*) BPMAV
      WRITE(*,103)
      WRITE(2,103)
      WRITE(6,103)
      READ(*,*) BPMEA
      IF (BPMEA.LE.0.0) BPMEA=BPMAV
      WRITE(*,104)
      WRITE(2,104)
      WRITE(6,104)
      READ(*,*) SULFUR
      WRITE(*,105)
      WRITE(2,105)
      WRITE(6,105)
      READ(*,*) WTMOL
      WRITE(*,106)
      WRITE(2,106)
      WRITE(6,106)
      READ(*,*) PCTMP
      WRITE(*,107)
      WRITE(2,107)
      WRITE(6,107)
      READ(*,*) PCTMN
      WRITE(*,108)
      WRITE(2,108)
      WRITE(6,108)
      READ(*,*) PCTMA
      WRITE(*,109)
      WRITE(2,109)
      WRITE(6,109)
      READ(*,*) RINGSN
      WRITE(*,110)
      WRITE(2,110)
      WRITE(6,110)
      READ(*,*) RINGSA
      WRITE(*,111) DEGAPI, BPMAV, BPMEA, SULFUR, WTMOL,
     1             PCTMP, PCTMN, PCTMA, RINGSN, RINGSA
      WRITE(2,111) DEGAPI, BPMAV, BPMEA, SULFUR, WTMOL,
     1             PCTMP, PCTMN, PCTMA, RINGSN, RINGSA
      WRITE(6,111) DEGAPI, BPMAV, BPMEA, SULFUR, WTMOL,
     1             PCTMP, PCTMN, PCTMA, RINGSN, RINGSA
      RETURN
101 FORMAT(' PROGRAM TO PERFORM THE ''DETAILED PROCEDURE'' OF',
     1 ' CHAPTER 24',/,' ENTER THE API GRAVITY OF THE PETROLEUM',
     2 ' FRACTION')
102 FORMAT(' ENTER THE MOLAR AVERAGE BOILING POINT, DEG R')
103 FORMAT(' ENTER THE MEAN AVERAGE BOILING POINT, DEG R',/,
     1 ' TO USE THE MOLAR AVERAGE BOILING POINT, ENTER ZERO')
104 FORMAT(' ENTER THE WEIGHT PERCENTAGE OF SULFUR')
105 FORMAT(' ENTER THE MOLECULAR WEIGHT')
106 FORMAT(' ENTER THE MOLE PERCENTAGE OF PARAFFINS')
107 FORMAT(' ENTER THE MOLE PERCENTAGE OF NAPHTHENES')
```

Table 24.9
Continued

```
  108 FORMAT(' ENTER THE MOLE PERCENTAGE OF AROMATICS')
  109 FORMAT(' ENTER THE NUMBER OF RINGS PER NAPHTHENE MOLECULE')
  110 FORMAT(' ENTER THE NUMBER OF RINGS PER AROMATIC MOLECULE')
  111 FORMAT(' SUMMARY OF INPUT:',/,'   API GRAVITY =',F7.3,/,
     1 ' MOLAR AVG BOILING PT =',F8.2,/,
     2 ' MEAN   AVG BOILING PT =',F8.2,/,
     3 ' WEIGHT PERCENT SULFUR=',F6.2,/,
     4 ' MOLECULAR WEIGHT =', F8.2,/,
     5 ' MOLE PERCENT PARAFFINS=',F7.3,/,
     6 ' MOLE PERCENT NAPHTHENES=',F7.3,/,
     7 ' MOLE PERCENT AROMATICS=',F7.3,/,
     8 ' NUMBER OF RINGS PER NAPHTHENE MOLECULE=',F7.3,/,
     9 ' NUMBER OF RINGS PER AROMATIC MOLECULE='.F7.3)
      END
      SUBROUTINE COMPUT
C ROUTINE TO CONTROL COMPUTATIONS
C OF CHAPTER 24
      DIMENSION XMOLF(10)
      COMMON /READS/ DEGAPI, BPMAV, BPMEA, SULFUR, WTMOL,
     1               PCTMP, PCTMN, PCTMA, RINGSN, RINGSA
      DATA R /1.98717E-3/
C STEP 1 - BOILING POINTS - DONE OUTSIDE THIS PROGRAM
C STEP 2 - SPECIFIC GRAVITY AND PSEUDOCRITICAL TEMPERATURE
      SG=141.5/(131.5+DEGAPI)
      IF (BPMEA.LE.0.0) BPMEA=BPMAV
      TCRIT=24.2787*(BPMEA**0.58848)*(SG**0.3596)
      WRITE(*,101) SG, TCRIT
      WRITE(2,101) SG, TCRIT
      WRITE(6,101) SG, TCRIT
C STEP 3 - HEAT OF VAPORIZATION AT THE MOLAR AVG BOILING PT
      HVMABP=21.5*BPMAV
      WRITE(*,102) HVMABP
      WRITE(2,102) HVMABP
      WRITE(6,102) HVMABP
C STEP 4 - HEAT OF VAPORIZATION AT 536.67 DEG R
      T=536.67
      HV537=HVMABP*HTFUNC(T,TCRIT,BPMAV)
      WRITE(*,103) HV537
      WRITE(2,103) HV537
      WRITE(6,103) HV537
C STEP 5 - FREE ENERGY OF VAPORIZATION AT 536.67 DEG R
      FV537=HVMABP*FTFUNC(T,TCRIT,BPMAV)
      WRITE(*,104) FV537
      WRITE(2,104) FV537
      WRITE(6,104) FV537
C STEP 6 - THE AVERAGE VALUE OF M
      BARM=0.02*PCTMP + 0.02*(1.-RINGSN)*PCTMN - 0.06*RINGSA*PCTMA
      WRITE(*,105) BARM
      WRITE(2,105) BARM
      WRITE(6,105) BARM
C STEP 7 - THE AVERAGE VALUE OF N
      BARN=(WTMOL-1.0079*BARM)/14.0268
      WRITE(*,106) BARN
      WRITE(2,106) BARN
      WRITE(6,106) BARN
C OMIT STEPS 8 - 10, WHICH WERE PRIMARILY FOR ILLUSTRATIVE PURPOSES
C STEP 11 - PART 1 - GRAM-ATOMS OF SULFUR PER MOLE
      XNS=WTMOL*SULFUR/3206.4
C STEP 11 - PART 2 - GRAM-ATOMS OF CARBON PER MOLE
      XNC=(WTMOL*(1.-0.01*SULFUR)-1.0079*BARM)/14.0268
C STEP 11 - PART 3 - GRAM-ATOMS OF HYDROGEN PER MOLE
      XNH=2.*XNC+BARM
      WRITE(*,107) XNC, XNH, XNS
      WRITE(2,107) XNC, XNH, XNS
      WRITE(6,107) XNC, XNH, XNS
C STEP 11 - EXTRA - C/H WEIGHT RATIO
      COVERH=(12.011*XNC)/(1.008*XNH)
      WRITE(*,108) COVERH
      WRITE(2,108) COVERH
```

Table 24.9
Continued

```
      WRITE(6,108) COVERH
C STEPS 12 THROUGH 14 - OBTAIN HEATS AND FREE ENERGIES OF
C PSEUDOCOMPONENT MIXTURES, USING THE AVERAGE XNC, RINGSN, AND RINGSA
      WRITE(*,109)
      WRITE(2,109)
      WRITE(6,109)
      READ(*,*) T
   10 FIDL=0.01*PCTMP*FPAR(BARN,T) + 0.01*PCTMN*FNAP(BARN,RINGSN,T)
     1                             + 0.01*PCTMA*FARO(BARN,RINGSA,T)
      HIDL=0.01*PCTMP*HPAR(BARN,T) + 0.01*PCTMN*HNAP(BARN,RINGSN,T)
     1                             + 0.01*PCTMA*HARO(BARN,RINGSA,T)
C STEP 15 - ADD IDEAL ENTROPY OF MIXING
C           DESIGNATE THE MOLE FRACTIONS AT EACH NUMBER OF CARBON
C           ATOMS AS XCL AND XCH, THE MOLE FRACTIONS AT EACH
C           NUMBER OF NAPHTHENIC RINGS AS XNL AND XNH, AND THE MOLE
C           FRACTIONS AT EACH NUMBER OF AROMATIC RINGS AS XAL AND XAH.
      UNITY=1.0
      XCL=AMOD(BARN,UNITY)
      XCH=1.-XCL
      XNAL=AMOD(RINGSN,UNITY)
      XNAH=1.-XNAL
      XAL=AMOD(RINGSA,UNITY)
      XAH=1.-XAL
C DESIGNATE THE TEN MOLE FRACTIONS OF PSEUDOCOMPONENTS AS XMOLF(I)
      XMOLF(1)=XCL*PCTMP/100.
      XMOLF(2)=XCH*PCTMP/100.
      XMOLF(3)=XCL*XNAL*PCTMN/100.
      XMOLF(4)=XCH*XNAL*PCTMN/100.
      XMOLF(5)=XCL*XNAH*PCTMN/100.
      XMOLF(6)=XCH*XNAH*PCTMN/100.
      XMOLF(7)=XCL*XAL*PCTMA/100.
      XMOLF(8)=XCH*XAL*PCTMA/100.
      XMOLF(9)=XCL*XAH*PCTMA/100.
      XMOLF(10)=XCH*XAH*PCTMA/100.
      FMIX=0.0
      DO 20  I=1,10
      X=XMOLF(I)
      IF (X.LE.0.0) GO TO 20
      FMIX=FMIX+R*T*X*ALOG(X)
   20 CONTINUE
      IF (XNS.GT.0.0) FMIX=FMIX+R*T*XNS*ALOG(XNS)+
     1 R*T*(1.-XNS)*ALOG(1.-XNS)
      WRITE(*,110) FMIX
      WRITE(2,110) FMIX
      WRITE(6,110) FMIX
      FIDL=FIDL+FMIX
C STEP 16 - ADD INCREMENTS FOR SULFUR
      FIDL=FIDL+FSUL(T)*XNS
      HIDL=HIDL+HSUL(T)*XNS
      WRITE(*,111) FIDL,HIDL,T
      WRITE(2,111) FIDL,HIDL,T
      WRITE(6,111) FIDL,HIDL,T
C REPEAT STEPS 12 THROUGH 16 FOR T=298.15 K (UNLESS T=298.15 ALREADY)
      IFLAG=0
      IF (T.NE.298.15) IFLAG=1
      IF (T.NE.298.15) T=298.15
      IF(IFLAG.NE.0) GO TO 10
C STEP 17 - PART 1 - CONVERT TO ENGLISH UNITS
      FIDL=1798.8*FIDL
      HIDL=1798.8*HIDL
      T=536.67
      WRITE(*,112) FIDL,HIDL,T
      WRITE(2,112) FIDL,HIDL,T
      WRITE(6,112) FIDL,HIDL,T
C STEP 17 -PART 2 - CALCULATE STANDARD FREE ENERGY AND HEAT OF
C                   FORMATION OF THE LIQUID FRACTION
      FLIQ=FIDL-FV537
      HLIQ=HIDL-HV537
      WRITE(*,113) FLIQ,HLIQ,T
      WRITE(2,113) FLIQ,HLIQ,T
      WRITE(6,113) FLIQ,HLIQ,T
```

Table 24.9
Continued

```
C STEP 18 - CALCULATE THE TOTAL HEAT OF COMBUSTION OF THE SEPARATED
C           ELEMENTS CONTAINED IN THE FRACTION
      HCELEM=-169179.*XNC-61442.5*XNH-127614.*XNS
      WRITE(*,114) HCELEM
      WRITE(2,114) HCELEM
      WRITE(6,114) HCELEM
C STEP 19 - COMPUTE STANDARD HEAT OF COMBUSTION OF THE LIQUID FRACTION
      HOCOMB=HCELEM-HLIQ
      WRITE(*,115) HOCOMB
      WRITE(2,115) HOCOMB
      WRITE(6,115) HOCOMB
C STEP 20 - COMPUTE THE CONSTANT VOLUME HEAT OF COMBUSTION
      HCOMB=HOCOMB+1000.*R*T*XNH/4.
C STEP 21 - COMPUTE THE HEAT OF COMBUSTION AS WEIGHED IN AIR
      HCOMB=HCOMB*(1.0+0.001184*((1./SG)-(1./7.76)))
      WRITE(*,116) HCOMB
      WRITE(2,116) HCOMB
      WRITE(6,116) HCOMB
C STEP 22 - NO WASHBURN CORRECTION USED
C STEP 23 - OBTAIN HEAT OF COMBUSTION PER POUND
      HCOMB=-HCOMB/WTMOL
      WRITE(*,117) HCOMB
      WRITE(2,117) HCOMB
      WRITE(6,117) HCOMB
  101 FORMAT(' SG=', F7.4,'; TCRIT=',F8.2)
  102 FORMAT(' HVMABP=',F7.0)
  103 FORMAT(' HV537=',F7.0)
  104 FORMAT(' FV537=',F7.0)
  105 FORMAT(' BARM=',F8.4)
  106 FORMAT(' BARN=',F8.4)
  107 FORMAT(' CHEMICAL FORMULA = C(',F7.3,') H(',F7.3,') S(',F7.3,')')
  108 FORMAT(' C/H WEIGHT RATIO = ',F7.4)
  109 FORMAT(' ENTER A TEMPERATURE FOR THE FREE ENERGY AND',
     1 ' ENTHALPY',/,' THE VALID RANGE IS 298.15 TO 1000 DEGREES K')
  110 FORMAT(' FREE ENERGY OF MIXING = ', F7.2,' KCAL/G-MOLE')
  111 FORMAT(' THE IDEAL GAS FREE ENERGY AND ENTHALPY OF FORMATION',
     1 /,' ARE ', F8.3,' AND ', F8.3,' KCAL/G-MOLE, RESPECTIVELY',/,
     2 ' AT ',F8.2, ' DEGREES K')
  112 FORMAT(' THE IDEAL GAS FREE ENERGY AND ENTHALPY OF FORMATION',
     1 /,' ARE ', F9.0,' AND ', F9.0,' BTU/LB-MOLE, RESPECTIVELY',/,
     2 ' AT ',F8.2, ' DEGREES R')
  113 FORMAT(' THE LIQUID FREE ENERGY AND ENTHALPY OF FORMATION',
     1 /,' ARE ', F9.0,' AND ', F9.0,' BTU/LB-MOLE, RESPECTIVELY',/,
     2 ' AT ',F8.2, ' DEGREES R')
  114 FORMAT(' HCELEM=',F10.0)
  115 FORMAT(' HOCOMB=',F10.0)
  116 FORMAT(' HCOMB=',F10.0,' BTU/LB-MOLE')
  117 FORMAT(' HCOMB=',F7.0,' BTU/LB')
      RETURN
      END
      FUNCTION HTFUNC(T,TCRIT,BPMAV)
C FUNCTION RETURNS A FACTOR TO CONVERT THE HEAT OF VAPORIZATION
C AT THE MOLAR AVERAGE BOILING POINT TO THAT AT TEMPERATURE T.
      HTFUNC=(1.-(T/TCRIT))/(1.-(BPMAV/TCRIT))
      HTFUNC=HTFUNC**0.38
      RETURN
      END
      FUNCTION FTFUNC(T,TCRIT,BPMAV)
C FUNCTION RETURNS A FACTOR TO CONVERT THE HEAT OF VAPORIZATION
C AT THE MOLAR AVERAGE BOILING POINT TO THE FREE ENERGY OF
C VAPORIZATION AT TEMPERATURE T
      FTFUNC=(T/TCRIT)*FSER(T,TCRIT,BPMAV)/((1.-(BPMAV/TCRIT))**0.38)
      RETURN
      END
      FUNCTION FSER(T,TCRIT,BPMAV)
C FUNCTION RETURNS THE VALUE OF A SERIES NEEDED IN FUNCTION FTFUNC
      COMMON /TERM/ T1,TCRIT1,BPMAV1
      SAVE /TERM/
      T1=T
```

Table 24.9
Continued

```
        TCRIT1=TCRIT
        BPMAV1=BPMAV
        FSER=TERMI(0)
        DO 10   I=1,99
        A=TERMI(I)
        FSER=FSER+A
C IF THE NEW TERM IS LESS THAN 1 PART IN 1.E6 OF FSER, CONSIDER
C THE SERIES COMPLETED
        A=ABS(A)
        B=ABS(FSER)
        IF((A/B).LT.1.0E-6) GO TO 100
C IF THE SERIES IS NOT COMPLETE AFTER 99 TERMS, WRITE AN ERROR MESSAGE
        IF(I.EQ.99) WRITE(*,101)
        IF(I.EQ.99) WRITE(2,101)
        IF(I.EQ.99) WRITE(6,101)
     10 CONTINUE
    100 RETURN
    101 FORMAT(' SERIES FSER NOT COMPLETED')
        END
        FUNCTION TERMI(I)
        COMMON /TERM/ T,TCRIT,BPMAV
C FUNCTION RETURNS THE VALUE OF A TERM IN THE SERIES IN FUNCTION FSER
        XI=FLOAT(I)
        A=XI/(XI+0.38)
        B=(1.-(T/TCRIT))**(XI+0.38)
        C=(1.-(BPMAV/TCRIT))**(XI+0.38)
        TERMI=A*(B-C)
        RETURN
        END
        FUNCTION HPAR(XNC,T)
C FUNCTION RETURNS THE ENTHALPY OF A PARAFFINIC PSEUDOCOMPONENT
C HAVING XNC CARBON ATOMS PER MOLECULE, AT TEMPERATURE T DEGREES K
        DIMENSION BASE(4), XINCR(4), TSTD(4)
C EACH BASE AND XINCR CORRESPOND TO A TEMPERATURE VALUE TSTD.
C THE ENTHALPY EQUALS THE BASE PLUS (THE INCREMENT TIMES THE NUMBER
C     OF CARBON ATOMS)
        DATA TSTD /298.15, 600., 800., 1000./
        DATA BASE / -12.04, -13.08, -13.67, -14.03/
        DATA XINCR / -5.227, -5.8, -6.022, -6.099/
        NLOW=0
        DO 10  I=1,4
        IF (T.GE.TSTD(I)) NLOW=I
     10 CONTINUE
        IF(NLOW.EQ.0) NLOW=1
        IF(NLOW.EQ.4) NLOW=3
        NHIGH=NLOW+1
        HLOW=BASE(NLOW)+XNC*XINCR(NLOW)
        HHIGH=BASE(NHIGH)+XNC*XINCR(NHIGH)
        TFAC=(T-TSTD(NLOW))/(TSTD(NHIGH)-TSTD(NLOW))
        HPAR=HLOW+TFAC*(HHIGH-HLOW)
        RETURN
        END
        FUNCTION HNAP(XNC,RINGSN,T)
C FUNCTION RETURNS THE ENTHALPY OF A NAPHTHENIC PSEUDOCOMPONENT
C HAVING XNC CARBON ATOMS PER MOLECULE AND RINGSN RINGS PER MOLECULE,
C AT TEMPERATURE T DEGREES K
        DIMENSION BASE(4), XINCR(4), XRING(4), TSTD(4)
C EACH BASE, XINCR, AND XRING CORRESPOND TO A TEMPERATURE VALUE TSTD.
C THE ENTHALPY EQUALS THE BASE PLUS (THE INCREMENT TIMES THE NUMBER
C     OF CARBON ATOMS) PLUS (THE XRING TIMES THE NUMBER OF EXTRA RINGS).
        DATA TSTD /298.15, 600., 800., 1000./
        DATA BASE / 21.33,14.41, 10.51, 6.6 /
        DATA XRING / 11.12, 7.2, 7.39, 8.19 /
        DATA XINCR / -8.331, -7.868, -7.48, -6.916/
        NLOW=0
        DO 10  I=1,4
        IF (T.GE.TSTD(I)) NLOW=I
     10 CONTINUE
        IF(NLOW.EQ.0) NLOW=1
```

Table 24.9
Continued

```
      IF(NLOW.EQ.4) NLOW=3
      NHIGH=NLOW+1
      HLOW=BASE(NLOW)+XNC*XINCR(NLOW)+(RINGSN-1.0)*XRING(NLOW)
      HHIGH=BASE(NHIGH)+XNC*XINCR(NHIGH)+(RINGSN-1.)*XRING(NHIGH)
      TFAC=(T-TSTD(NLOW))/(TSTD(NHIGH)-TSTD(NLOW))
      HNAP=HLOW+TFAC*(HHIGH-HLOW)
      RETURN
      END
      FUNCTION HARO(XNC,RINGSA,T)
C FUNCTION RETURNS THE ENTHALPY OF AN AROMATIC PSEUDOCOMPONENT
C HAVING XNC CARBON ATOMS PER MOLECULE AND RINGSA RINGS PER MOLECULE,
C AT TEMPERATURE T DEGREES K
      DIMENSION BASE(4), XINCR(4), XRING(4), TSTD(4)
C EACH BASE, XINCR, AND XRING CORRESPOND TO A TEMPERATURE VALUE TSTD.
C THE ENTHALPY EQUALS THE BASE PLUS (THE INCREMENT TIMES THE NUMBER
C      OF CARBON ATOMS) PLUS (THE XRING TIMES THE NUMBER OF EXTRA RINGS).
      DATA TSTD /298.15, 600., 800., 1000./
      DATA BASE / 61.76, 58.81, 56.08, 53.67 /
      DATA XRING / 40.76, 42.72, 43.63, 44.16 /
      DATA XINCR / -7.257, -7.397, -7.364, -7.173/
      IF (XNC.LT.8.0) HARO=HAROL(XNC,T)
      IF (XNC.LT.8.0) RETURN
      NLOW=0
      DO 10  I=1,4
      IF (T.GE.TSTD(I)) NLOW=I
   10 CONTINUE
      IF(NLOW.EQ.0) NLOW=1
      IF(NLOW.EQ.4) NLOW=3
      NHIGH=NLOW+1
      HLOW=BASE(NLOW)+XNC*XINCR(NLOW)+(RINGSA-1.0)*XRING(NLOW)
      HHIGH=BASE(NHIGH)+XNC*XINCR(NHIGH)+(RINGSA-1.)*XRING(NHIGH)
      TFAC=(T-TSTD(NLOW))/(TSTD(NHIGH)-TSTD(NLOW))
      HARO=HLOW+TFAC*(HHIGH-HLOW)
      RETURN
      END
      FUNCTION FPAR(XNC,T)
C FUNCTION RETURNS THE FREE ENERGY OF A PARAFFINIC PSEUDOCOMPONENT
C HAVING XNC CARBON ATOMS PER MOLECULE, AT TEMPERATURE T DEGREES K
      DIMENSION BASE(3), XINCR(3), TSTD(3)
C EACH BASE AND XINCR CORRESPOND TO A TEMPERATURE VALUE TSTD.
C THE FREE ENERGY EQUALS THE BASE PLUS (THE INCREMENT TIMES THE NUMBER
C      OF CARBON ATOMS)
      DATA TSTD /298.15, 600.,1000./
      DATA BASE / -12.55, -12.60, -11.94/
      DATA XINCR / 1.674, 8.909, 18.845/
      NLOW=0
      DO 10  I=1,3
      IF (T.GE.TSTD(I)) NLOW=I
   10 CONTINUE
      IF(NLOW.EQ.0) NLOW=1
      IF(NLOW.EQ.3) NLOW=2
      NHIGH=NLOW+1
      FLOW=BASE(NLOW)+XNC*XINCR(NLOW)
      FHIGH=BASE(NHIGH)+XNC*XINCR(NHIGH)
      TFAC=(T-TSTD(NLOW))/(TSTD(NHIGH)-TSTD(NLOW))
      FPAR=FLOW+TFAC*(FHIGH-FLOW)
      RETURN
      END
      FUNCTION FNAP(XNC,RINGSN,T)
C FUNCTION RETURNS THE FREE ENERGY OF A NAPHTHENIC PSEUDOCOMPONENT
C HAVING XNC CARBON ATOMS PER MOLECULE AND RINGSN RINGS PER MOLECULE,
C AT TEMPERATURE T DEGREES K
      DIMENSION BASE(3), XINCR(3), XRING(3), TSTD(3)
C EACH BASE, XINCR, AND XRING CORRESPOND TO A TEMPERATURE VALUE TSTD.
C THE FREE ENERGY EQUALS THE BASE PLUS (THE INCREMENT TIMES THE NUMBER
C      OF CARBON ATOMS) PLUS (THE XRING TIMES THE NUMBER OF EXTRA RINGS)
      DATA TSTD /298.15, 600.,1000./
      DATA BASE / 9.49, 0.52, -6.47/
      DATA XINCR / -0.3905, 7.447, 17.431/
```

Table 24.9
Continued

```
        DATA XRING /4.94, 0.39, -4.26/
        NLOW=0
        DO 10   I=1,3
        IF (T.GE.TSTD(I)) NLOW=I
   10 CONTINUE
        IF(NLOW.EQ.0) NLOW=1
        IF(NLOW.EQ.3) NLOW=2
        NHIGH=NLOW+1
        FLOW=BASE(NLOW)+XNC*XINCR(NLOW)+(RINGSN-1.)*XRING(NLOW)
        FHIGH=BASE(NHIGH)+XNC*XINCR(NHIGH)+(RINGSN-1.)*XRING(NHIGH)
        TFAC=(T-TSTD(NLOW))/(TSTD(NHIGH)-TSTD(NLOW))
        FNAP=FLOW+TFAC*(FHIGH-FLOW)
        RETURN
        END
        FUNCTION FARO(XNC,RINGSA,T)
C FUNCTION RETURNS THE FREE ENERGY OF AN AROMATIC PSEUDOCOMPONENT
C HAVING XNC CARBON ATOMS PER MOLECULE AND RINGSA RINGS PER MOLECULE,
C AT TEMPERATURE T DEGREES K
        DIMENSION BASE(3), XINCR(3), XRING(3), TSTD(3)
C EACH BASE, XINCR, AND XRING CORRESPOND TO A TEMPERATURE VALUE TSTD.
C THE FREE ENERGY EQUALS THE BASE PLUS (THE INCREMENT TIMES THE NUMBER
C      OF CARBON ATOMS) PLUS (THE XRING TIMES THE NUMBER OF EXTRA RINGS)
        DATA TSTD /298.15, 600.,1000./
        DATA BASE / 28.79, -3.30, -43.14/
        DATA XINCR / -0.1246, 7.157, 16.842/
        DATA XRING / 19.0, -3.89, -35.5/
        IF (XNC.LT.8.0) FARO=FAROL(XNC,T)
        IF (XNC.LT.8.0) RETURN
        NLOW=0
        DO 10   I=1,3
        IF (T.GE.TSTD(I)) NLOW=I
   10 CONTINUE
        IF(NLOW.EQ.0) NLOW=1
        IF(NLOW.EQ.3) NLOW=2
        NHIGH=NLOW+1
        FLOW=BASE(NLOW)+XNC*XINCR(NLOW)+(RINGSA-1.)*XRING(NLOW)
        FHIGH=BASE(NHIGH)+XNC*XINCR(NHIGH)+(RINGSA-1.)*XRING(NHIGH)
        TFAC=(T-TSTD(NLOW))/(TSTD(NHIGH)-TSTD(NLOW))
        FARO=FLOW+TFAC*(FHIGH-FLOW)
        RETURN
        END
        FUNCTION FSUL(T)
C FUNCTION RETURNS THE FREE ENERGY FOR SUBSTITUTION OF SULFUR FOR CH2,
C AT TEMPERATURE T DEGREES K
        DIMENSION BASE(3), TSTD(3)
C EACH BASE CORRESPONDS TO A TEMPERATURE VALUE TSTD.
C THE FREE ENERGY EQUALS THE BASE
        DATA TSTD /298.15, 600.,1000./
        DATA BASE / 4.5, -5.11, -15.49/
        NLOW=0
        DO 10   I=1,3
        IF (T.GE.TSTD(I)) NLOW=I
   10 CONTINUE
        IF(NLOW.EQ.0) NLOW=1
        IF(NLOW.EQ.3) NLOW=2
        NHIGH=NLOW+1
        FLOW=BASE(NLOW)
        FHIGH=BASE(NHIGH)
        TFAC=(T-TSTD(NLOW))/(TSTD(NHIGH)-TSTD(NLOW))
        FSUL=FLOW+TFAC*(FHIGH-FLOW)
        RETURN
        END
        FUNCTION HSUL(T)
C FUNCTION RETURNS THE ENTHALPY FOR SUBSTITUTION OF SULFUR FOR CH2,
C AT TEMPERATURE T DEGREES K
        DIMENSION BASE(4), TSTD(4)
C EACH BASE CORRESPONDS TO A TEMPERATURE VALUE TSTD.
C THE ENTHALPY EQUALS THE BASE
        DATA TSTD /298.15, 600.,800.,1000./
```

Table 24.9
Continued

```
      DATA BASE / 13.69, 11.39, 4.87, -1.65/
      NLOW=0
      DO 10  I=1,4
      IF (T.GE.TSTD(I)) NLOW=I
 10 CONTINUE
      IF(NLOW.EQ.0) NLOW=1
      IF(NLOW.EQ.4) NLOW=3
      NHIGH=NLOW+1
      HLOW=BASE(NLOW)
      HHIGH=BASE(NHIGH)
      TFAC=(T-TSTD(NLOW))/(TSTD(NHIGH)-TSTD(NLOW))
      HSUL=HLOW+TFAC*(HHIGH-HLOW)
      RETURN
      END
      FUNCTION HAROL(XNC,T)
      DIMENSION TSTD(7),HFORM(7,3)
      DATA TSTD /298.15, 300., 400., 500., 600., 800., 1000./
      DATA HFORM /19.82, 19.79, 18.56, 17.54, 16.71, 15.51, 14.82,
     1            11.95, 11.92, 10.34,  9.05,  8.02,  6.65,  6.01,
     2             4.24,  4.20,  2.37,  0.84, -0.39, -2.06, -2.86/
      NCLOW=INT(XNC+1.0E-5)
      NCLOW=NCLOW-5
      IF(NCLOW.LE.0) NCLOW=1
      IF(NCLOW.GE.3) NCLOW=2
      NCHIGH=NCLOW+1
      XNCFAC=XNC-FLOAT(5+NCLOW)
      NLOW=0
      DO 10  I=1,7
      IF (T.GE.TSTD(I)) NLOW=I
 10 CONTINUE
      IF(NLOW.EQ.0) NLOW=1
      IF(NLOW.EQ.7) NLOW=6
      NHIGH=NLOW+1
      HLOW=HFORM(NLOW,NCLOW)+XNCFAC*(HFORM(NLOW,NCHIGH)-
     1                               HFORM(NLOW,NCLOW))
      HHIGH=HFORM(NHIGH,NCLOW)+XNCFAC*(HFORM(NHIGH,NCHIGH)-
     1                               HFORM(NHIGH,NCLOW))
      TFAC=(T-TSTD(NLOW))/(TSTD(NHIGH)-TSTD(NLOW))
      HAROL=HLOW+TFAC*(HHIGH-HLOW)
      RETURN
      END
      FUNCTION FAROL(XNC,T)
      DIMENSION TSTD(7),FFORM(7,3)
      DATA TSTD /298.15, 300., 400., 500., 600., 800., 1000./
      DATA FFORM /30.99, 31.06, 35.01, 39.24, 43.66, 52.84, 62.27,
     1            29.23, 29.34, 35.39, 41.81, 48.48, 62.24, 76.32,
     2            28.10, 28.24, 36.55, 45.27, 54.28, 72.76, 91.59/
      NCLOW=INT(XNC+1.0E-5)
      NCLOW=NCLOW-5
      IF(NCLOW.LE.0) NCLOW=1
      IF(NCLOW.GE.3) NCLOW=2
      NCHIGH=NCLOW+1
      XNCFAC=XNC-FLOAT(5+NCLOW)
      NLOW=0
      DO 10  I=1,7
      IF (T.GE.TSTD(I)) NLOW=I
 10 CONTINUE
      IF(NLOW.EQ.0) NLOW=1
      IF(NLOW.EQ.7) NLOW=6
      NHIGH=NLOW+1
      FLOW=FFORM(NLOW,NCLOW)+XNCFAC*(FFORM(NLOW,NCHIGH)-
     1                               FFORM(NLOW,NCLOW))
      FHIGH=FFORM(NHIGH,NCLOW)+XNCFAC*(FFORM(NHIGH,NCHIGH)-
     1                               FFORM(NHIGH,NCLOW))
      TFAC=(T-TSTD(NLOW))/(TSTD(NHIGH)-TSTD(NLOW))
      FAROL=FLOW+TFAC*(FHIGH-FLOW)
      RETURN
      END
```

Notation

A	subscript referring to aromatic compounds
Btu	one Btu/lb-mole-$°R$ = 1 cal$_{IT}$/g-mole-$°K$
c	subscript referring to critical temperature
cal	thermochemical calorie (4.1840 absolute Joules)
cal$_{IT}$	International Steam Tables calorie (4.1868 absolute Joules)
C/H	weight ratio of carbon to hydrogen
d	density, g/cc
f	weight fraction sulfur
F	in ΔF_i, ΔF_j, or ΔF_k, the ideal gas standard free energy of formation of isomer i, j, or k, respectively
ΔF	the ideal gas standard free energy of formation of the mixture
$FOVT$	free energy of vaporization at T
ΔH	the ideal gas standard enthalpy of formation of the mixture
ΔH_i	the ideal gas standard enthalpy of formation of isomer i
HOV	heat of vaporization at atmospheric boiling point
$HOVT$	heat of vaporization at T
i	index in the mathematical series for $FOVT$; index indicating isomer number
IT	subscript indicating International Steam Tables
j	index indicating isomer number
k	index indicating isomer number
m	number of hydrogen atoms per molecule less twice the number of carbon atoms per molecule
m'	average value of m
$MABP$	molar average boiling point
$MeABP$	mean average boiling point
MW	molecular weight
n	unsubscripted, number of carbon atoms per molecule when sulfur is neglected
Δn	the change in the number of moles of gas during combustion of one mole of mixture, with water produced as a liquid
n_H, n_N, n_S	the numbers of hydrogen, carbon, and sulfur atoms, respectively, per molecule, when sulfur is considered
n'	average value of n
N	total number of isomers; subscript referring to naphthenes
P	pressure; subscript referring to paraffins
Q_A	number of aromatic rings per aromatic molecule
Q_A'	average value of Q_A
Q_N	number of naphthenic rings per naphthene molecule
Q_N'	average value of Q_N
R	gas constant
SG	specific gravity 60/60°F (ratio of density at 60°F to the density of water at 60°F)
T	absolute temperature
T_b	normal boiling point; $MABP$ of a mixture
T_c	pseudo critical temperature
TT	square of the temperature
$VABP$	volumetric average boiling point
X	mole fraction of compound of a particular type; as indicated by a subscript
y	mole fraction of an isomer; subscript indicates serial number assigned to the isomer

References

1. Adler, S. B., and K. R. Hall, *Hydrocarbon Processing* 64 (11), 71–75 (1985).
2. Antoine, A. C., "Use of Petroleum-Based Correlations and Estimation Methods for Synthetic Fuels," NASA Technical Memorandum No. 81533, NTIS No.N80-27509 (June 1980).
3. American Society for Testing and Materials, Procedure ANSI/ASTM D2382-76, "Standard Test Method for Heat of Combustion of Hydrocarbon Fuels by Bomb Calorimeter (High-Precision Method)," Federal Test Method Standard No. 791b.
4. Huang, P. K., "Characterization and Thermodynamic Correlations for Undefined Hydrocarbon Mixtures," Thesis, Pennsylvania State University, March 1977.
5. Huang, P. K., and Daubert, T. E., "Prediction of the Enthalpy of Petroleum Fractions—The Pseudocompound Method," *Ind. Eng. Chem. Proc. Des. Dev.* 13, 359 (1974).
6. Watson, K. M., *Ind. Eng. Chem.* 35, 398 (1943).
7. Hougen, O. A., Watson, K. M., and Ragatz, R. A., *Chemical Process Principles,* Volume 1, John Wiley & Sons, New York, NY (1947).
8. American Petroleum Institute, *Technical Data Book—Petroleum Refining,* 1980.
9. American Petroleum Institute Project 44.
10. Hubbard, W. N., Scott, D. W., and Waddington, G., Chapter 5 in Rossini, F. D., editor, *Experimental Thermochemistry,* Interscience Publishers, New York, NY (1956).
11. Stull, D. R. et al., *The Chemical Thermodynamics of Organic Compounds,* John Wiley & Sons, New York, NY (1969).
12. Smith, Buford D., *AIChE J.* 5, 26 (1959).

25

Alternate Datum States for Enthalpies of Petroleum Fractions

With contributions by R. L. Montgomery

Alternate zero enthalpy reference states for discrete components (pure hydrocarbons) were discussed in Chapter 18. The three zero enthalpy datum states described there can also be defined for pseudo components (fractions) of petroleum by analogy with the corresponding definitions for discrete components, as follows:

1. The first alternative is the API-44 datum—Ideal gas of the pseudo component at absolute zero temperature.
2. The second alternative is the API-TDB datum—Saturated liquid of the pseudo component at −200°F.
3. The third alternative is the absolute enthalpy datum—Standard states of the elements at absolute zero temperature, the standard states being graphite carbon, orthorhombic sulfur, and ideal gas for hydrogen, oxygen, and nitrogen.

These conversions can be made for hydrocarbon mixtures when molecular properties are known. Such detailed information is not available for petroleum fractions, so correlations must be used for heat capacities and other properties, which are obtained in CUTPRO of EQUIL where Equation 7.34 is used for heat (enthalpy) calculations. This same equation will be used in deriving the datum state conversion constants so as to mimimize errors.

Conversions between reference states 1 and 2 will not introduce errors in heat balance calculations because the conversion constants will cancel out. Conversions between reference states 1 and 3 might introduce errors in heat balances made on chemical reactors, where molecular forms are changed, and inconsistencies or errors in conversion constants would not cancel out.

A computer program is available for calculating the constants needed for converting datum state 1 to 2, or 1 to 3, state 1 being the basic datum state from which states 2 and 3 are derived.

Conversion of API-44 to API-TDB Enthalpies

Figure 18.2, which illustrated this conversion for discrete components, applies equally well to pseudo components of petroleum, as do Equations 18.1 through 18.3. *HOV*, the heat of vaporization at the atmospheric boiling point in Equation 18.2, will be found for the pseudo components by the Watson-Giacalone (2,5) equation, as had been suggested in API-TDB (1). This gives the following expression for *HOVT*, in FORTRAN notation:

$$HOVT = (1.986 * T_c * TT * ALOG(P_c/14.686)/$$
$$(T_c - TT)) * ((T_c - 259.7)/$$
$$(T_c - TT)) ** 0.38 \qquad (25.1)$$

230

where T_c = critical temperature of pseudo component, °R, Equation 7.16

TT = normal boiling point of pseudo component, °R

P_c = critical pressure of pseudo component, psia, Equation 7.17

AK = Watson's characterization factor, Equation 7.9

Equation 25.1 is applied to all pseudo components of petroleum, even though the terms are not indexed here. It may be applied to all the 11 or 21 pseudo components, having ID numbers between 1 and 50; or it may be applied to hypothetical components, having ID numbers between 51 and 100, as specified by the user.

HOVT values from Equation 25.1 are used in Equation 18.1 to calculate the values of *EE* for applying the −200°F liquid zero enthalpy datum state to petroleum components, the same procedure described previously for discrete components, the difference being the source of *HOVT* values and the ideal gas state heat capacities.

A FORTRAN program for calculating the values of *HOVT* and *EE* for pseudo components of petroleum is part of AZHPF, a program for making the 1 to 2 and the 1 to 3 conversions. This program is discussed later.

Conversion of API-44 to Absolute Enthalpies

In the third alternate zero enthalpy datum state the conversion constant is the enthalpy of formation of the pseudo component from the elements at absolute zero temperature, which has been named *HOFZ*. By including the heat of formation at absolute zero temperature in the enthalpy calculation for pseudo components of petroleum, absolute enthalpies are obtained.

The evaluation of *HOFZ* requires a knowledge of the amounts of each element in the pseudo component, the molecular weight, and the enthalpy of formation of the pseudo component at one temperature, which could be 77°F, the widely accepted standard, from which *HOFZ* can be calculated. This is usually accomplished by using the heat of combustion of the oil component.

There are optional procedures for computing *HOFZ*, the heat (enthalpy) of formation at absolute zero temperature. These methods either use the heat of combustion or are related to it. One method of finding *HOFZ* employs the modified Hougen-Watson-Ragatz correlation method presented in Chapter 24. This procedure is also used to calculate the heat of combustion, being the source of Figure 24.1.

Graphical Heat of Combustion Predictions

Figure 24.1 gives heats of combustion of petroleum fractions as a graphical function of API gravity (20 to 90) for lines of constant Watson Characterization Factor (11.2 to 12.8). A short FORTRAN Program named FIGONE reads and interpolates, or extrapolates Figure 24.1, and obtains numerical values of the heat of combustion and corrects for sulfur content. Input data for this program include wt% sulfur, API gravity, and the Watson *K* factor. The sulfur correction is + 120 Btu/lb for each wt% sulfur.

API-TDB Equation for Heat of Combustion

A simple analytical expression for the heat of combustion of petroleum fractions is the following equation from API-*Technical Data Book* (1).

$$- HOCA = (16,796. + APIG * (54.5 - APIG \\ * (0.217 + 0.0019 * APIG))) * WM \quad (25.2)$$

where $APIG$ = API gravity 60/60 of pseudo component

WM = molecular weight of pseudo component

$HOCA$ = heat removed during combustion to keep temperature at 60°F, with H_2O in gas state, in Btu/lbmole

Equation 25.2 estimates the net heat of combustion; therefore the heat of vaporization of the water formed in the combustion must be added to obtain the gross heat of combustion. (The difference between the heat of combustion at 60°F, the nominal temperature for Equation 25.2, and that at the chosen temperature of 77°F is neglected.)

The amounts of the elements are calculated, or estimated, the heat of combustion of the total elements is calculated, and the heat of formation is found by combining these two heats of combustion. These calculations are made at 77°F. A method for estimating the amounts of the elements carbon, hydrogen, and sulfur follows.

The carbon/hydrogen weight ratio has been correlated graphically by Winn (6) as a function of gravity and the Watson characterization factor. Moura (4) fitted the following equation to Winn's nomograph:

$$COH = AW(SG^{BW}) \quad (25.3)$$

where COH = ratio (lbs carbon)/(lbs hydrogen)

$AW = \exp(3.07546 - 0.0834254\ AK)$

SG = specific gravity, 60/60°F

$BW = 4.3214893 + (0.0269877AK \\ - 0.57975478)AK$

Values of *a* and *b* (in C_aH_b) are found from *COH* values by the following two expressions, which are in FORTRAN terms:

$$BB = WM/(1.008*(COH + 1.0)) \qquad (25.4)$$

$$AA = 1.008*BB*COH/12.011 \qquad (25.5)$$

where $AA = a$ in C_aH_b
$\qquad BB = b$ in C_aH_b

The computer program also permits sulfur in the molecule. Additions to the above lead to a small subroutine for evaluating a, b, and c in $C_aH_bS_c$.

Detailed Procedure of Chapter 24

This procedure for computing thermochemical properties was presented in Chapter 24. This calculation procedure permits calculating the heat of combustion of oil fractions and can also be used to calculate the enthalpy of formation directly, without calculating the heat of combustion.

When the heat of combustion is obtained by the Detailed Procedure of Chapter 24, the molecular weight of the pseudo component is available from that calculation. If the heat of combustion is from another source, the molecular weight is calculated by Equation 7.2.

When the heat of combustion is obtained by the Detailed Procedure of Chapter 24, standard enthalpy of formation of the pseudo petroleum component in the ideal gas state is available from the same procedure.

If the heat of combustion is measured or estimated by one of the other techniques, the standard enthalpy of formation of the pseudo component is computed by applying Steps 17 through 23 of the Detailed Procedure in reverse to the heat of combustion.

HOFZ from Heats of Combustion

The standard heat of formation, HOF, of a petroleum pseudo component is found, by subtracting the heat of combustion of the component from the heat of combustion of the component's elements.

The value of HOFZ is computed from HOF by the following equation:

$$HOFZ = HOF + HVST \qquad (25.6)$$
$$- (H^o - H^o{}_o)_{T=536.7°R}, \text{ for the petroleum}$$
pseudo component
$$+(H^o-H^o{}_o)_{T=536.7°R}, \text{ for the elements in the}$$
component

where $\qquad HOFZ =$ heat (enthalpy) of formation of the pseudo component, in the ideal gas state, from its elements, all at 0°R

$HOF =$ standard heat of formation of the pseudo component from its elements, all at 77°F

$HVST =$ heat of vaporization at 77°F

$(H^o - H^o{}_o)_{T=536.7°R} =$ enthalpy at $T = 536.7$°R relative to 0°R

Computer Programs for Pseudo Components

Two FORTRAN programs have been written for computing values of EE and HOFZ for pseudo components of petroleum. TST345 gives three choices of methods for estimating the heats of combustion, the enthalpies of formation, and the values of HOFZ. These three methods are (a) Equation 25.2 for the heat of combustion plus other calculations for the enthalpy of formation; (b) Figure 24.1 for the heat of combustion plus other calculations for the enthalpy of formation; and (c) The Detailed Procedure of Chapter 24 for both heat of combustion and the enthalpy of formation.

The other FORTRAN program, which is called AZHPF, finds EE values by the same procedure as does TST345, but uses only the Equation 25.2 method for the heat of combustion computations. The AZHPF calculations of the EE and HOFZ values are the same as those in TST345. With more options for the user, Program TST345 is more suitable for development studies. With no options for the user, Program AZHPF is more suitable for routine work and is intended for inclusion in EQUIL, which explains the COMMON statements. FORTRAN Program AZHPF is listed in Table 25.1.

Program AZHPF was prepared with addition to EQUIL in mind. As has been brought out previously, the properties of pseudo components are found in CUTPRO for use in EQUIL. Presently, there are no sulfur content data input in EQUIL. Therefore, the program for calculating HOFZ in AZHPF was based on average sulfur contents of liquid petroleum fuels given by Maxwell (3):

API Gravity	Sulfur, wt%	Inerts*, wt%
0	2.95	1.15
5	2.35	1.00
10	1.80	0.95
15	1.35	0.85
20	1.00	0.75
25	0.70	0.70
30	0.40	0.65
35	0.30	0.60

* Air and carbon dioxide

Equation 25.2 was developed for these average values of sulfur and inerts contents so heats of combustion from Equation 25.2 should be correct for these average num-

(text continued on page 235)

Table 25.1
Computer Program AZHPF for Calculating *HOVT, EE* and *HOFZ* for Pseudo Components of Petroleum

```
      COMMON /PRINT/ TITLE(20),CNAME(7,30),FD(30),WM(30),BP(30),
     1    SG(30),ICOD,KHF,KYF,KTL,IKV,IHS,IHD,IOUT,NC,
     2    TI,PI,EF,TO,PO,TTOL,PTOL,XTOL,SUMF,HF,SF
      COMMON /CRIT/ ID(30),TC(30),PC(30),WC(30)
      COMMON /HSTR/ HCF(9,30),HSI(30),HOF(30)
      DIMENSION AK(30),APIG(30),EE(30),RP(30),HOVT(30),BF(30),
     *  TCBR(30),PTEH(30),TCST(30),PTSH(30),HVST(30),TT(30),HV200(30),
     *  HOV(30),AW(30),BW(30),COH(30),AA(30),BB(30),CC(30),HEAT(30),
     *  HOCOMB(30),HOELEMS(30),HPF(30),HOFZ(30),HOFZPC(30)
      OPEN(5,FILE = 'AZHPF.DAT',STATUS = 'NEW')
      OPEN(6,FILE = '              ',STATUS = 'NEW')
      WRITE(*,7)
    7 FORMAT(//,' Enter number of pseudo components.')
      READ(*,*) NPF
      DO 10 I = 1, NPF
      WRITE(*,8) I
    8 FORMAT(//, ' Enter ID and two properties of component:',I3,/
     *' Options : Watson K with API or SpGr or with BP deg F;'/
     *' Specific Gravity and Boiling Point, degrees F.'/
     *' Reqd order : ID, WK, API, BP, SG, 1 integer + 4 real numbers'/
     *' Only two property entries, with zeros in other two spaces.'/)
      READ(*,*) ID(I),AK(I),APIG(I),BP(I),SG(I)
      WRITE(5,9) ID(I),AK(I),APIG(I),BP(I),SG(I)
    9 FORMAT(I3,7X,3F10.2,F10.4)
   10 CONTINUE
      REWIND 5
      OPEN(5,FILE = 'AZHPF.DAT',STATUS = 'OLD')
      DO 11 I = 1, NPF
      READ(5,9) ID(I),AK(I),APIG(I),BP(I),SG(I)
      IF(AK(I).NE.0.AND.BP(I).NE.0) GO TO 21
      IF(AK(I).NE.0.AND.SG(I).NE.0) GO TO 22
      IF(AK(I).NE.0.AND.APIG(I).NE.0) GO TO 23
      IF(SG(I).NE.0.AND.BP(I).NE.0) GO TO 24
   21 TT(I) = BP(I) + 459.7
      SG(I) = TT(I)**.333333/AK(I)
      APIG(I) = 141.5/SG(I) - 131.5
      GO TO 12
   22 APIG(I) = 141.5/SG(I) - 131.5
      TT(I) = (AK(I)*SG(I))**3.
      BP(I) = TT(I) - 459.7
      GO TO 12
   23 SG(I) = 141.5/(APIG(I) + 131.5)
      TT(I) = (AK(I)*SG(I))**3.
      BP(I) = TT(I) - 459.7
      GO TO 12
   24 APIG(I) = 141.5/SG(I) - 131.5
      TT(I) = BP(I) + 459.7
      AK(I) = TT(I)**.333333/SG(I)
   12 CONTINUE
      WM(I) = 204.38*EXP(.00218*TT(I))*EXP( - 3.07*SG(I))*
     * TT(I)**.118*SG(I)**1.88
      T2 = TT(I)*TT(I)
      TC(I) = 341.7 + 811.*SG(I) + (.4244 + .1174*SG(I))*TT(I)
     1  + (.4669 - 3.2623*SG(I))*1.E5/TT(I)
      PC(I) = EXP(8.3634 - .0566/SG(I)
```

Table 25.1
Continued

```
    1  − (.24244 + (2.2898 + .11857/SG(I))/SG(I))*1.E − 3*TT(I)
    2  + (1.4685 + (3.648 + 0.47227/SG(I))/SG(I))*1.E − 7*T2
    3  − (.42019 + 1.6977/SG(I)**2)*1.E − 10*T2*TT(I))
       BF(I) = ((12.8/AK(I) − 1.)*(10./AK(I) − 1.)*100.)**2
       IF(AK(I).GT.12.8.OR.AK(I).LT.10.) BF(I) = 0.0
       HCF(2,I) = ((.02678*AK(I) − .32646 + BF(I)*(.08081*SG(I) − .084773))*WM(I)
    1  )
       HCF(3,I) = ((( − 0.6946 + .6061*AK(I) − .01902*AK(I)**2
    1  + BF(I)*(1.0887 − 1.0413*SG(I))))*WM(I)*1.E − 4)/2.
       HCF(4,I) = ((( − 0.5131 − BF(I)*(.26216 − .23474*SG(I))))*WM(I)*1.E − 7)/3.
C  HOVT(at − 200F) and HVST(at 77F) by APITDB 1976 Method
       RP(I) = PC(I)/14.696
       HOV(I) = 1.986*TC(I)*TT(I)*ALOG(RP(I))/(TC(I) − TT(I))
       TCBR(I) = (TC(I) − 259.7)/(TC(I) − TT(I))
       PTEH(I) = TCBR(I)**0.38
       HOVT(I) = HOV(I)*PTEH(I)
       TCST(I) = (TC(I) − 536.7)/(TC(I) − TT(I))
       PTSH(I) = TCST(I)**0.38
       HVST(I) = HOV(I)*PTSH(I)
C  HV200,Ideal gas state H value at − 200 F; and EE, APITDB zero H factor
       HV200(I) = 259.7*(HCF(2,I) + 259.7*(HCF(3,I) + 259.7*HCF(4,I)))
       HCF(8,I) = HOVT(I) − HV200(I)
       HOVT(I) = HOVT(I)/WM(I)
       EE(I) = HCF(8,I)/WM(I)
C  CALCULATE EMPIRICAL FORMULA NUMBERS FOR CARBON, HYDROGEN AND SULFUR
       AW(I) = EXP(3.07546 − 0.0834254*AK(I))
       BW(I) = 4.321489 + (0.0269877*AK(I) − 0.579755)*AK(I)
       COH(I) = AW(I)*(SG(I)**BW(I))
       CALL CHSABC(COH(I),WM(I),AA(I),BB(I),CC(I),APIG(I))
C  CALCULATE HEAT OF COMBUSTION(of pseudo component) PLUS H2O CORRECTION
       HEAT(I) = 16796. + APIG(I)*(54.5 − APIG(I)*(0.217 + 0.0019*APIG(I)))
       HEAT(I) = HEAT(I) + 0.5*(BB(I)/WM(I))*18923.
C  CALCULATE STANDARD HEAT OF COMBUSTION AT 77 DEG F
       RT = 1065.744
       HOCOMB(I) = HEAT(I)/(1. + 1.184E − 3*((1./SG(I)) − (1./7.76)))
       HOCOMB(I) = ( − 1.)*WM(I)*HOCOMB(I) − RT*BB(I)/4.
C  CALCULATE HEAT OF COMBUSTION OF ELEMENTS, BTU/LB-MOLE
       HOELEMS(I) = − 169179.*AA(I) − 61442.5*BB(I) − 127614.*CC(I)
C  CALCULATE STANDARD HEAT OF FORMATION, BTU/LB-MOLE OF FRACTION
       HOF(I) = HOELEMS(I) − HOCOMB(I)
C  CALCULATE HPF = ENTHALPY OF THE FRACTION AT 536.67 DEG R ABOVE ZERO R
       T = 536.7
       HPF(I) = T*(HCF(2,I) + T*(HCF(3,I) + T*HCF(4,I)))
C  CALCULATE HOFZPC, HEAT OF FORMATION AT ZERO R, BTU/LB-MOLE
       HOFZPC(I) = HOF(I) + HVST(I) − HPF(I) + 453.25*AA(I) + 1820.11*BB(I) +
    1  1894.1*CC(I)
       HCF(9,I) = HOFZPC(I)
       HOFZ(I) = HCF(9,I)/WM(I)
   11  CONTINUE
C  Write Results
       WRITE(6,13)
   13  FORMAT(3X,'ID',1X,'WATSON K',2X,'APIG',4X,'BP F',3X,'SpGr',4X,'MW'
    *  ,6X,'HOVT',6X,' EE ',7X,'HOFZ',/,41X,2(5X,'BTU/LB',),4X,'BTU/LB')
       DO 15 I = 1,NPF
       WRITE(6,14) ID(I),AK(I),APIG(I),BP(I),SG(I),WM(I),HOVT(I),EE(I),
```

Table 25.1
Continued

```
      *HOFZ(I)
   14 FORMAT(2X,I3,3F8.2,F7.3,F8.2,3F10.4)
   15 CONTINUE
      REWIND 5
      REWIND 6
      STOP
      END
      SUBROUTINE CHSABC(COH,WM,AA,BB,CC,API)
      DIMENSION A(15), S(15)
      DATA A/
    1 0.,5.,10.,15.,20.,25.,30.,35.,40.,45.,50.,55.,60.,65.,70./
      DATA S/
    1 2.95,2.35,1.8,1.35,1.,.7,.4,.3,.2,.15,.1,.075,.05,.025,0.0/
      IF(API.LT.0.0) GO TO 10
      IF(API.EQ.0.0) GO TO 11
      IF(API.EQ.70.OR.API.GT.70.) GO TO 12
      DO 13 K = 1,15
      IF(API.GE.A(K).AND.API.LT.A(K + 1)) GO TO 15
      GO TO 13
   15 SUL = S(K) + (S(K) − S(K + 1))*(A(K) − API)/5.
   13 CONTINUE
      GO TO 14
   10 SUL = 3.0
      GO TO 14
   11 SUL = 2.95
      GO TO 14
   12 SUL = 0.0
   14 CONTINUE
      SR1 = 3206.4
      SR2 = 3206.4 − 32.064*SUL
      ONECOH = 1. + COH
      AA = WM*(COH/ONECOH)*(SR2/SR1)/12.011
      BB = WM*(SR2/SR1)/(1.008*ONECOH)
      CC = WM*SUL/SR1
      RETURN
      END
```

bers. It is necessary then to include these sulfur content values in the calculation of the heat of formation of the elements of the pseudo components. Subroutine CHSABC was prepared to calculate the values of *AA, BB,* and *CC,* i.e., carbon, hydrogen, and sulfur contents, using the previous data and Equations 25.3–25.5.

Note that five basic properties of the pseudo components are used in the Program AZHPF: Watson characterization factor, API gravity, boiling point, specific gravity, and molecular weight. Any two of the first four of these are sufficient to define the pseudo components, with one exception—only one property can be a gravity. Workable combinations are Watson K with either gravity or the boiling point; boiling point with either gravity is a viable option.

Results for Pseudo Components

For the calculations made to illustrate the use of AZHPF, given in Tables 25.2 and 25.3, boiling point and specific gravities were specified. When entering property data values, the user enters zeros for those two properties not being input.

Other properties, i.e., Watson K, API gravity, and molecular weight, were calculated and are given in Table 25.2. Two molecular weight values are shown for comparison of the values of WM from BLOCK DATA for these discrete components and the values calculated by Equation 7.2. Calculated Watson K's and API gravities are also included in Table 25.2.

Table 25.2
Properties of 10 Hydrocarbons for Enthalpy Datum State Correction Calculations as Discrete and Pseudo Components

ID	Hydrocarbon Name	Given Values		Calculated		Mole Weight	
		BP°F	SpGr	Watson-K	API-G	Discr.	Pseudo
138	3,3 Diethylpentene	295.1	0.7570	12.03	55.42	128.26	134.29
164	n-Dodecane	421.3	0.7530	12.73	56.42	170.33	180.49
231	1-Nonene	296.4	0.7330	12.43	61.54	136.24	136.48
235	1-Dodecene	416.0	0.7620	12.56	54.20	168.34	177.35
310	sec-Butylcyclopentane	309.9	0.7990	11.47	45.60	126.24	135.25
313	n-Hexylcyclopentane	397.2	0.8010	11.86	45.15	154.29	165.45
412	Isopropylcyclohexane	310.6	0.8070	11.36	43.84	126.24	134.68
415	n-Pentylcyclohexane	398.6	0.8080	11.76	43.62	154.29	165.14
508	Isopropylbenzene	306.3	0.8660	10.57	31.89	120.20	127.03
524	n-Pentylbenzene	401.7	0.8630	11.03	32.46	148.24	159.00

Table 25.3
Values of *HOVT*, *EE*, and *HOFZ* Calculated for 10 Hydrocarbons by "Discrete" and by "Pseudo" Component Procedures

ID	Hydrocarbon Name	HOVT, Btu/Lb		EE, Btu/Lb		HOFZ, Btu/Lb	
		Discr.	Pseudo	Discr.	Pseudo	Discr.	Pseudo
138	3,3 Diethylpentene	169.32	171.75	115.28	166.28	− 609.86	− 500.33
164	n-Dodecane	168.23	167.79	117.86	151.12	− 563.23	− 899.82
231	1-Nonene	178.08	170.65	126.82	158.04	− 197.57	− 855.33
235	1-Dodecene	166.83	168.21	117.36	153.56	− 266.45	− 754.61
310	sec-Butylcyclopentane	166.67	173.58	119.57	176.01	− 429.03	− 118.05
313	n-Hexylcyclopentane	169.67	170.33	122.67	166.86	− 411.06	− 281.94
412	Isopropylcyclohexane	165.73	174.05	130.47	177.40	− 506.09	− 62.05
415	n-Pentylcyclohexane	167.10	170.58	129.64	168.50	− 468.69	− 227.96
508	Isopropylbenzene	186.36	178.66	155.15	182.33	133.61	217.21
524	n-Pentylbenzene	173.31	173.52	135.29	176.92	22.71	71.76

Results of *HOVT*, *EE*, and *HOFZ* calculations from Program AZHPF are given in Table 25.3 for 10 hydrocarbons, treated as "pseudo" components. Also entered in Table 25.3 are the corresponding values of the same properties of these same hydrocarbons as "discrete" components. These "discrete" component results are from Tables 18.2–18.5.

Noting the comparisons of the "discrete" and "pseudo" values of *HOVT, EE,* and *HOFZ* values for these 10 hydrocarbons in Table 25.3, it can be seen that the agreement is better for the *HOVT* values and poor for the *EE* and *HOFZ* values. This is because Equation 25.1 for *HOVT* is valid for most hydrocarbons, while the cor-

relations and heats of combustion for *EE* and *HOFZ* are valid only for straight-run petroleum fractions.

References

1. American Petroleum Institute, *Technical Data Book—Petroleum Refining* (1980).
2. Giacalone, *Gazz. Chim. Ital.* 81, 180 (1951).
3. Maxwell, J. B. *Data Book on Hydrocarbons*, D. Van Nostrand Co., Princeton, N.J. (1950).
4. Moura, C. A. D. private communication, (1987).
5. Watson, K. M. *Ind. Eng. Chem.* 35, 398 (1943).
6. Winn, F. W. *Petroleum Refiner* 36, 2, 157 (1957).

Author Index

Subject Index